Cell Surface and Extracellular Glycoconjugates

Biology of Extracellular Matrix Series

Editor
ROBERT P. MECHAM

Robert P. Mecham: REGULATION OF MATRIX ACCUMULATION

Thomas N. Wight and Robert P. Mecham: BIOLOGY OF PROTEOGLYCANS

Richard Mayne and Robert E. Burgeson: STRUCTURE AND FUNCTION OF COLLAGEN TYPES

Deane F. Mosher: FIBRONECTIN

W. Steven Adair and Robert P. Mecham: ORGANIZATION AND ASSEMBLY OF PLANT AND ANIMAL EXTRACELLULAR MATRIX

Linda J. Sandell and Charles D. Boyd: EXTRACELLULAR MATRIX GENES

John A. McDonald and Robert P. Mecham: RECEPTORS FOR EXTRACELLULAR MATRIX

CELL SURFACE AND EXTRACELLULAR GLYCOCONJUGATES
Structure and Function

Edited by **DAVID D. ROBERTS**
Laboratory of Pathology
National Cancer Institute
National Institutes of Health
Bethesda, Maryland

ROBERT P. MECHAM
Department of Cell Biology and Physiology
Washington University School of Medicine
St. Louis, Missouri

ACADEMIC PRESS, INC.
A Division of Harcourt Brace & Company
San Diego New York Boston
London Sydney Tokyo Toronto

Front cover photographs: (Top) Immunofluorescent staining of microfibril-associated glycoprotein (MAGP) in the extracellular matrix of bovine pulmonary artery endothelial cells. (Center) Cross section of bovine hypertensive pulmonary artery stained with elastin Van Gieson stain to highlight the elastic lamellae. (Bottom) Intracellular staining of MAGP in permeabilized endothelial cells. Courtesy of Dr. Robert P. Mecham.

Academic Press, Inc.
1250 Sixth Avenue, San Diego, California 92101-4311

United Kingdom Edition published by
Academic Press Limited
24–28 Oval Road, London NW1 7DX

Library of Congress Cataloging-in Publication Data

Cell surface and extracellular glycoconjugates : structure and function / edited by David D. Roberts, Robert P. Mecham.
p. cm. -- (Biology of extracellular matrix)
Includes bibliographical references and index.
ISBN 0-12-589630-1
1. Glycoconjugates. 2. Membrane proteins. 3. Membrane lipids.
I. Roberts, David D. (David Duncan), Date . II. Mecham, Robert P. III. Series.
QP702.G577C45 1993
574.19'2483--dc20 93-7137
CIP

PRINTED IN THE UNITED STATES OF AMERICA
93 94 95 96 97 98 BB 9 8 7 6 5 4 3 2 1

Contents

The Use of Glycosylation Inhibitors to Study Glycoconjugate Function

ALAN D. ELBEIN

Use of Mammalian Cell Mutants to Study the Functions of N- and O-Linked Glycosylation

PAMELA STANLEY

The Structure and Function of Glycosphingolipid Receptors in Inflammation and Cancer

BRUCE A. MACHER and CHERYL L. M. STULTS

Biochemical and Molecular Biological Approaches to Modify Oligosaccharides: Analysis of Glycosylation Function in Human Choriogonadotropin

O. P. BAHL, N. R. THOTAKURA, and W. Y. CHEN

The Role of Glycosylation in Adhesive Protein Function

MARVIN L. TANZER, MARTIN S. GINIGER, JOHN W. DEAN III, and S. CHANDRASEKARAN

Contributors

Numbers in parentheses indicate the pages on which the authors' contributions begin.

O. P. BAHL (245), Department of Biological Sciences, State University of New York at Buffalo, Buffalo, New York 14260

S. CHANDRASERKARAN (271), Department of BioStructure and Function, School of Dental Medicine, University of Connecticut Health Center, Farmington, Connecticut 06030-3705

W. Y. CHEN (245), Department of Biological Sciences, State University of New York at Buffalo, Buffalo, New York 14260

JOHN R. COUCHMAN (33), Department of Cell Biology, University of Alabama at Birmingham, Birmingham, Alabama 35294

JOHN W. DEAN III (271), Department of Periodontology, University of Tennessee College of Dentistry, Memphis, Tennessee 38163

TAMARA L. DOERING (83),[1] Department of Biological Chemistry, Johns Hopkins Medical School, Baltimore, Maryland 21205

ALAN D. ELBEIN (119), Department of Biochemistry and Molecular Biology, University of Arkansas for Medical Sciences, Little Rock, Arkansas 72205

KIYOSHI FURUKAWA (1), Department of Biochemistry, Institute of Medical Science, University of Tokyo, Tokyo 108, Japan

MARTIN S. GINIGER (271), Department of BioStructure and Function, School of Dental Medicine, University of Connecticut Health Center, Farmington, Connecticut 06030-3705

AKIRA KOBATA (1),[2] Department of Biochemistry, Institute of Medical Science, University of Tokyo, Tokyo 108, Japan

BRUCE A. MACHER (223), Department of Chemistry and Biochemistry, San Francisco State University, San Francisco, California 94132

[1] *Present address*: Department of Molecular and Cell Biology, University of California at Berkeley, Berkeley, California 94720.

[2] *Present address*: Tokyo Metropolitan Institute of Gerontology, 35-2 Sakaecho, Itabashi-ku, Tokyo 173, Japan.

PAMELA STANLEY (181), Department of Cell Biology, Albert Einstein College of Medicine, New York, New York 10461

CHERYL L. M. STULTS (223), Department of Chemistry and Biochemistry, San Francisco State University, San Francisco, California 94132

MARVIN L. TANZER (271), Department of BioStructure and Function, School of Dental Medicine, University of Connecticut Health Center, Farmington, Connecticut 06030-3705

N. R. THOTAKURA (245), Molecular and Cellular Endocrinology Branch, NIDDK, National Institutes of Health, Bethesda, Maryland 20892

ANNE WOODS (33), Department of Cell Biology, University of Alabama at Birmingham, Birmingham, Alabama 35294

Preface

Over the past five years, the field of carbohydrates and glycoconjugates has been rediscovered by investigators in many disciplines. In immunology, specific carbohydrate structures and complementary carbohydrate binding proteins known as selectins have emerged as important mediators of lymphocyte recirculation, activation, and inflammatory responses. In tumor biology, alteration of glycosylation or expression of lectins by tumor cells has been demonstrated to profoundly influence their malignant behavior *in vitro* and *in vivo*. Biological activities of several hormones and growth factors have been shown to require proper glycosylation of the protein or to require direct interaction of the factor with glycoconjugates on the cell surface for expression of their biological activities. Many interactions of pathogens with the host involve specific recognition of glycoconjugates on the host epithelia. The biological significance of the complex temporal and spatial control of glycoconjugate expression during development is just beginning to be elucidated. Finally, the commercial development of recombinant hormones and enzymes as therapeutic agents has produced an urgent need to better understand how glycosylation affects the biological and pharmacological properties of these molecules.

Each of the common monosaccharides on the mammalian cell surface can be linked to a second monosaccharide to form up to eight positional and anomeric isomers, each possessing distinct chemical and biological properties. Structural analysis of glycoconjugates therefore requires determination not only of the sequence of monosaccharides, but also of the nature of each glycosidic linkage, the number and positions of branches, and the identity and position of modifications such as sulfation, phosphorylation, and acetylation. Because of the enormous diversity of carbohydrate structures and the greater technical difficulties in elucidating their structure and function relative to those of proteins and nucleic acids, reference works are needed to inform researchers in diverse disciplines of the approaches and resources that can be employed to address these important problems. This volume addresses the determination of structure and function of the glycoconjugates that occur on the cell surface and in molecules involved in cell–matrix interactions. It does not include a discussion of cytoplasmic and nuclear glycosylation nor a complete description of proteoglycans except those associated with the cell surface. An earlier volume in this series, *Biology of Proteoglycans*, and several other excellent recent monographs may be consulted for comprehensive discussion of the matrix and connective tissue proteoglycans.

This volume is divided into three sections to address structure and function and their specific applications, although the authors have made efforts to emphasize the connections between the structural and biological approaches to function. The first chapters introduce the structures of the major classes of glycoconjugates and address the applications of structural analyses to elucidation of function. The major classes of glycoconjugates discussed include the asparagine- and serine/threonine-linked complex oligosaccharides on glycoproteins, glycosyl phosphatidylinositol linkages, and the cell membrane proteoglycans. The second section includes descriptions of several powerful techniques to examine function, including the use of somatic mutations in biosynthesis or degradation of the O- and N-linked oligosaccharides and the use of specific inhibitors of biosynthetic enzymes to perturb structure and function. The final section discusses the integration of these techniques to address the function of specific glycoconjugates, including the glycoprotein hormones, the membrane glycolipids, and the cell adhesion proteins and their receptors.

The scope of this volume is purposefully broad, with the intent to provide basic background and technical information about this rapidly expanding field of biology. Our understanding of the functions of glycoconjugates is far from complete, and the authors present many unanswered questions and suggestions for future research. Our hope is that this book will provide guidance and inspiration for new investigators entering the field of glycobiology and for researchers in other disciplines attempting to define the role of glycoconjugates in their biological systems.

David D. Roberts
Robert P. Mecham

Use of Structural Analyses to Examine the Functions of N- and O-Linked Sugar Chains of Glycoproteins

Kiyoshi Furukawa and Akira Kobata

Department of Biochemistry, Institute of Medical Science, University of Tokyo, 4-6-1 Shirokanedai, Minato-ku, Tokyo 108, Japan

I. Introduction

From recent progress in vascular cell biology, a lock and key theory for leukocyte cell adhesion has emerged, in which the cell surface carbohydrate ligands (key) are recognized by lectin-like molecules (lock). These carbohydrate receptors are inducible on the surface of endothelial cells and platelets or constitutively expressed on the surface of certain types of leukocytes. Based on their cDNA nucleotide sequences, these cell adhesion molecules are grouped into a protein family named selectins (Bevilacqua *et al.*, 1989; Tedder *et al.*, 1989; Bowen *et al.*, 1989; Siegelman *et al.*, 1989; Lasky *et al.*, 1989; Camerini *et al.*, 1989; Johnston *et al.*, 1989). Activated vascular endothelial cells and platelets at the sites of inflammation and wound healing are induced to express E-selectin (ELAM-1) and P-selectin (PADGEM, GMP-140,

Cell Surface and Extracellular Glycoconjugates

or CD62) at the cell surface and bind neutrophils, monocytes, and/or a subset of T lymphocytes through their cell surface carbohydrate ligands, including sialyl X [Neu5Acα 2→3Galβ1→4(Fucα1→3)GlcNAc] and/or X [Galβ1→4(Fucα1→3)GlcNAc], respectively (Phillips *et al.*, 1990; Goelz *et al.*, 1990; Walz *et al.*, 1990; Lowe *et al.*, 1990; Larsen *et al.*, 1990). The tissue-specific interaction of lymphocyte subpopulations with high endothelial venules in lymphoid tissues is also mediated by the lymphocyte cell surface molecule, L-selectin (formerly Mel-14, LAM-1), which binds to sialylated or sulfated fucosyl oligosaccharides expressed on lymphoid tissue (Rosen *et al.*,1985; Imai *et al.*, 1990, 1991). Detailed analyses identified additional ligands for E-selectin, including sialyl Lewis^a [Neu5Acα2→3Galβ1→3(Fucα1→4)GlcNAc] (Berg *et al.*, 1991) and Vim2 antigen [Neu5Acα2→3Galβ1→4GlcNAcβ1→3Gal-β1→4(Fucα1→3) GlcNAc] (Tiemeyer *et al.*, 1991). Since the structural characterization of selectin ligands was done primarily by use of monoclonal antibodies, whose carbohydrate epitope structures, in most cases, are not well elucidated, the real structures of the functional ligands *in vivo* are still obscure. Of course, one of the possible ligands for E-selectin is undoubtedly related to sialyl X. Therefore, structural determination of these biologically active oligosaccharides is an important objective in cell biology.

This chapter describes the methods for the structural analyses of N- and O-linked sugar chains of glycoproteins in order to examine their functions. With use of these methods, the oligosaccharide structures of small quantities of glycohormones, enzymes, cell adhesion molecules, recombinant products, and extracellular matrix components have been elucidated (Kawano *et al.*, 1988; Yamashita *et al.*, 1986; Furukawa *et al.*, 1989; Asada *et al.*, 1991; Takeuchi *et al.*, 1989; Takamoto *et al.*, 1989). Structural information thus obtained has contributed to elucidating the biological roles of the sugar chains of glycoproteins (Tsuchiya *et al.*, 1989; Takeuchi *et al.*, 1989).

II. Analysis of Sugar Composition

Analytical data of monosaccharide composition are necessary to choose a strategy for the structural study of the sugar chains of glycoproteins. So far, the major sugar chains of glycoproteins can be classified into two groups according to their sugar–peptide linkage regions (Fig. 1). Those that are linked to the asparagine (Asn) residues of polypeptides are called N-linked sugar chains. Others, which are linked to either serine (Ser) or threonine (Thr) residue, are called O-linked sugar chains. Since the O-linked sugar chains do not contain mannose residues, which are always found in the N-linked sugar chains, the

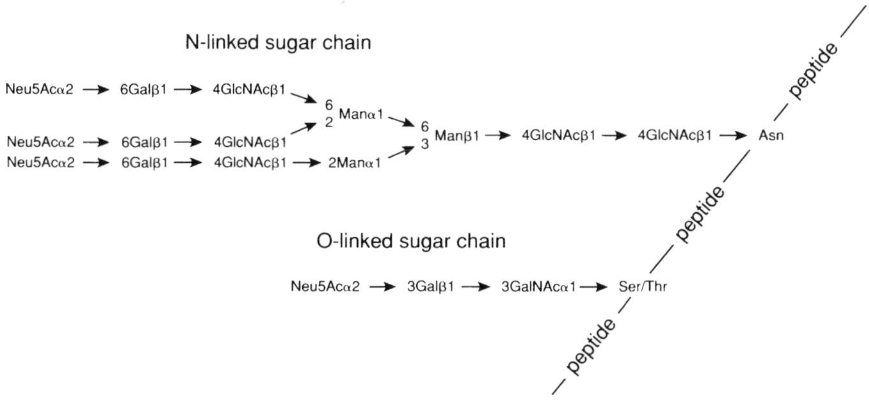

FIG. 1. Two major groups of glycoprotein sugar chains.

distribution of the two sugar chain groups in a glycoprotein can be inferred from monosaccharide analysis. Furthermore, the types of sugar chains released from glycoprotein samples can be estimated by the molar ratio of composing monosaccharides. However, the presence of *N*-acetylgalactosamine, which is a typical monosaccharide component of O-linked sugar chains, does not always indicate that the glycoprotein contains O-linked oligosaccharides, because some of the N-linked sugar chains also contain that sugar residue (Green and Baenziger, 1988; Nyame *et al.*, 1989; Weisshaar *et al.*, 1991; Chan *et al.*, 1991; Nakata *et al.*, 1991). Therefore, the data of monosaccharide composition are important for the structural analysis of N- and O-linked sugar chains whose structural studies are performed with use of different procedures as will be described later.

Since susceptibilities of the glycosidic linkages of different sugars to acid hydrolysis vary, it is impossible to obtain complete data for monosaccharide composition from a single acid hydrolysis. Reliable data can be obtained only by performing a study of the time course of monosaccharide release by acid hydrolysis and extraporating the data of hydrolysis to 0 hr to compensate for the degradation of each monosaccharide residue in acidic solution.

All monosaccharide residues, obtained by acid hydrolysis of sugar chains, contain an aldehyde group. This aldehyde group can be quantitatively converted to a tritium-labeled carbinol residue by the following reaction

$$\text{—CHO} \xrightarrow{\text{NaB}^3\text{H}_4} \underset{\displaystyle |\atop {}^3\text{H}}{\text{—CHOH.}}$$

Since the radioactivity incorporated into each monosaccharide is proportional to its molar amount, data of monosaccharide composition can be obtained by determining the radioactivity of monosaccharide alcohols separated by high-performance liquid chromatography. As shown in Fig. 2, sugar alcohols derived from human transferrin sugar chains were separated by the Shodex SUGAR SP-1010 column. This method will allow analysis of sugar compositions of glycoproteins from bacteria and plants as well as animals, because many additional sugar alcohols can be separated (Takeuchi *et al.*, 1987).

III. Release of Oligosaccharides from Polypeptide Backbones

Many glycoproteins contain more than one sugar chain per molecule. In most cases, these multiple sugar chains have different structures. Even in the case of ovalbumin, which carries only one sugar chain, microheterogeneity of the sugar chain structure has been observed (Huang *et al.*, 1970). In the early development of structural studies, analysis of N-linked sugar chains of glycoproteins was conducted using each glycopeptide, separated by gel filtration column chromatography after exhaustive pronase digestion. However, this method is compli-

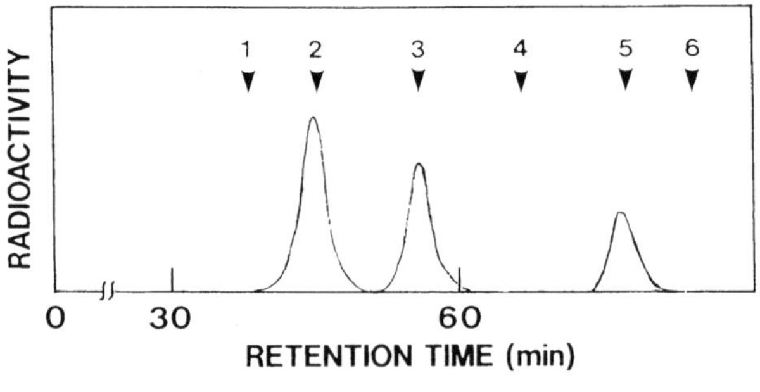

FIG. 2. Carbohydrate composition analysis of the sugar chains obtained from human transferrin. The sugar chains released from the glycoprotein by hydrazinolysis (100°C, 10 hr) were hydrolyzed with 4 N HCl at 100°C for 3 hr. The hydrolysate was N-acetylated and then reduced with NaB^3H_4. Both were mixed with standard sugar alcohols and injected into the HPLC system with a Shodex SUGAR-1010 column. The panel shows the elution pattern of labeled alditols and aminoalditols from the sample. The numbers at the top of the figure indicate the elution positions of standard sugar alcohols: (1) *N*-acetylgalactosaminitol; (2) *N*-acetylglucosaminitol; (3) mannitol; (4) fucitol; (5) galactitol; and (6) glucitol. The results indicate that the transferrin sugar chains are composed of galactose, *N*-acetylglucosamine, and mannose in a ratio of 1 : 1.8 : 1.4 as determined by calculating the radioactivity incorporated into each peak, indicating that the sugar chains are of the biantennary complex type.

cated by the structural diversity of the peptide portion and often leads to erroneous conclusions about oligosaccharide structures. Such a dilemma, encountered during oligosaccharide structural analysis, prompted us to develop methods for releasing sugar chains from polypeptide backbones without any amino acids. To date, several enzymatic and chemical methods for releasing N- and O-linked sugar chains from glycoproteins as intact oligosaccharides have been developed.

A. *Enzymatic Methods*

Enzymes that release sugar chains from polypeptides are found in animals, plants, and microorganisms and are classified into three group by the difference in their cleavage regions. The first group hydrolytically cleaves the *N,N'*-diacetylchitobiose moieties of N-linked sugar chains by the following reaction and is called endo-β-*N*-acetylglucosaminidases.

$$R\rightarrow \text{GlcNAc}\beta1\rightarrow4\text{GlcNAc}\rightarrow\text{Asn (peptide)} \rightarrow$$
$$R\rightarrow\text{GlcNAc} + \text{GlcNAc}\rightarrow\text{Asn (peptide)}$$
$$(R, \text{sugar chains})$$

Among them, endo D, H, C_{II}, and F were satisfactorily purified and have been used for the structural study of N-linked sugar chains of glycoproteins. Their substrate specificities are illustrated in Table I.

The basic glycon structure required by endo D is the trisaccharide Manα1→3Manβ1→4GlcNAc (Tai *et al.*, 1975; Ito *et al.*, 1975b). The most important part is the α-mannosyl residue of the trisaccharide. Substitution at the C-4 position of this mannose residue with other sugars shows no effect (Mizuochi *et al.*, 1984), but substitution at the C-2 position inhibits the action of endo D (Tai *et al.*, 1975; Ito *et al.*, 1975b). The aglycon specificity of endo D is rather broad, and oligosaccharides that are linked to ($\pm$Fucα1→6)GlcNAc or ($\pm$Fucα1→6)-GlcNAc-Asn (peptide) can be cleaved.

In the case of endo H, the basic glycon structure required is the tetrasaccharide Manα1→3Manα1→6Manβ1→4GlcNAc (Arakawa and Muramatsu, 1974). The nonreducing terminal α-mannose residue can be substituted with other sugars at its C-2 position. Therefore, the enzyme can hydrolyze all high-mannose-type sugar chains and most of the hybrid-type sugar chains. Since the complex-type sugar chains do not contain the Manα1→3 residue linked to the Manα1→6Man arm, they are not substrates of endo H, even after removal of their outer chain moieties. The enzyme hydrolyzes sugar chains linked to ($\pm$Fucα1→6)GlcNAc or ($\pm$Fucα1→6)GlcNAc-Asn (peptide) (Foddy *et al.*, 1986).

TABLE I

STRUCTURAL REQUIREMENTS OF VARIOUS ENDO-β-N-ACETYLGLUCOSAMINIDASES (ENDO) FOR THEIR SUBSTRATES

Enzymes	Structures required for substrates[a]	Sources
endo D	R↘6 Manβ1→4GlcNAcβ1→4R_1 R→4Manα1↗3	*Diplococcus pneumoniae*
endo H	R↘6 Manα1↘ R→2Manα1↗3 6 R→4Manβ1→4GlcNAcβ1→4R_1 R↗3	*Streptomyces griseus* and *S. plicatus*
endo C_{II}	R→2Manα1→3Manα1↘6 R→4Manβ1→4GlcNAcβ1→4R_2 R→2Manα1↗3	*Clostridium perfringens*
endo F	The structural requirement is similar to that of endo H, but endo F slowly cleaves the biantennary complex-type sugar chains.	*Flavobacterium meningosepticum*

[a] R, sugar or hydrogen; R_1, (±Fucα1→6)GlcNAc→Asn–(peptides), (±Fucα1→6)GlcNAc, or (±Fucα1→6)GlcNAc$_{OH}$; R_2, GlcNAc→Asn, GlcNAc, or GlcNAc$_{OH}$. The subscript **OH** indicates NaB^{3}H$_4$-reduced oligosaccharides.

The basic glycon structure required by endo C_{II} is the pentasaccharide Manα1→3Manα1→6(Manα1→3)Manβ1→4GlcNAc (Ito *et al.*, 1975a). The two terminal α-mannosyl residues must be either unsubstituted or substituted only at the C-2 position by other sugars. However, substitution at the C-4 position of the α-mannosyl residue of the Manα1→-3Manβ1→group inhibits the action of endo C_{II} (Tai *et al.*, 1977). The aglycon specificity of the enzyme is the same as that of endo H.

Endo F is somewhat similar to endo H, which hydrolyzes high-mannose-type and hybrid-type sugar chains as described above. However, it also slowly cleaves the biantennary-complex-type but not the tri- and tetraantennary-complex-type sugar chains (Tarentino *et al.*, 1985). Since endo-type enzymes have strict substance specificities, these specificities are quite useful in determining the structures of the sugar chains of glycoproteins.

The second group cleaves the β-aspartylglycosylamine linkage of N-linked sugar chains, releasing oligosaccharide–glycosylamine and peptide. The glycosylamine is then spontaneously cleaved to an oligosaccharide and ammonia (Fig. 3). They are called N-Glycanase, glycopeptidase, or peptide-N^4-(N-acetyl-β-glucosaminyl) asparagine amidase. An enzyme that belongs to this group, first found in almond

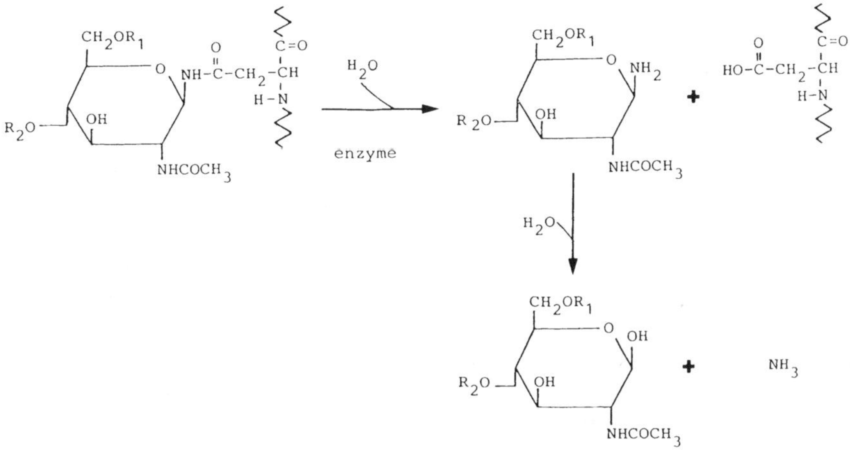

FIG. 3. Mechanism of hydrolysis of glycopeptidase. R_1 represents either hydrogen or a fucose residue and R_2 represents sugar chains.

emulsin (Takahashi, 1977), has a broad substrate specificity and can release all types of N-linked sugar chains. However, since the hydrolytic activity of the enzyme is influenced by the size and conformation of the peptide as well as the crowdedness of the sugar chains on a polypeptide (Takahashi and Nishibe, 1978; Plummer and Tarentino, 1981), the enzyme often cannot quantitatively release N-linked sugar chains from polypeptides (Tarentino and Plummer, 1982). To avoid such incomplete hydrolysis, glycoproteins are usually converted to glycopeptides by limited protease digestion or denatured by boiling in the presence of β-mercaptoethanol and sodium dodecyl sulfate or by treatment with chaotropic reagents such as KSCN and $NaClO_4$, prior to the enzyme digestion (Tarentino and Plummer, 1982; Hirani *et al.*, 1987).

The third group hydrolytically cleaves the α-*N*-acetylgalactosaminyl linkages to Ser or Thr residues, found in mucin-type glycoproteins, and releases the O-linked sugar chains by the following reaction.

Galβ1→GalNAcα1→Ser/Thr-peptide→
Galβ1→3GalNAc + Ser/Thr-peptide

This class of enzymes is called endo-α-*N*-acetylgalactosaminidases or *O*-glycanases. The enzyme was purified from *Diplococcus pneumoniae* and has been well characterized (Endo and Kobata, 1976; Umemoto *et al.*, 1977). Since its aglycon specificity is very strict, the enzyme releases only a disaccharide with the Galβ1→3GalNAc structure from various

O-linked sugar chains, but does not act on glycoprotein sugar chains with the GalNAc, ±Neu5Acα2→3Galβ1→3(±Neu5Acα2→6)GalNAc, and Fucα1→2Galβ1→3(±Neu5Acα2→6)GalNAc structures. Therefore, the enzyme can be used only for the study of the desialylated mucins. Recently, a new endoglycosidase, which releases various O-linked sugar chains as oligosaccharides, has been partially purified from the culture medium of *Streptomyces* sp. OH-11242 (Iwase *et al.*, 1988, 1992). The enzyme can release O-linked oligosaccharides as large as hepta- to octaoligosaccharides. Purification of this enzyme to a level that is free from other exoglycosidases will provide a strong tool for the study of O-linked oligosaccharide structures.

Since the glycon specificities of these endo-enzymes range from very narrow to relatively wide, use of the proper enzymes can be quite effective for preparation of the sugar chains, discrimination of the types of the sugar chains released, and estimation of the carbohydrate structures of the released sugar chains.

Enzymes that cleave the β-galactosyl linkages within sugar chains have been found in culture media of various bacteria and are called endo-β-galactosidases. The enzyme purified from *D. pneumoniae* (Takasaki and Kobata, 1976) releases GalNAc/Galα1→3(Fucα1→2)Gal groups from type 2 but not from type 1 A and B blood group determinants by the following reaction.

GalNAc/Galα1→3(Fucα1→2)Galβ1→4GlcNAc/Glc→*R*→
GalNAc/Galα1→3(Fucα1→2)Gal + GlcNAc/Glc→*R*
(*R*, hydrogen or sugar chain)

The enzyme from *Escherichia freundii,* which was initially found to act on keratan sulfate (Kitamikado *et al.*, 1970), cleaves the β-galactosyl linkages within poly-*N*-acetyllactosamine groups of glycoproteins and glycolipids by the following reaction.

R→GlcNAcβ1→3Galβ1→4*R*′→*R*→GlcNAcβ1→3Gal + *R*′

The glycon (*R*) and aglycon (*R*′) specificities of the enzyme have been studied, and the structures of oligosaccharides released from human granulocyte and erythrocyte membrane glycoproteins by digestion with the *E. freundii* enzyme are listed in Table II (Fukuda *et al.*, 1984a, 1984b). These data indicated that the enzyme acts on site I but not on site II on the following fucose containing structure.

II   I
↓   ↓
—GlcNAcβ1→3Galβ1→4GlcNAcβ1→3Galβ1→4GlcNAc—
3
↗
Fucα1↗

TABLE II

STRUCTURES OF OLIGOSACCHARIDES RELEASED BY ENDO-β-GALACTOSIDASE (*E. freundii*)

Structures
GlcNAcβ1→3Gal
Galβ1→4GlcNAcβ1→3Gal
Fucα1→2Galβ1→4GlcNAcβ1→3Gal
Galβ1→4GlcNAcβ1→3Gal 3 ↑ Fucα1
GlcNAcβ1→3Galβ1→4GlcNAcβ1→3Gal 3 ↑ Fucα1
Galβ1→4GlcNAcβ1→3Galβ1→4GlcNAcβ1→3Gal 3 ↑ Fucα1
Galβ1→4GlcNAcβ1→3Galβ1→4GlcNAcβ1→3Gal 3 3 ↑ ↑ Fucα1 Fucα1
Neu5Acα2→6Galβ1→4GlcNAcβ1→3Gal
Neu5Acα2→6Galβ1→4GlcNAcβ1→3Gal 3 ↑ Fucα1
Neu5Acα2→3Galβ1→4GlcNAcβ1→3Galβ1→4GlcNAcβ1→3Gal 3 ↑ Fucα1
Galβ1→4GlcNAcβ1↘6 Gal GlcNAcβ1↗3
Galβ1→4GlcNAcβ1↘6 Gal Galβ1→4GlcNAcβ1↗3
Galβ1→4GlcNAcβ1↘6 Gal Fucα1→2Galβ1→4GlcNAcβ1↗3
Galβ1→4GlcNAcβ1↘6 Gal GlcNAcβ1→3Galβ1→4GlcNAcβ1↗3
Galβ1→4GlcNAcβ1↘6 Gal Galβ1→4GlcNAcβ1→3Galβ1→4GlcNAcβ1↗3
Galβ1→4GlcNAcβ1↘6 Galβ1→4GlcNAcβ1→3Gal Galβ1→4GlcNAcβ1↗3

(*continues*)

TABLE II—*Continued*

Structures
Galβ1→4GlcNAcβ1↘6 Gal Fucα1→2Galβ1→4GlcNAcβ1→3Galβ1→4GlcNAcβ1↗3
Fucα1→2Galβ1→4GlcNAcβ1↘6 Gal Fucα1→2Galβ1→4GlcNAcβ1→3Galβ1→4GlcNAcβ1↗3

Furthermore, the enzyme can cleave the β-galactosyl linkage of I antigen, in which the C-3 and C-6 positions of the galactose residue are substituted with *N*-acetylglucosamine residues. Quite interestingly, the enzyme acts only on those of glycoproteins but not those of glycolipids (Scudder *et al.*, 1984).

Another endo-β-galactosidase obtained from *Clostridium perfringens* (Fukushi *et al.*, 1987) catalyzes the following reaction.

Galα1→3Galβ1→4GlcNAcβ1→*R*→Galα1→3Gal + GlcNAcβ1→*R*
(*R*, sugar chain)

Since the enzyme specifically removes the Galα1→3Gal groups in the outer chain moieties of N- and O-linked sugar chains, it might be effectively used to examine the function of this group. Because human and old-world monkeys possess natural antibody capable of binding to the α-galactose residues (Galili *et al.*, 1987), the physiological role of the Galα1→3Gal groups expressed on the sugar chains of microorganisms, human erythrocytes, and a variety of recombinant products with pharmaceutical purpose can be examined using this enzyme.

Taking the specificity of each enzyme described above into consideration, one can examine the functions of N- and O-linked sugar chains of glycoproteins by digestion with these enzymes, because the enzymes leave the polypeptide portion intact. However, the accessibility of the enzymes to the sugar chains of glycoproteins is limited by protein conformation. Therefore, release of the sugar chains from glycoproteins must be confirmed by some other method such as a lectin blot analysis (Takeuchi *et al.*, 1990).

B. Chemical Methods

Although the enzymatic method to release sugar chains from glycoproteins is mild, the critical point of the above methods is the accessibility of the enzymes to the glycoprotein sugar chains. Most chemical methods cannot save the protein part intact but can release sugar chains of glycoproteins quantitatively. Therefore, for structural study of the sugar chains of glycoproteins, these methods are considered to be more suitable.

The best chemical method for releasing N-linked sugar chains from glycoproteins as oligosaccharides is probably hydrazinolysis. Hydrazine was initially used for determining the carboxyl terminal amino acid, and later this reagent was also employed to cleave the GlcNAc–Asn linkage and *N*-acetylglucosaminyl linkages (Bayard *et al.*, 1974). However, this method included several serious problems such as modification of the reducing terminal of the released oligosaccharides. Subsequently, specific cleavages of the GlcNAc–Asn linkage and quantitative recovery of the released N-linked sugar chains by hydrazinolysis were accomplished by careful treatment (Mizuochi *et al.*, 1978).

The principle of hydrazinolysis is that the glycosylamine linkage is cleaved by heating in anhydrous hydrazine at 100°C for 10 hr while glycosidic linkages remain untouched. By this procedure, all N-linked sugar chains are released from the polypeptide backbone. Although the liberated sugar chains remain linked to hydrazine and all N-acetyl amino-sugar residues, including N-acetylneuraminic acid residues, are de-N-acetylated, the modified oligosaccharide derivatives can be recovered as original oligosaccharides by exhaustive N-acetylation (Takasaki *et al.*, 1982) as follows.

$$\text{—CH=N—NH}_2 \xrightarrow{N\text{-acetylation}} \text{—CH=N—NHAc} \xrightarrow{H_2O} \text{CH=O} + \text{H}_2\text{NNHAc}$$

However, the glycolyl group of *N*-glycolyl neuraminic acid is also susceptible to hydrazinolysis. Therefore, identification of the sialic acid residues must be undertaken using the intact glycoprotein. After N-acetylation, there is no danger of degradation of the reducing termini of the released sugar chains, and they can be used for structural studies. The presence of O-linked sugar chains in the sample may cause problems, because a part of the O-linked sugar chains is also released by an unknown mechanism during hydrazinolysis. Therefore, special care must be taken when the glycoprotein sample contains such sugar chains.

Oligosaccharide mixtures, obtained by hydrazinolysis, are converted to tritium-labeled alcohols by reduction with NaB^3H_4. Since NaB^3H_4 with very high specific activity is commercially available, even picomole quantities of oligosaccharide can be labeled and used for structural analyses.

O-linked sugar chains are released from polypeptides by mild heating in alkaline solution. In order to protect released oligosaccharides from the peeling reaction, a high concentration of $NaBH_4$ is included in the reaction mixture. The most widely used condition to release O-linked sugar chains is to incubate glycoproteins in 0.05 *N* NaOH solution containing 1 *M* $NaBH_4$ at 60°C for 48 hr (Carlson, 1968). Under these conditions, approximately 70–80% of the O-linked sugar chains of most glycoproteins are released as oligosaccharides.

IV. Fractionation of Oligosaccharides

The released oligosaccharide mixture obtained by hydrazinolysis and alkaline-borohydride treatment can be fractionated either by paper electrophoresis or by ion-exchange column chromatography.

High-voltage paper electrophoresis is routinely conducted with pyridine acetate buffer, pH 5.4, at 89 V/cm for 1.5 hr. The radioactive peaks are detected by the radiochromatoscanning of the paper, and the radioactive oligosaccharides included in each radioactive peak are recovered by eluting with water.

Anion-exchange column chromatography is carried out using a Mono Q HR5/5 column. Oligosaccharide samples applied to the column are eluted with 5 m*M* sodium acetate buffer, pH 4.0, then with a linear gradient from 5 to 300 m*M* sodium acetate buffer, pH 4.0, and finally with 500 m*M* sodium acetate buffer, pH 4.0. Radioactive fractions are pooled, passed through a Bio-Rad 50W-X12 (H^+ form) column to remove the salt, and evaporated to dryness.

Neutral and desialylated-neutral oligosaccharides can be fractionated routinely by Bio-Gel P-4 (under 400 mesh) column chromatography (Yamashita *et al.*, 1982b). Paper chromatography, in some cases, is very effective for separating isomeric oligosaccharides but is time-consuming (Kobata, 1972; Kobata *et al.*, 1978). Inclusion of a dextran hydrolysate as an internal standard in the oligosaccharide mixture gives information about the size of each oligosaccharide in a mixture. The sizes of oligosaccharides, expressed as glucose units, are important for estimating the oligosaccharide structure by the sequential exoglycosidase digestion method as will be described next.

V. Structural Analysis of Oligosaccharides

A. *Sequential Exoglycosidase Digestion*

In contrast to polypeptides and polynucleotides, which are linear molecules, sugar chains can be branched. This situation makes the sequencing of sugar chains more complicated than sequencing of proteins and nucleic acids. The most widely used method of determining the monosaccharide sequence of sugar chains is the sequential exoglycosidase digestion. Exoglycosidases are hydrolases that remove monosaccharide residues from the nonreducing termini of sugar chains. The enzymes have two kinds of specificities, the glycon specificity and the aglycon specificity, twoard their substrates. The glycon specificity is directed to the monosaccharide to be removed, including its anomeric configuration. Therefore, the enzymes that hydrolyze β-galactosyl linkages are called β-galactosidases. Although the glycon specificity toward the monosaccharide moiety is strict in many cases, some enzymes show wider specificities. For example, the enzymes that cleave β-*N*-acetylglucosaminyl linkages also hydrolyze β-*N*-acetylgalactosaminyl linkages and are therefore called β-*N*-acetylhexosaminidases.

The principle of sequencing of monosaccharides in sugar chain depends on the glycon specificities of exoglycosidases. For example, the Galα1→3Galβ1→4GlcNAcβ1→ group releases a galactose when incubated with α-galactosidase. No other exoglycosidases can act on this trisaccharide sequence. After removal of the α-galactose residue, however, the newly exposed β-galactose residue can be removed by β-galactosidase digestion. Accordingly, the trisaccharide group can be removed only by sequential digestion with α-galactosidase, β-galactosidase, and β-*N*-acetylhexosaminidase. In many cases, a sugar chain may contain more than two nonreducing termini because of branching as follows.

```
Galβ1→4GlcNAcβ1↘6
                 Manα1↘6
Galβ1→4GlcNAcβ1↗2       Manβ1→4GlcNAcβ1→4GlcNAc→Asn–peptide
Galβ1→4GlcNAcβ1→ 2Manα1↗3
```

Therefore, the amount of each monosaccharide released must be determined after each exoglycosidase digestion by an appropriate method.

Oligosaccharides with reducing termini can be labeled by reduction with NaB^3H_4. Since the tritium label is located at the reducing end of the oligosaccharide and exoglycosidase digestion releases a monosaccharide from the other end of the sugar chain, the amount of monosaccharide released by each exoglycosidase digestion can be estimated by analyzing the change in the effective size of the radioactive oligosaccharide by gel permeation chromatography before and after the enzyme digestion. For this purpose, column chromatography with use of Bio-Gel P-4 (under 400 mesh), which can separate most of the sugar chains of glycoproteins, has been developed (Yamashita *et al.*, 1982b). The radioactive digestion product can be then used for the next exoglycosidase treatment.

For example, one of the complex-type sugar chains released from recombinant factor VIII by hydrazinolysis was sequenced as follows

(Hironaka *et al.*, 1992). After reduction with NaB^3H_4 and separation on a column containing immobilized *Griffonia simplicifolia* isolectin-I, the oligosaccharide was eluted at the position equivalent to 15.2 glucose units by Gio-Gel P-4 column chromatograph (Fig. 4A). When the oligosaccharide in Fig. 4A was digested with coffee bean α-galactosi-

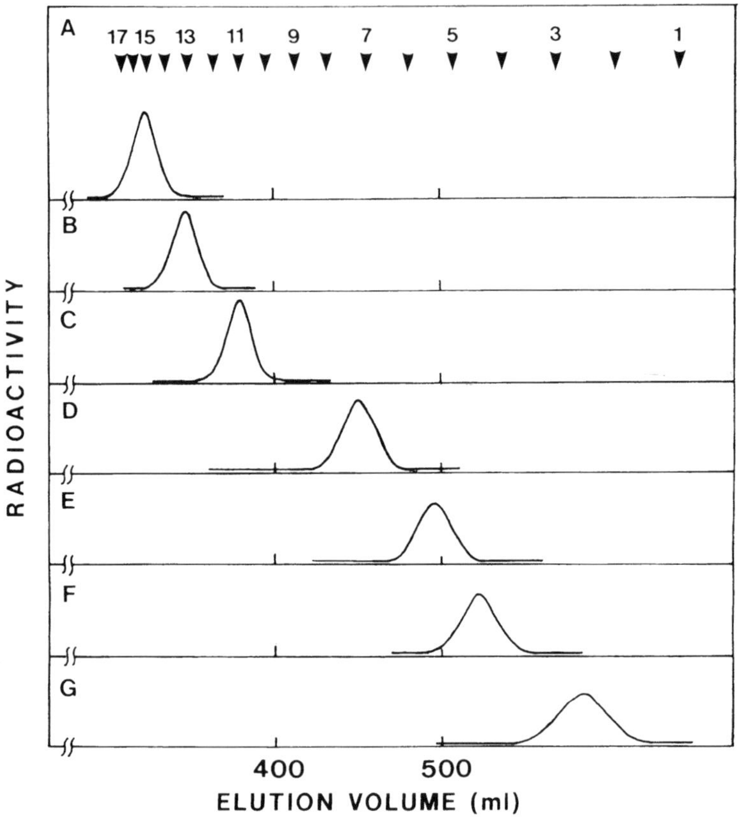

FIG. 4. Sequential exoglycosidase digestion of the oligosaccharides of recombinant human blood coagulation factor VIII (Hironaka *et al.*, 1992, *J. Biol. Chem.* **267,** in press). The oligosaccharide and its digestion products were analyzed by Bio-Gel P-4 column chromatography: (A) the radiolabeled oligosaccharide recovered from a GS-I–agarose column in the bound fraction; (B) the oligosaccharide in A digested with coffee bean α-galactosidase; (C) the oligosaccharide in B digested with diplococcal β-galactosidase; (D) the oligosaccharide in C digested with diplococcal β-*N*-acetylhexosaminidase; (E) the oligosaccharide in E digested with jack bean α-mannosidase; (F) the oligosaccharide in E digested with snail β-mannosidase; and (G) the oligosaccharide in F digested with jack bean β-*N*-acetylhexosaminidase. Arrowheads at the top of the figure indicate the elution positions of glucose oligomers used as internal standards, and the numbers indicate the glucose units.

dase, a radioactive product eluting at 13.2 glucose units was obtained (Fig. 4B). Since the effective size of most hexoses is approximately 1.0 glucose unit, the result indicates that two galactose residues were removed by the α-galactosidase digestion. The oligosaccharide in Fig. 4A was totally resistant to diplococcal β-galactosidase. However, the radioactive oligosaccharide in Fig. 4B was hydrolyzed by the enzyme and converted to a radioactive product with the mobility of 11.2 glucose units (Fig. 4C), indicating that two galactose residues were removed. The radioactive oligosaccharide in Fig. 4C was hydrolyzed by diplococcal β-*N*-acetylhexosaminidase and converted to a radioactive product with the mobility of 7.2 glucose units (Fig. 4D). Since one *N*-acetylglucosamine residue behaves in many cases as approximately 2.0 glucose units, the result indicates that two *N*-acetylglucosamine residues were removed by the β-*N*-acetylhexosaminidase digestion. The radioactive oligosaccharide was then sequentially digested by incubating with jack bean α-mannosidase (Fig. 4E), snail β-mannosidase (Fig. 4F), and jack bean β-*N*-acetylhexosaminidase (Fig. 4G), releasing two mannoses, one mannose, and one *N*-acetylglucosamine, respectively, and was finally converted to *N*-acetyl-[^{3}H]glucosaminitol. These results indicated that the structure of the oligosaccharide can be written as (Galα1→Galβ1→GlcNAc)$_2$(Manα1→)$_2$Man β1→GlcNAcβ 1→GlcNAc$_{\mathbf{OT}}$, where the subscript **OT** indicates NaB^{3}H$_4$-reduced oligosaccharides.

Sequencing of the oligosaccharides by the approach described above can be done with using 10 μg or even less of sample. However, since the principle of this method is based solely on the substrate specificity of each enzyme, special care must be taken to exclude contamination of other exoglycosidases in the enzyme preparations.

Since O-linked sugar chains obtained from the polypeptide backbone by β-elimination have been already reduced, they cannot be labeled as in the case of N-linked sugar chains released by hydrazinolysis. In order to overcome this bottleneck, NaB^{3}H$_4$ was added in the β-elimination reaction mixture in order to obtain labeled oligosaccharides (Aminoff *et al.*, 1980). This method, however, required an extremely high concentration of NaB^{3}H$_4$ to protect the released oligosaccharides from the peeling reaction. Production of large amounts of ^{3}H$_2$ gas has also hampered the frequent use of this method for labeling O-linked sugar chains. Amano *et al.* (1988) devised a new method to label the oligosaccharides released from glycoproteins by alkaline-NaBH$_4$ treatment. The oligosaccharide alcohol mixture was de-N-acetylated by heating in anhydrous hydrazine at 106°C for 24 hr. Hydrazine was removed by repeated evaporation with toluene, and the residue was N-acetylated with either ^{3}H acetic anhydride or ^{14}C acetic anhydride in 0.1 *M*

Na_2–CO_3–$NaHCO_3$ buffer, pH 10.7. Since all O-linked sugar chains contain at least one N-acetylhexosamine residue and the total amount of amino-sugars in an oligosaccharide sample can be estimated by analysis of its sugar composition, the de-N-acetylation method can be used for radiolabeling O-linked sugar chains without producing excessive radioactive waste. Structural analysis of the labeled O-linked oligosaccharides thus obtained can be performed by sequential exoglycosidase digestion in a manner similar to that described for N-linked oligosaccharides.

B. Aglycon Specificity of Exoglycosidases

Many exoglycosidases exhibit aglycon specificity for the sugar chain structures to which the monosaccharide to be hydrolyzed is linked. This specificity is quite useful for the structural study of sugar chains, because the enzymatic digestion will afford information as to the structure of the sugar chains to which the target monosaccharide is linked. Aglycon specificities of exoglycosidases, which are useful for the structural study of the sugar chains, will be described.

1. α-MANNOSIDASES

Four types of α-mannosyl linkages, the Manα1→2Man, Manα1→3Man, Manα1→4Man, and Manα1→6Man linkages, in the sugar chains of glycoproteins are currently known. *Aspergillus saitoi* contains two α-mannosidases with different aglycon specificities (Amano and Kobata, 1986). α-Mannosidase I cleaves only the Manα1→2Man linkage (Yamashita *et al.*, 1980). Therefore, it has been used as an effective tool to identify high-mannose-type sugar chains, because all of them will be converted to a heptasaccharide, Manα1→6(Manα1→3)Manα1→6(Manα1→3)Manβ1→4GlcNAcβ1→4GlcNAc, by incubation with this enzyme. Since the heptasaccharide can be separable from most of the complex-type and hybrid-type oligosaccharides, the total amount of the high-mannose-type oligosaccharides in a very complex mixture of oligosaccharides released from glycoproteins can be easily determined with use of this α-mannosidase. Another enzyme, α-mannosidase II from the same source, releases one mannose residue from the *R*→Manα1→6(Manα1→3)Manβ1→4GlcNAcβ1→4$GlcNAc_{OT}$ group but not from the isomeric Manα1→6(*R*→Manα1→3)Manβ1→4GlcNAcβ1→4$GlcNAc_{OT}$ group, where *R* represents sugars (Amano and Kobata, 1986). Therefore, this enzyme is useful for assigning the position of an outer branch in monoantennary-complex-type sugar chains. Since jack bean α-mannosidase cleaves all α-mannosyl linkages, it can be used to determine the number of all α-mannosyl residues in a sugar chain.

2. β-GALACTOSIDASES

The enzymes purified from two sources have been widely used for the structural study of sugar chains of glycoproteins. The enzyme from jack bean meal cleaves the Galβ1→4GlcNAc linkage 50 times faster than the Galβ1→3GlcNAc linkage when the enzyme concentration is limited to 0.6 unit/ml. At higher enzyme concentration, it cleaves both linkages readily (Arakawa *et al.,* 1974). The aglycon specificity of the β-galactosidase from *D. pneumoniae* is narrower. It cleaves the Galβ1→4GlcNAc linkage but not the Galβ1→3GlcNAc linkage even at high enzyme concentration (Paulson *et al.,* 1978). Therefore, jack bean enzyme is suitable for determining the total β-galactosyl residues at the nonreducing termini of sugar chains and diplococcal enzyme is used to determine the amount of the Galβ1→4GlcNAc groups.

3. β-*N*-ACETYLHEXOSAMINIDASES

β-*N*-Acetylhexosaminidase purified from jack bean meal has wide aglycon specificity and cleaves all β-*N*-acetylglucosaminyl linkages. In contrast, the enzyme from *D. pneumoniae* has a very narrow aglycon specificity. Among the five β-*N*-acetylglucosaminyl linkages shown below, which are often found in the complex-type oligosaccharides, the enzyme cleaves only the GlcNAcβ1→2Man linkage (Yamashita *et al.,* 1981).

```
GlcNAcβ1↘6
          Manα1↘
GlcNAcβ1↗2       6
          GlcNAcβ1→4Manβ1→4GlcNAcβ1→4GlcNAc
GlcNAcβ1↘4       3
          Manα1↗
GlcNAcβ1↗2
```

However, the enzyme cannot cleave the GlcNAcβ1→2Man linkage of the GlcNAcβ1→6(GlcNAcβ→2)Man group and of the GlcNAcβ1→2Manα1→6(GlcNAcβ1→4)Man group, although it cleaves the GlcNAcβ1→2 linkage of the GlcNAcβ1→4(GlcNAcβ1→2)Man group (Table III). Therefore, these two β-*N*-acetylhexosaminidases are very useful tools for determining the structures of complex-type oligosaccharides.

4. α-L-FUCOSIDASES

α-L-Fucosidases purified from *Aspergillus niger* (Bahl, 1970), *C. perfringens* (Aminoff and Furukawa, 1970), and *Bacillus fulminans* (Kochibe, 1973) do not hydrolyze synthetic substrates. They cleave

TABLE III

ACTION OF DIPLOCOCCAL β-*N*-ACETYLHEXOSAMINIDASE ON VARIOUS OLIGOSACCHARIDES

Structures of oligosaccharides[a]
GlcNacβ1→2Manα1→3Manβ1→4GlcNAc$_{OT}$
GlcNAcβ1→2Manα1→6Manβ1→4GlcNAc$_{OT}$
GlcNAcβ1→4Manα1→3Manβ1→4GlcNAc$_{OT}$
GlcNAcβ1→6Manα1→6Manβ1→4GlcNAc$_{OT}$
GlcNAcβ1→2Manα1↘6 Manβ1→4GlcNAc$_{OT}$ GlcNAcβ1→2Manα1↗3
GlcNAcβ1↘6 Manα1↘6 GlcNAcβ1↗2 Manβ1→4GlcNAc$_{OT}$ GlcNAcβ1→2Manα1↗3
GlcNAcβ1↘6 Manα1↘ GlcNAcβ1↗2 6 Manβ1→4GlcNAc$_{OT}$ GlcNAcβ1↘4 3 Manα1↗ GlcNAcβ1↗2
GlcNAcβ1↘6 Manα1→6Manβ1→4GlcNAc$_{OT}$ GlcNAcβ1↗2
Manα1↘6 Manα1↘ Manα1↗3 6 GlcNAcβ1→4Manβ1→4GlcNAc$_{OT}$ 3 GlcNAcβ1→2Manα1↗
Manα1→3Manα1↘ 6 GlcNAcβ1→4Manβ1→4GlcNAc$_{OT}$ GlcNAcβ1↘4 3 Manα1↗ GlcNAcβ1↗2
GlcNAcβ1→2Manα1↘ 6 GlcNAcβ1→4Manβ1→4GlcNAc$_{OT}$ 3 GlcNAcβ1→2Manα1↗

[a] The underlined *N*-acetylglucosamine residues are removed by the enzyme.

the Fucα1→2Gal linkage but not the Fucα1→3GlcNAc, Fucα1→4GlcNAc, and Fucα1→6GlcNAc linkages. Two α-L-fucosidases with different aglycon specificities were purified from almond emulsin (Ogata *et al.*, 1977). α-L-Fucosidase I cleaves the Fucα1→3GlcNAc and the Fucα1→4GlcNAc linkages, but not the Fucα1→2Gal and

Fucα1→6GlcNAc linkages and synthetic substrates. α-L-Fucosidase II does not act on synthetic substrates, the Fucα1→3GlcNAc, Fucα1→4GlcNAc, and Fucα1→6GlcNAc linkages, but cleaves the fucosyl linkage of Fucα1→2Galβ1→4Glc. However, the enzyme cannot release fucose from Fucα1→2Galβ1→3GlcNAcβ1→3Galβ1→4Glc. This evidence indicated that the aglycon specificity of α-L-fucosidase II is similar to but not the same as those of bacterial α-L-fucosidases described above. α-L-Fucosidases purified from *Turbo cornutus, Charonia lampas,* and *Chamelea gallina* cleave all α-fucosyl linkages so far known (Nishigaki *et al.*, 1974; Reglero *et al.*, 1976).

5. Sialidases

The enzymes from *C. perfringens* (Cassidy *et al.*, 1965), *Vibrio chorela* (Schramm and Mohr, 1959), *Arthrobacter ureafacience* (Uchida *et al.*, 1974), and influenza virus (Gottschalk, 1957) are commercially available. Among them, the enzyme from *A. ureafacience* is the most useful for the purpose of sequential exoglycosidase digestion because it cleaves all sialyl linkages so far reported. Sialidase from Newcastle disease virus (NDV) was reported to cleave the Siaα2→3Gal linkage but not the Siaα2→6Gal/GalNAc linkage of glycoproteins (Paulson *et al.*, 1982) as long as the latter structure is unsubstituted (Amano *et al.*, 1988). Therefore, NDV sialidase can be used to assign the Siaα2→3Gal group in a sugar chain.

VI. Methylation Analysis

Methylation analysis is an essential procedure for the structural study of the sugar chains of glycoproteins. As described in the previous section, some of the glycosidic linkages in an oligosaccharide can be deduced using the aglycon specificities of exoglycosidases. Yet methylation analysis is the most reliable method for the elucidation of the whole structure of an oligosaccharide.

The principle of methylation analysis of sugar chains is to completely methylate the free hydroxyl groups of an oligosaccharide by an appropriate method. The permethylated oligosaccharide will then be hydrolyzed, and the partially O-methylated monosaccharides are quantitatively determined after being separated by an appropriate method. Therefore, effective and complete methylation is essential for obtaining reliable data. A mixture of methylsulfinyl carbanion and methyl iodide as a methylation reagent can convert any oligosaccharides effectively to permethylated oligosaccharides (Hakomori, 1964). Inclusion of labeled oligosaccharides reduced with NaB^3H_4 will help to estimate the recovery of permethylated samples during the purification.

The permethylated oligosaccharide is hydrolyzed in 93% acetic acid containing 0.5 *N* H_2SO_4 at 76°C for 4 hr (Li *et al.*, 1978). In some cases, the hydrolysis time must be extended up to 16 hr because of the relative resistance of the highly branched complex-type oligosaccharides (Yamashita *et al.*, 1982a). The hydrolyzed products are then converted to partially O-methylated alditol acetates and partially O-methylated 2-*N*-methylacetamido-2-deoxyalditol acetates by reduction with $NaBH_4$ followed by acetylation with acetic anhydride (Sawardeker *et al.*, 1965).

Separation and quantitation of various partially O-methylated alditol acetates can be performed by gas chromatography using a capillary column coated with bonded-phase fused silica such as DB-1 and DB-5 (J & W Scientific, Rancho Cordova, CA), Ultra 1 and Ultra 2 (Hewlett–Packard, CA), and SPB-1 and SPB-5 (Supelco, Bellefonte, PA). In Fig. 5, the data of methylation analysis of the oligosaccharide described in Fig. 4 are shown. It is worthy to note that the fractionation pattern was obtained by using only 0.2–1 nmol of sample.

VII. Fractionation and Structural Estimation of Oligosaccharides by Lectin Column Chromatography

The use of immobilized lectin columns is becoming an indispensable tool for group separation of the N- and O-linked sugar chains of glycoproteins. The first trial of the lectin column for the structural study of the sugar chains of glycoproteins was carried out by Ogata *et al.* (1975). Study of the interaction of various radiolabeled N-linked oligosaccharides and oligomannosides to a concanavalin A (Con A)–Sepharose column revealed that the presence of at least two unsubstituted or 2-O-substituted α-mannosyl residues in a sugar chain are required for its retention on the lectin column. All complex-type sugar chains contain a pentasaccharide, Manα1→6(Manα1→3)Manβ1→4GlcNAcβ1→4GlcNAc, as a common core that has two α-mannose residues, and the core itself binds to a Con A–Sepharose column. In the case of a biantennary-complex-type sugar chain, each outer chain is linked at the C-2 position of each α-mannose residue and still interacts with Con A. In tri- and tetraantennary-complex-type sugar chains, however, the α-mannose residue(s) is substituted with two outer chains, and they do not interact with Con A. Therefore, biantennary-complex-type sugar chains bind to a Con A column, but tri- and tetraantennary-complex-type sugar chains pass through the column. High-mannose-type sugar chains, which have many Manα1→ residues and →2Manα1→ residues, will bind more tightly to a Con A–Sepharose column. Furthermore,

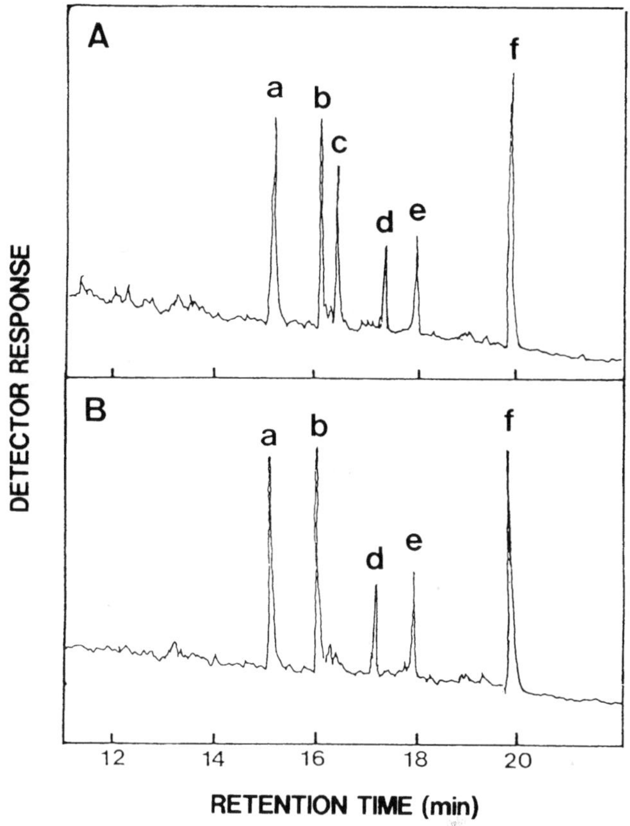

FIG. 5. Total ion chromatograms of partially O-methylated alditol acetates and amino-alditol acetates obtained by methylation analysis of the oligosaccharide described in Fig. 4 before (A) and after (B) coffee bean α-galactosidase digestion: (a) 2,3,4,6-tetra-*O*-methylgalactitol; (b) 3,4,6-tri-*O*-methylmannitol; (c) 2,4,6-tri-*O*-methylgalactitol; (d) 1,3,5,6-tetra-*O*-methyl-2-*N*-methylacetamido-2-deoxyglucitol; (e) 2,4-di-*O*-methyl-mannitol; and (f) 3,6-di-*O*-methyl-2-*N*-methyl-acetamido-2-deoxyglucitol. The chromatography was performed with use of a 30-m DB-5 capillary column programmed from 90 to 260°C at 8°C per minute.

Narasimhan *et al.* (1986) reported that bisected biantennary sugar chains do not bind to a Con A–Sepharose column.

Bisected biantennary oligosaccharides, with the sequence Galβ1→4GlcNAcβ1→, bind to an erythroagglutinin (E-PHA)–agarose column (Irimura *et al.*, 1981), whereas nonbisected sugar chains do not (Cummings and Kornfeld, 1982). Study of the interaction of various bisected oligosaccharides to an E-PHA–agarose column at room temperature (Yamashita *et al.*, 1983) revealed that the minimum structure required for binding to the column is as follows.

Galβ1→4GlcNAcβ1→2Manα1↘
                                              6
                              GlcNAcβ1→4Manβ1→4GlcNAcβ1→4R_3
                   R_2↘4   ↗3
                              Manα1
        R_1→4GlcNAcβ1↗2

In the structure, R_1 and R_2 represent either hydrogen or sugars and R_3 represents either (±Fucα1→6)GlcNAc→Asn or (±Fucα1→6)-GlcNAc$_{OT}$. With this basic information, the immobilized lectin column can be used as an effective tool to fractionate and assign the structures of bisected complex-type sugar chains.

Complex-type and hybrid-type sugar chains can be classified by the presence or absence of an α-fucosyl residue linked at the C-6 position of the proximal *N*-acetylglucosamine residue of the trimannosyl core. These fucosylated sugar chains can be easily separated from their non-fucosylated counterparts by passing them through an immobilized *Aleuria aurantia* lectin (AAL) column (Yamashita *et al.*, 1985). Since the binding of oligosaccharides to an AAL–Sepharose column is not affected by the branching or the presence of bisecting *N*-acetylglucosamine residue, the fractionation can be much more effective than immobilized lentil and pea lectin columns (Kornfeld *et al.*, 1981).

Complex-type sugar chains often show variation in their branching. Although mono- and biantennary-complex-type sugar chains can be separated from highly branched oligosaccharides by a Con A–Sepharose column, tri- and tetraantennary oligosaccharides cannot be separated by this column. An immobilized *Datura stramonium* agglutinin (DSA) column can be effectively used to fractionate these highly branched oligosaccharides. Study of the behavior of various radiolabeled oligosaccharides (Yamashita *et al.*, 1987) revealed that oligosaccharides that contain the Galβ1→4GlcNAcβ1→4(Galβ1→4GlcNAcβ1→2)Man group are retarded in the column as long as the pentasaccharide group is not substituted with other sugars. Oligosaccharides containing an unsubstituted Galβ1→4GlcNAcβ1→6(Galβ1→4GlcNAcβ 1→2)Man group bind to the column and are eluted with the buffer containing *N*-acetylglucosamine oligomers. The binding specificity is not affected by the structure of the inner core portion or by the presence of a bisecting *N*-acetylglucosamine residue. Therefore, the column can fractionate a mixture of complex-type sugar chains into three fractions: the pass-through fraction contains mono- and biantennary-complex-type oligosaccharides, the retarded fraction contains triantennary-complex-type oligosaccharides with the Galβ1→4-GlcNAcβ1→4(Galβ1→4GlcNAcβ1→2)Man group, and the bound fraction contains triantennary oligosaccharides with the Galβ1→4GlcNAcβ1→6(Galβ1→4GlcNAcβ1→2)Man group and tetraantennary oligosaccharides. Complex-type oligosaccharides with at least one Gal-β1→4GlcNAcβ1→3Galβ1→4GlcNAc group in their outer chain moieties also bind to the column and are eluted with the buffer containing *N*-acetylglucosamine oligomers. They can be descriminated from the bound tri- and tetraantennary oligosaccharides, however, by their change in behavior on the column after digestion with *E. freundii* or *F. keratolyticus* endo-β-galactosidase.

The structural requirements for binding of oligosaccharides to lectin columns enable us to utilize lectin column chromatography in the structural analyses of oligosaccharides. For example, the structures of N-linked oligosaccharides can be estimated as follows (Fig. 6A). A mixture of neutral oligosaccharides can be fractionated by Con A–Sepharose column chromatography into three fractions: the pass-through fraction (Con A^-), the bound fraction eluted with 5 m*M* α-methyl-D-glucopyranoside (Con A^+ or fraction II), and the bound fraction eluted with 100 m*M* α-methyl-D-mannopyranoside (Con A^{++} or fraction I). Since the Con A^- fraction contains highly branched and/or bisected oligosaccharides, this fraction can be further fractionated by sequential column chromatography with an E-PHA–agarose column and a DSA–Sepharose column into three fractions: E-PHA^r (fraction III), E-PHA^-·DSA^r (fraction V), and E-PHA^-·DSA^+ (fraction IV) as shown in Fig. 6A. The structures of oligosaccharides in each fraction can be estimated based on their binding specificities toward the three lectins (Fig. 6B). These oligosaccharides can be further separated into oligosaccharides with and without the Fucα1→6$GlcNAc_{OT}$ group by passing them through an AAL–Sepharose column.

Serial lectin column chromatography is also applicable to the separation and structural analysis of O-linked sugar chains. For example, the separation of O-linked sugar chains can be performed by the following procedure (Fig. 7A). A mixture of neutral oligosaccharides i–v shown in Fig. 7B, released from human milk galactosyltransferase by alkaline-borohydride treatment and radiolabeled with ^{3}H acetic anhydride (Amano *et al.*, 1991), was successfully fractionated by sequential chromatography of columns immobilized with *Ulex europaeus* agglutinin-I (UEA-I), AAL, and *Ricinus communis* agglutinin-I (RCA-I). Since oligosaccharides with the Fucα1→2Gal group interact with UEA-I, oligosaccharides ii, iv, and v in Fig. 7B can interact with UEA-I. However, the substitution with the α-galactose at the C-3 position of the fucosylated β-galactose residue as found in oligosaccharide v prevents

A

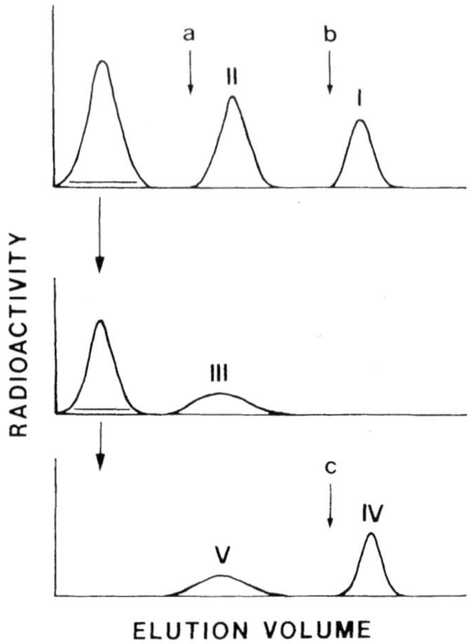

B

Fractions	types of oligosaccharides
I	high mannose-type, hybrid-type
II	biantennary complex-type
III	bisected bi-and 2,4-branched tri-antennary complex-type
IV	non-bisected tri- and tetra-antennary complex-type and bisected 2,6-branched tri- and tetra-antennary complex-type
V	non-bisected 2,4-branched triantennary complex-type

FIG. 6. Structural estimation of an N-linked oligosaccharide mixture by serial lectin column chromatography. A neutral oligosaccharide mixture can be separated on a Con A–Sepharose column. The fraction that passes through the column can be further fractionated by sequential chromatography on an E-PHA–agarose column and a DSA–Sepharose column. The five fractions thus obtained are named fractions I–V (A) and contain oligosaccharides with the structures listed in B. Arrows a, b, and c indicate the positions where the elution buffers are changed to buffer containing 5 m*M* α-methyl-D-glucopyranoside (a), 100 m*M* α-methyl-D-mannopyranoside (b), and 1% *N*-acetylglucosamine oligomers (c). The ordinate of each figure represents radioactivity.

A

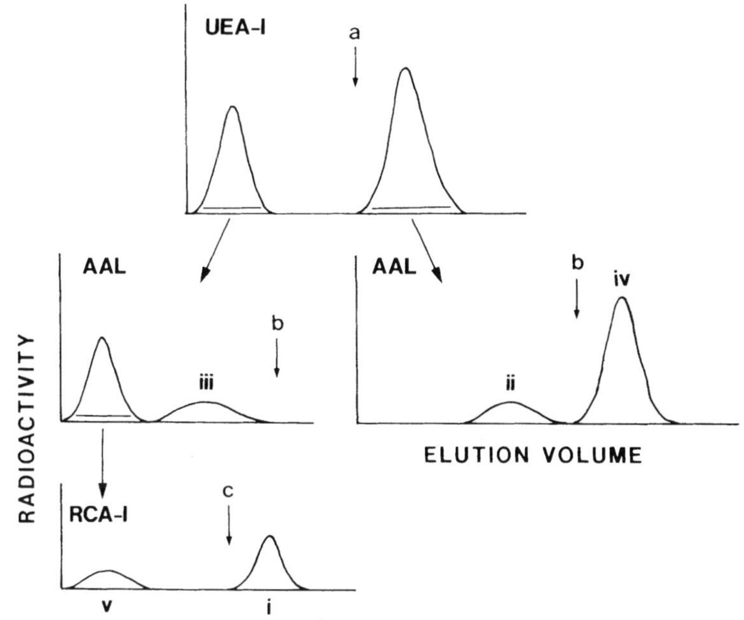

B

Oligosaccharides	structures
i	Galβ1→4GlcNAcβ1→6(Galβ1→3)GalNAc
ii	Fucα1→2Galβ1→4GlcNAcβ1→6(Galβ1→3)GalNAc
iii	Galβ1→4(Fucα1→3)GlcNAcβ1→6(Galβ1→3)GalNAc
iv	Fucα1→2Galβ1→4(Fucα1→3)GlcNAcβ1→6(Galβ1→3)GalNAc
v	Fucα1→2(Galα1→3)Galβ1→4GlcNAcβ1→6(Galβ1→3)GalNAc

FIG. 7. Fractionation of the O-linked oligosaccharides from human milk galactosyltransferase by serial lectin column chromatography. Oligosaccharides i–v, obtained from the sample whose structures are shown in B, were applied to an UEA-I–agarose column. The fractions that passed through and bound to the column were further fractionated by an AAL–Sepharose column. The fraction that passed through the two columns was finally applied to an RCA-I–agarose column. The detailed conditions of each chromatography and the carbohydrate structures are described in *J. Biol. Chem.*, 1991, **266,** 11461.

binding of the oligosaccharide to the column. On the other hand, oligosaccharides with type 2H (Fucα1→2Galβ1→4GlcNAc/Glc) and X [Galβ1→4(Fucα1→3)GlcNAc] determinants such as oligosaccharides ii and iii interact weakly with AAL. Oligosaccharides with the Galβ1→4GlcNAc group, which occurs in oligosaccharide i, interact with RCA-I. Based on the lectin binding specificities of these oligosaccharides, oligosaccharide i was recovered in fraction UEA-I^-·AAL^-·RAC-I^+; oligosaccharide ii, in fraction UEA-I^+·AAL^-; oligosaccharide iii, in UEA-I^-·AAL^+; oligosaccharide iv, in fraction UEA-I^+·AAL^+; and oligosaccharide v, in fraction UEA-I^-·AAL^-·RCA-I^- (Fig. 7A). Conversely, the structures of oligosaccharides recovered in each fraction separated by a serial lectin column chromatography can be estimated by their binding specificities to the lectins.

Although the structural information obtained by lectin column chromatography is preliminary, such information can be obtained using only 500–1000 cpm of each oligosaccharide and is helpful for planning further structural analyses, such as sequential exoglycosidase digestion.

In recent years, high-resolution NMR spectroscopy and fast-atom bombardment mass spectrometry have been introduced for the structural study of sugar chains (Montreuil and Vliegenthart, 1979; Vliegenthart *et al.*, 1981). However, they still require significantly large amounts of materials to be analyzed. The use of methods described in this chapter in various combinations will complement the NMR and mass spectroscopic approaches and enable us to obtain structural information for N- and O-linked sugar chains even from small quantities of glycoproteins.

Acknowledgments

The authors acknowledge the excellent secretarial assistance of Yumiko Kimizuka. The present work was supported by a Grant-in-Aid for Specially Promoted Research from the Ministry of Education, Science, and Culture of Japan.

References

Amano, J., and Kobata, A. (1986). Purification and characterization of a novel α-mannosidase from *Aspergillus saitoi*. *J. Biochem.* (*Tokyo*) **99,** 1645–1654.

Amano, J., Nishimura, R., Mochizuki, M., and Kobata, A. (1988). Comparative study of the mucin-type sugar chains of human chorionic gonadotropin present in the urine of patients with trophoblastic diseases and healthy pregnant women. *J. Biol. Chem.* **263,** 1157–1165.

Amano, J., Staehl, P., Berger, E. G., Kochibe, N., and Kobata, A. (1991). Structures of mucin-type sugar chains of the galactosyltransferase purified from human milk. *J. Biol. Chem.* **263,** 11461–11477.

Aminoff, D., and Furukawa, K. (1970). Enzymes that destroy blood group specificity. I. Purification and properties of α-L-fucosidase from *Clostridium perfringens*. *J. Biol. Chem.* **245,** 1659–1669.

Aminoff, D., Gathman, W. D., McLean, C. M., and Yodomae, T. (1980). Quantitation of oligosaccharides released by the β-elimination reaction. *Anal. Biochem.* **101,** 44–53.

Arakawa, M., and Muramatsu, T. (1974). Endo-β-*N*-acetylglucosaminidases acting on the carbohydrate moieties of glycoproteins: The differential specificities of the enzymes from *Streptomyces griseus* and *Diplococcus pneumoniae*. *J. Biochem.* (*Tokyo*) **76,** 307–317.

Arakawa, M., Ogata, S., Muramatsu, T., and Kobata, A. (1974). β-Galactosidases from jack bean meal and almond emulsin: Application in the enzymatic distinction of Galβ1→4GlcNAc and Galβ1→3GlcNAc linkages. *J. Biochem.* (*Tokyo*) **75,** 703–714.

Asada, M., Furukawa, K., Kantor, C., Gahmberg, C. G., and Kobata, A. (1991). Structural study of the sugar chains of human leukocyte cell adhesion molecules CD11/CD18. *Biochemistry* **30,** 1561–1571.

Bahl, O. P. (1970). Glycosidases of *Aspergillus niger*. II. Purification and general properties of 1,2-α-L-fucosidase. *J. Biol. Chem.* **245,** 299–304.

Bayard, B., and Montreuil, J. (1974). Methodologie de la structure et du metabolisme des glycoconjugues. *CNRS* (*Paris*) 208–218.

Berg, E. L., Robinson, M. K., Masson, O., Butcher, E. C., and Magnani, J. L. (1991). A carbohydrate domain common to both sialyl Le^a and Le^x is recognized by the endothelial cell leukocyte adhesion molecule ELAM-1. *J. Biol. Chem.* **266,** 14869–14872.

Bevilacqua, M. P., Stengelin, S., Gimbrone, M. A., Jr., and Seed, B. (1989). Endothelial leukocyte adhesion molecule 1: An inducible receptor for neutrophils related to complement regulatory protein and lectin. *Science* **243,** 1160–1165.

Bowen, B. R., Nguyen, T., and Lasky, L. A. (1989). Characterization of a human homologue of the murine peripheral lymph node homing receptor. *J. Cell Biol.* **109,** 421–427.

Camerini, D., James, S. P., Stamenkovic, I., and Seed, B. (1989). Leu-8/TQ1 is the human equivalent of the Mel-14 lympho node homing receptor. *Nature* (*London*) **342,** 78–82.

Carlson, D. M. (1968). Structures and immunological properties of oligosaccharides isolated from pig submaxillary mucins. *J. Biol. Chem.* **243,** 616–626.

Cassidy, T. T., Jourdian, G. W., and Roseman, S. (1965). The sialic acid. VI. Purification and properties of sialidase from *Clostridium perflingens*. *J. Biol. Chem.* **240,** 3501–3506.

Chan, A. L., Morris, H. R., Panico, M., Etienne, A. T., Rogers, M. E., Gaffney, P., Creighton-Kempsford, L., and Dell, A. (1991). A novel sialylated *N*-acetylgalactosamine-containing oligosaccharide is the major complex-type structure present in Bowes melanoma tisue plasminogen activator. *Glycobiology* **1,** 173–185.

Cummings, R. D. and Kornfeld, S. (1982). Characterization of the structural determinants required for the high affinity interaction of asparagine-linked oligosaccharides with immobilized *Phaseolus vulgaris* leukoagglutinating and erythroagglutinating lectins. *J. Biol. Chem.* **257,** 11230–11234.

Elder, T. H., and Alexander, S. (1982). Endo-β-*N*-acetylglucosaminidase F: Endoglycosidase from *Flavobacterium meningosepticum* that cleaves both high-mannose and complex glycoproteins. *Proc. Natl. Acad. Sci. USA* **79,** 4540–4544.

Endo, Y., and Kobata, A. (1976). Partial purification and characterization of an endo-α-*N*-acetylgalactosaminidase from the culture medium of *Diplococcus pneumoniae*. *J. Biochem.* (*Tokyo*) **80,** 1–8.

Foddy, L., Feeney, J., and Hughes, R. C. (1986). Properties of baby-hamster kidney (BHK) cells treated with Swainsonine, an inhibitor of glycoprotein processing. *Biochem. J.* **233,** 697–706.

Fukuda, M., Dell, A., Oates, J. E., and Fukuda, M. M. (1984a). Structures of branched lactosaminoglycan, the carbohydrate moiety of band 3 isolated from adult human erythrocytes. *J. Biol. Chem.* **259,** 8260–8273.

Fukuda, M., Spooncer, E., Oates, J. E., Dell, A., and Klock, J. C. (1984b). Structure of sialylated fucosyl lactosaminoglycan isolated from human granulocytes. *J. Biol. Chem.* **259,** 10925–10935.

Fukushi, N., Muramatsu, H., Uezono, M. M., and Muramatsu, T. (1987). A new endo-β-galactosidase releasing Galα1→3Gal from carbohydrate moieties of glycoproteins and from a glycolipid. *J. Biol. Chem.* **262,** 10086–10092.

Furukawa, K., Roberts, D. D., Endo, T., and Kobata, A. (1989). Structural study of the sugar chains of human platelet thrombospondin. *Arch. Biochem. Biophys.* **270,** 302–312.

Galili, U., Clark, M. R., Shohet, S. B., Buehler, J., and Macher, B. A. (1987). Evolutionary relationship between the natural anti-Gal antibody and the Galα1→3Gal epitope in primates. *Proc. Natl. Acad. Sci. USA* **84,** 1369–1373.

Goelz, S. E., Hession, C., Goff, D., Griffiths, B., Tizard, R., Newman, B., Chi-Rosso, G., and Lobb, R. (1990). ELFT: A gene that directs the expression of an ELAM-1 ligand. *Cell* **63,** 1349–1356.

Gottschalk, A. (1957). Neuraminidase: The specific enzyme of influenza virus and *Vibrio cholerae*. *Biochim. Biophys. Acta* **23,** 645–646.

Green, E. D., and Baenziger, J. U. (1988). Asparagine-linked oligosaccharides on leutropin, follitropin, and thyrotropin. *J. Biol. Chem.* **263,** 25–35.

Hakomori, S. (1964). A rapid permethylation of glycolipid, and polysaccharide catalyzed by methylsulfinyl carbanion in dimethyl sulfoxide. *J. Biochem.* (*Tokyo*) **55,** 205–208.

Hirani, S., Bernasconi, R. J., and Rasmussen, J. R. (1987). Use of *N*-glycanase to release asparagine-linked oligosaccharides for structural anlysis. *Anal. Biochem.* **162,** 485–492.

Hironaka, T., Furukawa, K., Esmon, P. C., Fournel, M. A., Sawada, S., Kato, M., Minaga, T., and Kobata, A. (1992). Comparative studies of the sugar chains of factor VIII purified from human plasma and from the culture media of recombinant baby hamster kidney cells. *J. Biol. Chem.* **267,** 8012–8020.

Huang, C. C., Mayer, H. E., Jr., and Montgomery, R. (1970). Microheterogeneity and paucidispersity of glycoproteins. I. The carbohydrate of chicken ovalbumin. *Carbohydr. Res.* **13,** 127–137.

Imai, Y., Singer, M. S., Fennie, C., Lasky, L. A., and Rosen, S. D. (1991). Identification of a carbohydrate-based endothelial ligand for a lymphocyte homing receptor. *J. Cell Biol.* **113,** 1213–1221.

Imai, Y., True, D. D., Singer, M. S., and Rosen, S. D. (1990). Direct demonstration of the lectin activity of $gp90^{MEL}$, a lymphocyte homing receptor. *J. Cell Biol.* **111,** 1225–1232.

Irimura, T., Tsuji, T., Tagami, S., Yamamoto, K., and Osawa, T. (1981). Structure of a complex-type sugar chain of human glycophorin A. *Biochemistry* **20,** 560–566.

Ito, S., Muramatsu, T., and Kobata, A. (1975a). Endo-β-*N*-acetylglucosaminidase acting on carbohydrate moieties of glycoproteins: Purification and properties of the two enzymes with different specificities from *Clostridium perfringens*. *Arch. Biochem. Biophys.* **171,** 78–86.

Ito, S., Muramatsu, T., and Kobata, A. (1975b). Release of galactosyl oligosaccharides by endo-β-*N*-acetylglucosaminidase D. *Biochem. Biophys. Res. Commun.* **63,** 938–944.

Iwase, H., Ishii, I., Ishihara, K., Tanaka, Y., Omura, S., and Hotta, K. (1988). Release of oligosaccharides possessing reducing-end *N*-acetylgalactosamine from mucus glycoprotein in *Streptomyces* sp. OH-11242 culture medium through action of endo-type glycosidase. *Biochem. Biophys. Res. Commun.* **151,** 422–428.

Iwase, H., Ishii-Karakasa, I., Hotta, K., Tanaka, Y., and Omura, S. (1992). Analysis of porcine gastric mucus glycoprotein added to a culture medium of *Streptomyces* sp. OH-11242 as the only source of carbon. *Comp. Biochem. Phys.* **101,** 651–655.

Johnston, G. I., Cook, R. G., and McEver, R. P. (1989). Cloning of GMP-140, a granule membrane protein of platelets and endothelium: Sequence similarity to protein involved in cell adhesion and inflammation. *Cell* **56,** 1033–1044.

Kawano, T., Endo, T., Nishimura, R., Mochizuki, M., Kochibe, N., and Kobata, A. (1988).

Structural differences found in the sugar chains of eutopic and ectopic free α-subunits of human glycoprotein hormone. *Arch. Biochem. Biophys.* **267,** 787–796.

Kitamikado, M., Ueno, R., and Nakamura, T. (1970). Enzymatic degradation of whale cartilage keratan sulfate. I. Isolation of keratan sulfate degrading bacterium. *Bull. Jpn. Soc. Sci. Fish.* **36,** 592–596.

Kobata, A. (1972). Isolation method of human milk oligosaccharides. *In* "Methods in Enzymology" (V. Ginsburg, ed.), Vol. 28, pp. 262–271. Academic Press, San Diego.

Kobata, A., Yamashita, K., and Tachibana, Y. (1978). Oligosaccharides from human milk. *In* "Methods in Enzymology" (V. Ginsburg, ed.), Vol. 50, pp. 216–220. Academic Press, San Diego.

Kochibe, N. (1973). Purification and properties of α-L-fucosidase from *Bacillus fulminas. J. Biochem.* (*Tokyo*) **74,** 1141–1149.

Kornfeld, S., Reitman, M. L., and Kornfeld, R. (1981). The carbohydrate-binding specificity of pea and lentil lectins. *J. Biol. Chem.* **256,** 6633–6640.

Larsen, E., Palabrica, T., Sajer, S., Gilbert, G. E., Wagner, D. D., Furie, B. C., and Furie, B. (1990). PADGEM-dependent adhesion of platelets to monocytes and neutrophils is mediated by a lineage-specific carbohydrate, LNF III (CD15). *Cell* **63,** 467–474.

Lasky, L. A., Singer, M. S., Yednock, T. A., Dowbenko, D., Fennie, C., Rodriguez, H., Hguyen, T., Stachel, S., and Rosen, S. D. (1989). Cloning of a lymphocyte homing receptor reveals a lectin domain. *Cell* **56,** 1045–1055.

Li, E., Tabas, I., and Kornfeld, S. (1978). The synthesis of complex-type oligosaccharides. *J. Biol. Chem.* **253,** 7762–7770.

Lowe, J. B., Stoolman, L. M., Nair, R. P., Larsen, R. D., Berhend, T. L., and Marks, R. M. (1990). ELAM-1-dependent cell adhesion to vascular endothelium determined by a transfected human fucosyltransferase cDNA. *Cell* **63,** 475–484.

Mizuochi, T., Amano, J., and Kobata, A. (1984). A new evidence about the substrate specificity of endo-β-*N*-acetylhexosaminidase D. *J. Biochem.* (*Tokyo*) **95,** 1209–1213.

Mizuochi, T., Yonemasu, K., Yamashita, K., and Kobata, A. (1978). The asparagine-linked sugar chains of subcomponent C1q of the first component of human complement. *J. Biol. Chem.* **253,** 7404–7409.

Montreuil, J., and Vliegenthart, J. F. K. (1979). Primary structure and conformation of glycans *N*-glycosidically linked to peptide chains. *In* "Glycoconjugate Research" (J. D. Gregory and R. W. Jeanloz, eds.), Vol. I, pp. 35–78. Academic Press, New York.

Nakata, N., Furukawa, K., Greenwalt, D. E., and Kobata, A. (1991). Bovine milk epithelial glycoprotein IV (CD36) contains bi-, tri- and tetra-antennary complex-type sugar chains with the GalNAcβ1→4GlcNAc group. *Glycoconjugate J.* **8,** 250.

Narasimhan, S., Freed, J. C., and Schachter, H. (1986). The effect of a "bisecting" *N*-acetylglucosaminyl group on the binding of biantennary complex oligosaccharides to concanavalin A, *Phaseolus vulgaris* erythroagglutinin (E-PHA), and *Ricinus communis* agglutinin (RCA-120) immobilized on agarose. *Carbohydr. Res.* **149,** 65–83.

Nishigaki, M., Muramatsu, T., Kobata, A., and Maeyama, K. (1974). The broad aglycon specificity of α-L-fucosidase from marine gastropds. *J. Biochem.* (*Tokyo*) **75,** 509–517.

Nyame, K., Smith, D. F., Damian, R. T., and Cummings, R. D. (1989). Complex-type asparagine-linked oligosaccharides in glycoproteins synthesized by *Schistosoma mansoni* adult males contain terminal β-linked *N*-acetylgalactosamine. *J. Biol. Chem.* **264,** 3235–3243.

Ogata, A.-M., Muramatsu, T., and Kobata, A. (1977). α-L-Fucosidase from almond emulsin: Characterization of the two enzymes with different specificities. *Arch. Biochem. Biophys.* **181,** 353–358.

Ogata, S., Muramatsu, T., and Kobata, A. (1975). Fractionation of glycopeptides by affinity column chromatography on concanavalin A–Sepharose. *J. Biochem.* (*Tokyo*) **78,** 687–696.

Paulson, J. C., Prieels, J.-P., Glasgow, L. R., and Hill, R. L. (1978). Sialyl- and fucosyltransferases in the biosynthesis of asparaginyl-linked oligosaccharides in glycoproteins. *J. Biol. Chem.* **253,** 5617–5624.

Paulson, J. C., Weinstein, J., Dorland, L., van Halbeek, H., and Vliegenthart, J. F. G. (1982). Newcastle disease virus contains a linkage-specific glycoprotein sialidase. *J. Biol. Chem.* **257,** 12734–12738.

Phillips, M. L., Nudelman, E., Gaeta, F. C. A., Perez, M., Singhal, A. K., Hakomori, S.-I., and Paulson, J. C. (1990). ELAM-1 mediates cell adhesion by recognition of a carbohydrate ligand, sialyl-Le^x. *Science* **250,** 1130–1132.

Plummer, T. H., Jr., and Tarentino, A. L. (1981). Facile cleavage of complex oligosaccharide from glycopeptides by almond emulsin peptide: *N*-Glycosidase. *J. Biol. Chem.* **256,** 10243–10246.

Reglero, A., and Cabegas, J. A. (1976). Glycosidases from mollusca: Purification and properties of α-L-fucosidase from *Chamelea gallina. Eur. J. Biochem.* **66,** 379–387.

Rosen, S. D., Singer, M. S., Yednock, T. A., and Stoolman, L. M. (1985). Involvement of sialic acid on endothelial cells in organ-specific lymphocyte recirculation. *Science* **228,** 1005–1007.

Sawardecker, J. S., Sloneker, J. H., and Jeanes, A. R. (1965). Quantitative determination of monosaccharides as their acetates by gas–liquid chromatography. *Anal. Chem.* **37,** 1602–1604.

Schramm, G., and Mohr, E. (1959). Purification of neuraminidase from *Vibrio cholerae. Nature* (*London*) **183,** 1677–1678.

Scudder, P., Hanfland, P., Uemura, K., and Feizi, T. (1984). Endo-β-D-galactosidase of *Bacteroides fragilis* and *Escherichia freundii* hydrolyze linear but not branched oligosaccharide domains of glycolipids of the neolacto series. *J. Biol. Chem.* **259,** 6588–6592.

Siegelman, M. H., van de Rijn, M., and Weissman, I. L. (1989). Mouse lympho node homing receptor cDNA clone encodes a glycoprotein revealing tandem interaction domain. *Science* **243,** 1165–1172.

Tai, T., Yamashita, K., and Kobata, A. (1977). The substrate specificities of endo-β-*N*-acetylglucosaminidase C_{II} and H. *Biochem. Biophys. Res. Commun.* **78,** 434–441.

Tai, T., Yamashita, K., Ogata, A. M., Koide, N., Muramatsu, T., Iwashita, S., Inoue, Y., and Kobata, A. (1975). Structural studies of two ovalbumin glycopeptides in relation to the endo-β-*N*-acetylglucosaminidase D. *J. Biol. Chem.* **250,** 8569–8575.

Takahashi, N. (1977). Demonstration of a new amidase acting on glycoproteins. *Biochem. Biophys. Res. Commun.* **76,** 1194–1201.

Takahashi, N., and Nishibe, H. (1978). Some characterization of a new glycopeptidase acting on aspartylglycosylamine linkages. *J. Biochem.* (*Tokyo*) **84,** 1467–1473.

Takamoto, M., Endo, T., Isemura, M., Kochibe, N., and Kobata, A. (1989). Structures of asparagine-linked oligosaccharides of human placental fibronectin. *J. Biochem.* (*Tokyo*) **105,** 742–750.

Takasaki, S., and Kobata, A. (1976). Purification and characterization of an endo-β-galactosidase produced by *Diplococcus pneumoniae. J. Biol. Chem.* **251,** 3603–3609.

Takasaki, S., Mizuochi, T., and Kobata, A. (1982). Hydrazinolysis of asparagine-linked sugar chains to produce free oligosaccharides. *In* "Methods in Enzymology" (V. Ginsburg, ed.), Vol. 83, pp. 263–268. Academic Press, San Diego.

Takeuchi, M., Takasaki, S., Inoue, N., and Kobata, A. (1877). Sensitive method for carbohydrate composition analysis of glycoproteins by high performance liquid chromatography. *J. Chromatogr.* **400,** 207–213.

Takeuchi, M., Takasaki, S., Miyazaki, H., Kato, H., Hoshi, S., Kochibe, N., and Kobata, A. (1988). Comparative study of the asparagine-linked sugar chains of human erythropoietins purified from urine and the culture medium of recombinant Chinese hamster ovary cells. *J. Biol. Chem.* **263,** 3657–3663.

Takeuchi, M., Takasaki, S., Shimada, M., and Kobata, A. (1990). Role of sugar chains in the *in vitro* biological activity of human erythropoitin produced in recombinant Chinese hamster ovary cells. *J. Biol. Chem.* **265,** 12127–12130.

Tarentino, A. L., Gomez, C. M., and Plummer, T. H., Jr. (1985). Deglycosylation of asparagine-linked glycans by peptide: *N*-Glycosidase F. *Biochemistry* **24,** 4665–4671.

Tarentino, A. L., and Plummer, Jr., T. H. (1982). Oligosaccharides accessibility to peptide: *N*-Glycosidase as promoted by protein-unfolding reagents. *J. Biol. Chem.* **257,** 10776–10780.

Tedder, T. F., Issacs, C. M., Ernst, T. J., Demetri, G. D., Adler, D. A., and Disteche, C. M. (1989). Isolation and chromosomal localization of cDNA encoding a novel human lymphocyte cell surface molecule, LAM-1: Homology with the mouse lymphocyte homing receptor and other adhesion proteins. *J. Exp. Med.* **170,** 123–133.

Tiemeyer, M., Swidler, S. J., Ishihara, M., Moreland, M., Schweingruber, H., Hirtzer, P., and Brandley, B. K. (1991). Carbohydrate ligands for endothelial-leukocyte adhesion molecule 1. *Proc. Natl. Acad. Sci. USA* **88,** 1138–1142.

Tsuchiya, N., Endo, T., Matsuta, K., Yoshinoya, S., Aikawa, T., Kosuge, E., Takeuchi, F., Miyamoto, T., and Kobata, A. (1989). Effects of galactose depletion from oligosaccharide chains on immunological activities of human IgG. *J. Rheumatol.* **16,** 285–290.

Uchida, Y., Tsukada, Y., and Sugimori, T. (1974). Production of microbial neuraminidase induced by colominic acid. *Biochim. Biophys. Acta* **350,** 425–431.

Umemoto, J., Bhavanandan, V. P., and Davidson, E. A. (1977). Purification and properties of an endo-α-*N*-acetyl-D-galactosaminidase from *Diplococcus pneumoniae*. *J. Biol. Chem.* **252,** 8609–8614.

Vliegenthart, J. F. G., van Halbeek, H., and Dorland, L. (1981). The application of 500-MHz high-resolution ^{1}H-NMR spectroscopy for the structural determination of carbohydrates derived from glycoproteins. *Pure Appl. Chem.* **53,** 45–77.

Walz, G., Aruffo, A., Kolanus, W., Bevilacqua, M., and Seed, B. (1990). Recognition by ELAM-1 of the sialyl-Le^x determinant on myeloid and tumor cells. *Science* **250,** 1132–1135.

Weisshaar, G., Hiyama, J., Renwick, A. G., and Nimtz, M. (1991). NMR investigation of the *N*-linked oligosaccharides at individual glycosylation sites of human leutropin. *Eur. J. Biochem.* **195,** 257–268.

Yamashita, K., Hitoi, A., and Kobata, A. (1983). Structural determinants of *Phaseolus vulgaris* erythroagglutinating lectin for oligosaccharides. *J. Biol. Chem.* **258,** 14753–14756.

Yamashita, K., Hitoi, A., Matsuda, Y., Miura, T., Katsunuma, N., and Kobata, A. (1986). Structures of sugar chains of human kidney γ-glutamyltranspeptidase. *J. Biochem.* (*Tokyo*) **99,** 55–62.

Yamashita, K., Ichishima, E., Arai, M., and Kobata, A. (1980). An α-mannosidase purified from *Aspergillus saitoi* is specific for α-1,2-linkage. *Biochem. Biophys. Res. Commun.* **96,** 1335–1342.

Yamashita, K., Kamerling. J. P., and Kobata, A. (1982a). Structural study of the carbohydrate moiety of hen ovomucoid. *J. Biol. Chem.* **257,** 12809–12814.

Yamashita, K., Kochibe, N., Ohkura, T., Ueda, I., and Kobata, A. (1985). Fractionation of L-fucose-containing oligosachharides on immobilized *Aleuria aurantia* lectin. *J. Biol. Chem.* **260,** 4688–4693.

Yamashita, K., Mizuochi, T., and Kobata, A. (1982b). Analysis of oligosaccharides by gel filtration. *In* "Methods in Enzymology (V. Ginsburg, ed.), Vol. 83, pp. 105–126. Academic Press, San Diego.

Yamashita, K., Ohkura, T., Yoshima, H., and Kobata, A. (1981). Substrate specificity of diplococcal β-*N*-acetylhexosaminidase, a useful enzyme for the structural studies of the complex-type asparagine-linked sugar chains. *Biochem. Biophys. Res. Commun.* **100,** 226–232.

Yamashita, L., Totani, K., Ohkura, T., Takasaki, S., Goldstein, I. J., and Kobata, A. (1987). Carbohydrate binding properties of complex-type oligosaccharides on immobilized *Datura stramonium* lectin. *J. Biol. Chem.* **262,** 1602–1607.

Structure and Biology of Pericellular Proteoglycans

John R. Couchman and Anne Woods

Department of Cell Biology, University of Alabama at Birmingham, Birmingham, Alabama 35294

I. Introduction

Proteoglycans are receiving ever-increasing attention. They have been implicated in a wide variety of processes, ranging from cell adhesion and migration to matrix assembly, growth factor sequestration and regulation, involvement in many aspects of immune function, binding of plasma proteins, and control of thrombogenesis. A basic definition of proteoglycans is that they are complex molecules comprising a core protein to which are attached one or more glycosaminoglycan chains. The composition and structure of glycosaminoglycans have been reviewed (Jackson *et al.*, 1991; Kjellén and Lindahl, 1991). Within the broad but simple definition of proteoglycans, there is extreme polymorphism. Many core proteins have now been described, some of which are related and can be placed in families; others appear to be quite distinct. In addition, the type and number of glycosaminoglycan chains

Cell Surface and Extracellular Glycoconjugates

attached to each core, the length of the chains, the extent of sulfation, and modification of uronic acid by C5 epimerization are subject to variation. Other post-translational modifications such as N- and O-linked oligosaccharide substitution, fatty acylation, and linkage to phosphatidylinositol can all add to the complexity. The mechanisms by which cells control these variations are not yet clear. Furthermore, hybrid proteoglycans have been known for some time (e.g., the large cartilage proteoglycan aggrecan, which bears both chondroitin and keratan sulfate chains) and it has now been demonstrated that several proteoglycans may be hybrids, bearing both heparan and chondroitin or dermatan sulfate chains. As will be discussed, some proteoglycan core proteins can exist without the addition of glycosaminoglycan chains (e.g., CD44, invariant γ chain) and these will be termed facultative proteoglycans. In contrast, most proteoglycans identified to date always have glycosaminoglycan chains attached (e.g., syndecan). We term these constitutive proteoglycans, even though their glycosaminoglycan chains may vary with cell or tissue type and during development. Only one type of glycosaminoglycan appears to be normally expressed without a core protein. This is hyaluronan, which is synthesized at the cell membrane and released as free chains (reviewed in Toole, 1990).

Interest in proteoglycans has increased in parallel with molecular cloning techniques, which have facilitated understanding of their structure and potential function. For example, at least 10 distinct cell surface proteoglycans have now been sequenced through cDNA techniques. Attempts at proteoglycan classification have become necessary since many distinct core proteins may be substituted with the same type of glycosaminoglycan chain. Broadly, proteoglycans identified to date can be conveniently classed into five groups, based largely on topographical distribution, but also incorporating familial relationships of core protein structure. These classes are the cell surface proteoglycans, basement membrane proteoglycans, small leucine-rich proteoglycans, large aggregating proteoglycans interacting with hyaluronan, and secretory granule proteoglycans. In this chapter, we concentrate on the proteoglycans of the cell surface and pericellular matrix and their possible roles in the control of cell behavior. While all proteoglycans, at some stage in their lives, may be pericellular in distribution, we shall concentrate on those for which it appears that their major and determining role is at the cell membrane or resident within the immediate pericellular matrix. Other excellent reviews discuss the properties of proteoglycans not covered here (Bernfield *et al.*, 1992; Esko, 1991; Gallagher, 1989; Jalkanen *et al.*, 1991, Kjellén and Lindahl, 1991; Kolset and Gallagher, 1990; Noonan and Hassell, 1993). The biology of hyaluronan, which may be an important contributor to cell behavior, has also been reviewed by Toole (1990).

II. Cell Surface Proteoglycans

This class of proteoglycans is at present the subject of much research directed at core protein and glycosaminoglycan structure, and their possible biological effects. Table I lists the cell surface proteoglycans identified by sequencing of their core proteins, together with previously suggested nomenclature, type of membrane association, and selected references. Cell surface proteoglycans vary extensively in size and structure, from the free glycosaminoglycan hyaluronan to NG2, which has a core protein of approximately 250 kDa with two to three attached chondroitin sulfate chains. In addition, the method of membrane association varies. Some association may occur through cell surface receptors for the glycosaminoglycan component [e.g., hyaluronan binding to membrane proteins (Aruffo *et al.,* 1990; Toole, 1990)] or for the core protein [e.g., basement membrane HSPG (Clément *et al.,* 1989)]. A second form of association is through a phosphatidylinositol linkage to core protein and insertion of this lipid moiety into the membrane [e.g.,

TABLE I
Cell Surface Proteoglycans

Name	GAG	Association	Reference
Syndecan 1	HS + CS	Transmembrane	Kiefer *et al.* (1990); Mali *et al.* (1990); Saunders *et al.* (1989a)
Syndecan 2, fibroglycan	HS	Transmembrane	Marynen *et al.* (1989); Pierce *et al.* (1992)
Syndecan 3, *N*-syndecan	HS + CS	Transmembrane	Carey *et al.* (1992); Gould *et al.* (1992)
Syndecan 4, ryudocan/amphiglycan	HS	Transmembrane	Kojima *et al.* (1992b); David *et al.* (1992)
Glypican	HS	PI-linked	David *et al.* (1990)
CD44, Pgp-1, Hermes-1 Ag, Hutch-1, p85	HS + CS	Transmembrane	Jalkanen *et al.* (1991)
NG2	CS	Transmembrane	Nishiyama *et al.* (1991b)
Betaglycan, Type III TGFβ receptor	HS + CS	Transmembrane	López-Casillas *et al.* (1991); Wang *et al.* (1991)
Thrombomodulin	CS + DS	Transmembrane	Jackman *et al.* (1986); Wen *et al.* (1987)
Invariant chain	CS	Transmembrane	Claesson *et al.* (1983)

glypican (David *et al.*, 1990)]. The third form of cell surface proteoglycans is those with protein core sequence that permits transmembrane orientation, thus producing intercalated proteoglycans with cytoplasmic, transmembrane, and external domains [e.g., syndecans (Bernfield *et al.*, 1992)].

A. *The Syndecan Family of Intercalated Proteoglycans*

Of these three types of cell surface proteoglycans, the intercalated proteoglycans are the best characterized in terms of protein core sequence. The first integral cell surface proteoglycan to be sequenced was termed syndecan (Saunders *et al.*, 1989a), and this will be referred to here as syndecan 1, following the nomenclature in Bernfield *et al.* (1992). This was first identified in mouse mammary epithelial cells and was primarily considered the epithelial form of this class of cell surface proteoglycans (Bernfield and Sanderson, 1990; Bernfield *et al.*, 1992). It is, however, expressed in other cell types, particularly during development (Bernfield and Sanderson, 1990; Bernfield *et al.*, 1992). Syndecan 1 is remarkably conserved between species within the transmembrane and cytoplasmic domains (close to 100% homology at the amino acid level among rat, human, hamster, and mouse; Bernfield *et al.*, 1992), suggesting that the cytoplasmic portion of the molecule may have specific biological roles, particularly in cell adhesion processes (discussed later). Other proteoglycans with sequence homology to syndecan 1 in the cytoplasmic and transmembrane domains have been identified. These include fibroglycan (syndecan 2), which was characterized first in human lung fibroblasts (Marynen *et al.*, 1989) and later as the major cell surface proteoglycan of rat liver cells (Pierce *et al.*, 1992); *N*-syndecan (syndecan 3), which was first identified in neural tissue (Carey *et al.*, 1992) but is also present in developing cartilage (Gould *et al.*, 1992); and ryudocan/amphiglycan (syndecan 4), which is present in endothelial, epithelial, and smooth muscle cells and skin fibroblasts (David *et al.*, 1992; Kojima *et al.*, 1992b). The profile of syndecan-like proteoglycans made by specific cell types appears to be quite complicated. Many cell types studied so far have mRNA transcripts for more than one syndecan species, and the level of expression can vary with developmental or differentiation stage (Bernfield and Sanderson, 1990; Bernfield *et al.*, 1992; Elenius *et al.*, 1991; Lories *et al.*, 1992; Solursh *et al.*, 1990; Trautman *et al.*, 1991). For example, rat aortic smooth muscle cells appear to have a higher expression of syndecan 4 mRNA than that of syndecan 1 mRNA, whereas cells of a rat endothelial line appear to have equal amounts of both (Kojima *et al.*, 1992b). In addition, cells of B lymphocyte lineage express syndecan at the cell surface during the pre-B and immature stages, lose it when they become circulating or peripheral B lymphocytes, and regain expression on differentiation into plasma cells (Sanderson *et al.*, 1989).

Again, because of its early identification, more is known about the glycosaminoglycan content of syndecan 1 than the other syndecans and how this can vary with cell type and location (Bernfield and Sanderson, 1990; Bernfield *et al.*, 1992; Jalkanen *et al.*, 1991; Salmivirta *et al.*, 1991; Sanderson and Bernfield, 1988). Syndecan 1 originally isolated from NMuMG cells was found to be a hybrid proteoglycan, having both chondroitin and heparan sulfate glycosaminoglycans (Rapraeger *et al.*, 1985), and localized over the entire cell surface of subconfluent cells, but redistributed to the basolateral surfaces at confluency (Rapraeger *et al.*, 1986). Three distinct isoforms of syndecan 1 exist (reviewed in Bernfield and Sanderson, 1990; Bernfield *et al.*, 1992), which have the same core protein size (69 kDa) on SDS–gel electrophoresis but vary in glycosylation pattern. The smallest (approximately 100 kDa) predominates in stratified epithelia and plasma cells and has one chain each of chondroitin and heparan sulfate. In both locations, it surrounds the whole cell surface. A slightly larger form of approximately 160 kDa is found in simple epithelia and vascular endothelia at basolateral surfaces. This contains two chondroitin and heparan sulfate chains. The third form, with an apparent molecular mass of 300 kDa has been isolated from cultured fibroblasts and also has two chains of each type of glycosaminoglycan, the increased size being due to larger heparan sulfate chains. This is primarily intracellular, with approximately 1% of the level of expression on the cell surface compared with epithelial cells. The expressed isoform varies with cellular organization. For example, primary keratinocytes in monolayer culture express the 160-kDa form, but convert to the 100-kDa form after induction of stratification. In addition, treatment of NMuMG cells with transforming growth factor β (TGF-β) increases the size of chondroitin sulfate chains on syndecan 1, thus increasing the apparent molecular mass (Rapraeger, 1989). In contrast, 3T3 cells increase synthesis of mRNA and surface expression of syndecan 1 of unaltered properties in response to TGF-β and bFGF (Elenius *et al.*, 1992). This is consistent with a difference in regulation of mesenchymal and epithelial syndecans. It is interesting that bFGF can, therefore, augment the synthesis of a potential receptor for itself.

Most studies on the tissue distribution of syndecans have concerned syndecan 1 and have used antibodies against the core protein. Syndecan 1 is detected early in embryogenesis, at the four-cell stage (Sutherland *et al.*, 1991), and shows temporal changes in distribution during development. Details of these developmental changes are outside the scope of this review, and the reader is referred to reviews in Bernfield and Sanderson (1990), Bernfield *et al.* (1992), and Jalkanen *et al.* (1991). There are some major points of note during organ development. Syndecan 1 expression in epithelial cells appears to decrease when the epithelium is undergoing remodeling, reappears when this is complete, and persists until lost in some terminally differentiated epithelia. The epithelial loss of syndecan is often accompanied by a new expression of syndecan 1 by condensing mesenchymal cells adjacent to the epithelium. These mesenchymal cells then lose syndecan 1 expression with differentiation. Future studies comparing the distribution of syndecan species through development may reveal much about the function of this family of cell surface proteoglycans. Indeed, data have indicated that, in developing cartilage, there is a reciprocal pattern of expression of syndecans 1 and 3, with syndecan 1 decreasing with differentiation (Solursh *et al.*, 1990) and syndecan 3 increasing (Gould *et al.*, 1992).

Table IIA shows the reported sequences of the transmembrane and cytoplasmic domains of the four syndecan-like molecules for the rat species. These are all class I intercalated proteins with an extracellular N terminus. Close similarity can be noted and the high overall degree of homology is shown in Table IIB. The similarity of structure indicates a biological role for the cytoplasmic domains of these molecules. There are, however, some differences among the cytoplasmic domains of the four forms, particularly with respect to the tyrosine residues (Table IIA). One of the tyrosines in the cytoplasmic domain of syndecans 1, 3, and 4 is contained within the sequence KKKDEGSY, which is one consensus sequence (KXXDXXXY) for tyrosine phosphorylation (Gould *et al.*, 1992). This is not the case for fibroglycan (syndecan 2), of either human (Marynen *et al.*, 1989) or rat (Pierce *et al.*, 1992) origin. Phosphorylation events could, therefore, control the interactions of the cytoplasmic domains of syndecans 1, 3, and 4, and may indicate a fundamental difference between these syndecans and fibroglycan. In addition, syndecan 3 contains a threonine residue just after this tyrosine, which is in a consensus sequence TXXE for phosphorylation by casein kinase II (Gould *et al.*, 1992). This could allow for further control of intracellular interactions. Finally, the sequences indicate that the cytoplasmic domain may contain a tyrosine internalization signal found in proteins endocytosed through coated pits (Ktistakis *et al.*, 1990; reviewed in Bernfield and Sanderson, 1990; Bernfield *et al.*, 1992; Jalkanen *et al.*, 1991).

In contrast to the transmembrane and intracellular portions of the syndecan-like molecules, the extracellular domains show little overall homology except in three regions. All except syndecan 4 (ryudocan/

TABLE II

AMINO ACID SEQUENCES AND HOMOLOGY OF THE TRANSMEMBRANE AND CYTOPLASMIC DOMAINS OF FOUR RAT SYNDECAN-RELATED PROTEOGLYCANS

A. Alignment of amino acid sequences[a]

Transmembrane domain

N-Syndecan (syndecan 3)	V L V A V I V G G V V G A L F A A F L V T L L I Y
Syndecan (syndecan 1)	V L G G V I A G G L V G L I F A V C L V A F M L Y
Fibroglycan (syndecan 2)	V L A A V I A G G V I G F L F A I F L I L L L V Y
Ryudocan (syndecan 4)	V L A A L I V G G V V G I L F A V F L I L L L V Y

Cytoplasmic domain

N-Syndecan (syndecan 3)	R M K K K D E G S Y T L E E P K Q A — S V T Y Q K — P D K Q E E F Y A
Syndecan (syndecan 1)	R M K K K D E G S Y S L E E P K Q A N G G A Y Q K — P T K Q E E F Y A
Fibroglycan (syndecan 2)	R M R K K D E G S Y D L G E R K — P S S A A Y Q K A P T K — — E F Y A
Ryudocan (syndecan 4)	R M K K K D E G S Y D L G — — K K P I — — — Y K K A P T — — N E F Y A

B. Sequence identity[b]

	N-Syndecan	Syndecan	Fibroglycan	Ryudocan
N-Syndecan	—	—	—	—
Syndecan	70	—	—	—
Fibroglycan	65	58	—	—
Ryudocan	62	53	73	—

[a] Most closely matching sequences are juxtaposed.
[b] Figures are percentages.

amphiglycan) have a dibasic sequence immediately external to the transmembrane domain. This provides a protease-sensitive site, possibly responsible for the reported cleavage and subsequent release of the ectodomain of syndecan 1 (Jalkanen *et al.*, 1987). A second similarity in sequence is the presence of potential glycosaminoglycan addition sites. The core protein sequences indicate different numbers of possible attachment sequences in syndecans 1, 2, 3, and 4 (reviewed in Bernfield *et al.*, 1992; Jalkanen *et al.*, 1991). It is unclear which of these sites are actually substituted with glycosaminoglycan chains in each syndecan, and whether these are chondroitin or heparan sulfate. Indeed, the glycosaminoglycan content of one syndecan type can vary, and although syndecan 1 is mostly reported as a hybrid proteoglycan, substitution solely with heparan sulfate chains is seen in developing mesenchyme (Salmivirta *et al.*, 1991). Thus, the primary structure of the core protein may not be sufficient to determine the glycosylation pattern. The type of glycosaminoglycan substitution on syndecan core proteins needs further biochemical analysis of isolated proteoglycans, but fibroglycan (syndecan 2) from human lung fibroblasts (Marynen *et al.*, 1989) and from rat liver (Pierce *et al.*, 1992) appears to contain only heparan sulfate. In addition, syndecans 1 and 4 from endothelial cells contain mainly heparan sulfate glycosaminoglycans (David *et al.*, 1992; Kojima *et al.*, 1992a,b).

Apart from these three areas of homology, the extracellular domains of the syndecan-like molecules so far identified are quite diverse. This is particularly evident with syndecan 3, which in chick (Gould *et al.*, 1992) and rat (Carey *et al.*, 1992) contains a threonine-, serine-, and proline-rich domain central in the core protein. This is similar to other sequences that encode the addition of O-linked oligosaccharides and is prevalent in mucin-like molecules. The differences in the extracellular domains of the syndecans may be helpful in generating core-protein-specific antibodies to enable determination of the distribution of these molecules *in vivo,* and their isolation for biochemical analysis. To date, most information on the distribution of syndecans concerns that of syndecan 1 (see earlier). Much of this has come from the use of monoclonal antibodies, which appear to be highly specific for syndecan 1 (e.g., antibody 281-2 against an epitope near the N terminus). In contrast, antibodies against conserved regions of syndecans, particularly the cytoplasmic or transmembrane domains, may be polyspecific for the syndecan family. Immunohistochemical analysis of syndecan-like molecular distribution may be aided by future polyspecific (against cytoplasmic peptides) and monospecific (against variable regions of the ectodomains) antibodies.

B. Phosphatidylinositol-Linked Proteoglycans

Proteoglycans that appear to be inserted into the cell membrane through a covalent phosphatidylinositol linkage have also been identified. These include heparan sulfate proteoglycans (HSPGs) from hepatocytes (Ishihara *et al.*, 1987), ovarian granulosa cells (Yanagashita and McQuillan, 1989), Schwann cells (Carey and Evans, 1989), embryonic heart cells (Chajek-Shaul *et al.*, 1989), human lung fibroblasts (David *et al.*, 1990), embryonic skin fibroblasts (Schmidtchen *et al.*, 1990), bone marrow cultures (Brunner *et al.*, 1991), and melanoma cells (Drake *et al.*, 1992). Of this class of proteoglycans, only one molecule, glypican, has yet been sequenced (David *et al.*, 1990). Glypican lacks the transmembrane sequence typical of the syndecans, but has a hydrophobic sequence at the C terminus that is covalently linked to phosphatidylinositol (David *et al.*, 1990). Other proteoglycans have been identified by release from the membrane by treatment with phospholipase C to cleave the phospholipid anchor, and/or metabolic labeling with radioactive inositol or ethanolamine. The first to be identified was in rat hepatocytes (Ishihara *et al.*, 1987), for which it was proposed that an endogenous phospholipase C cleaved the linkage between the core-protein-attached inositol and the membrane lipid, releasing a proteoglycan that had a myoinositol phosphate group still attached. The phospholipase was proposed to be insulin-activated, and the released proteoglycan endocytosed through a cell surface receptor recognizing the exposed myoinositol phosphate group. The endocytosed heparan sulfate chains of the proteoglycan from confluent hepatoma cells were reported to be cleaved from the core protein in endosomes and transported to the nucleus, where they blocked G1 to S phase transition (Ishihara and Conrad, 1989; Fedarko *et al.*, 1989). Binding and internalization of lung fibroblast glypican have not been noted, and there are no data on any possible role of this molecule in control of cell division. The divergent paths taken by phosphatidylinositol-bound HSPGs and those with transmembrane core proteins have been recently reviewed (Yanagashita and Hascall, 1992).

The phosphatidylinositol-linked proteoglycans appear to be a subset of total proteoglycans in all cell types in which they have been identified. Recent studies (Lories *et al.*, 1992) indicate that glypican is expressed in mammary epithelial cells in addition to lung fibroblasts. This study also demonstrated that traces of another syndecan-like molecule, which is a hybrid proteoglycan reactive with antibodies against syndecan 1, were synthesized in lung fibroblasts in addition to glypican, fibroglycan (syndecan 2), and another HSPG of 125-kDa core protein.

The expression of this hybrid was confirmed by amplifying syndecan cDNA sequences from fibroblast mRNA extracts and demonstrating antibody cross-reactivity of the recombinant products. The level of expression of the hybrid proteoglycan mRNA was always lower in fibroblasts than in the mammary epithelial cells, although confluent cultures of fibroblasts had increased expression in comparison with subconfluent cells. Analysis of proteoglycan profiles at the mRNA level may prove very fruitful in the future, since biochemical or immunohistochemical analysis may not detect proteoglycans that are rapidly released from the cell surfaces or degraded. Schmidtchen *et al.* (1990) has highlighted the possible problems in detecting glypican-like proteoglycans. A phosphatidylinositol-linked HSPG was detectable only in skin fibroblasts maintained under serum-free conditions, and the authors postulate that an endogenous phospholipase C activated by serum, or a phosphatidylinositol-specific phospholipase D present in serum, could release inositol-bound proteoglycans, thus preventing their detection. Indeed, Lories *et al.* (1992) showed the presence of glypican mRNA in most cell lines examined. Similarly, a syndecan 1 message appeared to be present in most cell lines tested. It is, therefore, important to determine the selective roles of these two proteoglycan types (see later).

C. The CD44 Family

CD44 is a polymorphic family of integral membrane glycoproteins, originally identified as Hermes antigens on lymphocytes and involved in lymphocyte homing and lymphohemopoiesis (reviewed in Jalkanen *et al.*, 1991; Shimizu and Shaw, 1991). Hermes antibodies, which block lymphocyte binding to high endothelial venules (Jalkanen *et al.*, 1986), recognize an 80- to 90-kDa glycoprotein and a higher molecular weight form (200 kDa) on lymphocyte surfaces. This higher molecular weight form has chondroitin sulfate attached to the protein (Jalkanen *et al.*, 1988) and hence is a facultative proteoglycan. CD44 expression is not limited to lymphocytes, since antigenically related proteins are found on other leukocytes and nonhematopoietic cells. Epithelial cells, fibroblasts, glial cells, Schwann cells, smooth and skeletal muscle cells, Kupffer cells, and kidney mesangial cells all synthesize CD44 to various extents (Brown *et al.*, 1991; Jalkanen *et al.*, 1991). Keratinocyte CD44 may also contain heparan sulfate (Brown *et al.*, 1991), and CD44 can, therefore, also be a facultative hybrid proteoglycan.

The 80- to 90-kDa form of CD44 has been the most studied and was previously referred to by different names. These include extracellular matrix receptor III, p80, Hermes-1 antigen, Hutch-1, phagocytic glycoprotein 1 (Pgp-1), and p85 (reviewed in Brown *et al.*, 1991; Jalkanen *et al.*, 1991). Studies with lymphocytes have implicated CD44 in cell–cell and cell–matrix interactions. Antibodies against CD44 inhibit lymphocyte adhesion to high endothelial venules (Berg *et al.*, 1989), T cell attachment to erythrocytes (Haynes *et al.*, 1989), and B cell binding to stromal cells (Miyake *et al.*, 1990). CD44 is also implicated in T and B cell activation (Murakami *et al.*, 1990; Shimizu *et al.*, 1989).

Although not homologous with other cell surface proteoglycans, CD44 shares some features of structural organization. As with the syndecans, there is a large N-terminal extracellular domain, a highly conserved transmembrane domain, and a short C-terminal cytoplasmic domain (reviewed in Jalkanen *et al.*, 1991). The cytoplasmic domain is normally well conserved between species and has a consensus sequence for protein kinase C phosphorylation. This sequence is, however, unlike that reported for NG2 proteoglycan (Nishiyama *et al.*, 1991b). One CD44 variant has a cytoplasmic domain consisting of just 3 amino acids in comparison with 72 in other forms (Goldstein *et al.*, 1989).

The extracellular domain of CD44 appears to have a region that is hypervariable between species in addition to more conserved domains. One feature is the presence of a dibasic sequence close to the external face of the plasma membrane, reminiscent of some syndecans. This dibasic sequence is common to mouse, bovine, and rat forms of CD44, but not to human forms of CD44. The basic structure of CD44 is complicated by alternate splicing and variable glycosylation. One variant isolated from human epithelial cells (CD44E; Brown *et al.*, 1991; Stamenkovic *et al.*, 1991) contains an inserted 132-amino-acid domain. At the C terminus of this insert is an SGD sequence, which by homology with perlecan may be substituted with heparan sulfate. This is consistent with the known hybrid nature of CD44E glycanation. Another splice variant from rat tumor cells that may confer metastic potential has been described (Günthert *et al.*, 1991). Additionally the molecule and its splice variants are heavily substituted with N- and O-linked oligosaccharides.

The conserved N-terminal extracellular domain has three disulfide-bonded loops and bears some homology with link protein and the hyaluronan-binding regions of aggrecan and versican. Blocking antibody studies provide further evidence that this is the hyaluronan-binding region of CD44. The presence of chondroitin sulfate chains is not needed for binding to endothelial venules, but facilitates interactions with fibronectin (Jalkanen and Jalkanen, 1992) and collagen (Faassen *et al.*, 1992). The significance of additional heparan sulfate (Brown *et al.*, 1991) is not currently understood, but may be important since the binding site on fibronectin for CD44 chondroitin sulfate chains is the high affinity heparin (and HSPG)-binding domain (Jalkanen and Jalkanen, 1992). The impact of variation in core protein structure on the interactive capabilities of CD44 also awaits determination. The promiscuous nature of CD44 interactions is underscored by earlier observations that the unglycanated molecule binds lymphocytes to high endothelial venules, possibly through vascular addressin. The addressin-binding activity is additional to N-terminal domain interactions with hyaluronan on the surface of high endothelial venules (reviewed in Shimizu and Shaw, 1991). This region appears to lie closer to the cell membrane than the hyaluronan binding site and the presence of the insert in CD44E may prevent this interaction (Stamenkovic *et al.*, 1991).

In common with the syndecans there is indirect evidence for transmembrane signaling through CD44 interactions. Indeed, the hyaluronan-binding activity of CD44 appears to be decreased when the cytoplasmic domain is truncated (Lesley *et al.*, 1992). Evidence points to a codistribution with either the vimentin or the actin-based cytoskeleton, for example with ankyrin (reviewed in Jalkanen *et al.*, 1991). These data are in agreement with a rapid redistribution of CD44 in migrating cells (Jacobson *et al.*, 1984) and a possible role in melanoma invasion of collagen gels (Faassen *et al.*, 1992). It has also been noted that CD44E is absent from points of stable cell–matrix interaction, such as focal adhesions, but is enriched in filopodia (Brown *et al.*, 1991).

D. NG2 Proteoglycan

Another type I transmembrane proteoglycan is NG2, which is found predominantly on neuronal cells (Stallcup, 1981), but also in developing mesenchyme (Nishiyama *et al.*, 1991a) and human melanoma cells (Harper and Reisfeld, 1987). The cDNA sequence for this proteoglycan (Nishiyama *et al.*, 1991b) indicates a large proteoglycan of 2325 amino acids, which is the largest core protein of a transmembrane proteoglycan to date. It contains a short cytoplasmic domain of 76 amino acids, a single transmembrane domain of 25 amino acids, and a large extracellular domain of 2224 amino acids, and appears to be quite distinct from other proteoglycan core proteins. Notable in the C-terminal cytoplasmic domain are three threonine residues whose flanking sequences conform to a motif for protein kinase C phosphorylation (Nishiyama *et al.*, 1991b). The threonine six residues from the cytoplasmic end of the putative transmembrane domain most closely matches the consensus sequence. Unlike syndecans and CD44, there is no dibasic sequence close to the external face of the putative transmembrane sequence. The extracellular domain consists of three regions, with two cysteine-rich domains separated by a serine–glycine-rich domain. The core protein is substituted with two to three chondroitin sulfate glycosaminoglycans of approximately 100 kDa (Nishiyama *et al.*, 1991b). Although the functions of this proteoglycan are not known, it does bind collagen type VI (Stallcup *et al.*, 1990), possibly through the central portion of the core protein.

E. Betaglycan

Five members of the transforming growth factor β (TGF-β) family have been identified. The receptors are complex, and three transmembrane classes, types I, II, and III, have been identified (reviewed in López-Casillas *et al.*, 1991; Ruoslahti and Yamaguchi, 1991; Wang *et al.*, 1991). Types I and II have high affinities for the different TGF-β isoforms, whereas the type III receptor appears to bind all isoforms with slightly lower affinity. The former two types also appear to be involved in transmembrane signaling following TGF-β binding, whereas type III interactions with ligand do not. The type III TGF-β receptor has been shown to be a hybrid proteoglycan, bearing both heparan and chondroitin sulfate chains, and is now termed betaglycan. Binding of ligand to betaglycan is through the core protein and the glycosaminoglycan chains are not necessary for TGF-β binding or expression at the cell surface (Cheifetz and Massagué, 1989). In addition to binding TGF-β, betaglycan can also bind bFGF through the heparan sulfate moiety, and the heparan sulfate content can be modified in response to bFGF (Andres *et al.*, 1992). It has been postulated that betaglycan acts as an immobilizer of TGF-β or presents the growth factor to other high-affinity receptors, similar to postulated roles for other membrane proteoglycans (Ruoslahti and Yamaguchi, 1991; Kjellén and Lindahl, 1991).

Two papers have documented the sequence of the core protein of rat betaglycan (López-Casillas *et al.*, 1991; Wang *et al.*, 1991). It is a type I transmembrane protein of 853 amino acids, with a C-terminal cytoplasmic domain of 41 (Wang *et al.*, 1991) or 43 (López-Casillas *et al.*, 1991) amino acids. The cytoplasmic domain is rich in serine and threonine, some of which may be sites for phosphorylation by protein kinase C. The extracellular domain has six SG sequences, two of which are surrounded by acidic amino acids, and thus could be potential glycosaminoglycan attachment sites. In addition, the extracellular domain has two dibasic sequences near the plasma membrane, one of which is identical to a sequence in TGF-α that is cleaved to generate soluble forms of this molecule. A soluble form of betaglycan is known to be released into the media of several cell types (Andres *et al.*, 1989).

Betaglycan does not bear homology to that of other proteoglycan core proteins sequenced so far. The only homology found (López-Cassilas *et al.*, 1991; Wang *et al.*, 1991) is that of the transmembrane and cytoplasmic domains of betaglycan to a major protein of vascular endothelium, termed endoglin (Gougos and Letarte, 1990).

F. Thrombomodulin

Thrombomodulin is an integral membrane proteoglycan, which bears a single chondroitin/dermatan sulfate chain (reviewed in Kjellén and Lindahl, 1991; Gallagher, 1989). It binds thrombin and acts as a cofactor in thrombin-catalyzed activation of protein C to convert this to a form with anticoagulant activity (reviewed in Kjellén and Lindahl, 1991). Activation of protein C does not appear to require the glycosaminoglycan chains. In contrast, the glycosaminoglycan content appears to be essential to prevent thrombomodulin-bound thrombin from cleaving fibrinogen and activating Factor V, and for the increase in thrombomodulin-bound thrombin sensitivity to antithrombin.

Sequencing of the core protein of human (Wen *et al.*, 1987) and bovine (Jackman *et al.*, 1986) thrombomodulin indicates it is a class I integral membrane protein with a short cytoplasmic tail that resembles the LDL receptor with tyrosine, serine, and threonine residues as possible substrates for phosphorylation (Wen *et al.*, 1987). In the LDL receptor, mutation of the tyrosine residue prevents endocytosis through coated pits (Davis *et al.*, 1986), and since thrombomodulin internalization is regulated by both thrombin and protein C, this tyrosine may be important in regulating the availability of thrombomodulin and thrombin (Wen *et al.*, 1987). The extracellular portion of the molecule is complex and contains an N-terminal domain, a domain with EGF-like repeats, and a serine-/threonine-rich region (Wen *et al.*, 1987). Thrombomodulin can thus be considered to have five functional domains (Kjellén and Lindahl, 1991). Recent deletion studies have shown that the glycosaminoglycan attachment site is within the serine-/threonine-rich domain of the molecule, whereas thrombin binding and cofactor activity are in separate regions of the EGF-like domain (Tsiang *et al.*, 1992). In addition, these studies indicate that the serine-/threonine-rich domain may act as a spacer from the membrane needed for efficient protein C activation.

G. Invariant γ Chain

During intracellular transport, human class II antigen α and β subunits are associated with an invariant transmembrane γ chain. This polypeptide can bear chondroitin sulfate glycosaminoglycan and is therefore a facultative proteoglycan (reviewed in Kjellén and Lindahl, 1991). The sequence of human invariant γ chain (Claesson *et al.*, 1983) indicates that membrane insertion is opposite that of the proteoglycans described above, in that the N teminus is cytoplasmic. The molecule consists of 216 amino acids, 30 of which are cytoplasmic. The extracellular domain is 160 amino acids and contains two sites for N-glycosylation in addition to one putative glycosaminoglycan attachment site of the SGXG sequence. The biological role of the invariant γ chain, or how this is modified by glycosaminoglycan addition, is not clear at present, but it has been suggested that the molecule could regulate intracellular transport of class II antigens (Kvist *et al.*, 1982; Sung and Jones, 1982).

H. Biological Roles of the Cell Surface Proteoglycans

As might be expected from the variety of structures of the proteoglycans, postulated roles of these molecules are numerous (reviewed in Bernfield, 1992; Esko, 1991, Gallagher, 1989; Kjellén and Lindahl, 1991). Participation of proteoglycans in growth factor binding and presentation of bound factors to higher affinity ligands are important issues which have been reviewed in detail (Gallagher, 1989; Jalkanen *et al.*, 1991; Kjellén and Lindahl, 1991; Ruoslahti and Yamaguchi, 1991) and will not be considered further here. In addition, proteoglycans or isolated glycosaminoglycans appear to be involved in blood coagulation regulation; heparin and heparan sulfate bind to antithrombin III and thrombomodulin binds to both thrombin and protein C (reviewed in Kjellén and Lindahl, 1991). Roles in lipid homeostasis and control of proteolysis have also been indicated (Bernfield, 1992; Esko, 1991; Kjellén and Lindahl, 1991), and even the binding of herpes simplex virus to cells appears to involve cell surface HSPG (Shieh *et al.*, 1992). Proteoglycans, especially HSPGs, appear to also be involved in cell–cell and cell–matrix interactions (reviewed in Bernfield and Sanderson, 1990; Bernfield *et al.*, 1992). Earlier work indicated that a balance among chondroitin/dermatan sulfate, hyaluronan, and heparan sulfate could control cell adhesion and migration (reviewed in Culp *et al.*, 1986; Höök *et al.*, 1986; Rosenberg *et al.*, 1986; Woods and Couchman, 1988). In general, it appears that the first two glycosaminoglycan species are antiadhesive and hence promote more migratory behavior, whereas heparan sulfate-containing molecules promote more stable adhesion (Boudreau *et al.*, 1991; Funderberg and Markwald, 1986; Kinsella and Wight, 1986; Lane and Solursh, 1991; Lewandowska *et al.*, 1987; Perris and Johansson, 1990). Current understanding of the biology of CD44 and syndecans supports this.

Studies on fibroblasts *in vitro* indicate that interactions of integrins (Hynes, 1990) with fibronectin matrix molecules through the RGD adhesive motif were sufficient to promote cell attachment and spreading but that this resulted in a migratory phenotype (Izzard *et al.*, 1986; Woods *et al.*, 1986; reviewed in Woods and Couchman, 1988). This migratory phenotype was characterized by a lack of specialized focal adhesions (Burridge *et al.*, 1988; Woods and Couchman, 1988) between the cell and the matrix, concomitant with a lack of organization of internal microfilaments into stress fibers. In contrast, if additional interactions occurred with heparin-binding domains of fibronectin, cells formed both focal adhesions and stress fibers (Woods *et al.*, 1986), and changed to a stationary phenotype. Indirect evidence implicated cell surface HSPG as the "receptor" for heparin-binding domains of fibronectin. Mutant cells lacking HSPG (LeBaron *et al.*, 1988), or having HSPG with reduced sulfation and binding to fibronectin, and having a shorter half-life at the cell surface (Couchman *et al.*, 1988), lacked the ability to form focal adhesions. In addition, immunohistochemical studies with antibodies against liver-derived HSPG demonstrated an enrichment of cell surface HSPG at focal adhesions and along the membrane overlying stress fibers, indicative of some linkage between the microfilament cytoskeleton and cell surface HSPG (Woods *et al.*, 1984). This possible linkage was highlighted by the fact that a hydrophobic HSPG was retained in detergent-resistant cytoskeleton-matrix preparations (Woods *et al.*, 1985). The nature of the HSPG involved could not be determined at the time these studies were performed, but future use of antibodies specific for the different core proteins of cell surface HSPGs may elucidate the species.

Specific antibodies against syndecan 1 have been used to try to determine the role of this proteoglycan in cell adhesion (reviewed in Bernfield and Sanderson, 1990; Bernfield *et al.*, 1992). Because of its location, correlated distribution with underlying microfilament bundles, loss from the cell surface on detachment of cells, and binding to actin, this is a good candidate for linking intracellular cytoskeleton to matrix molecules. In addition, studies (Leppä *et al.*, 1991) have shown that the testosterone-dependent expression of syndecan 1 can control epithelial-fibroblastic morphological changes in S115 mouse mammary tumor cells. Furthermore, when NMuMG cells are made syndecan 1 deficient by transfection with the antisense syndecan-1 coding region, they lose their normal epithelial-like microfilament organization, redistribute their $\beta 1$ integrins, and become more mesenchymal in behavior (Saunders *et al.*, 1989b; Kato and Bernfield, 1990).

Syndecan 1 binds types I, III, and V collagen, thrombospondin, and fibronectin (reviewed in Bernfield *et al.*, 1992); tenascin (Salmivirta *et al.,* 1991); and neurite growth-promoting protein (Salmivirta *et al.,* 1992). The binding is through the heparan sulfate chains, and the role of chondroitin sulfate on hybrid molecules is not known, although the interaction with tenascin seems to be reduced when chondroitin sulfate is present (Salmivirta *et al.,* 1991). Syndecan 1 does not show binding to the basement membrane molecule laminin (Koda *et al.,* 1985; Elenius *et al.,* 1990). The lack of binding to laminin and to vitronectin (Elenius *et al.,* 1990) illustrates that not all matrix molecules with heparin-binding activity can bind syndecan and that some specificity may be involved. Interestingly, laminin, like fibronectin, can promote focal adhesion and stress fiber formation (Woods and Couchman, 1988). This may indicate that another form of cell surface HSPG, in addition to the syndecans, may participate in cell–matrix interactions. Indeed, a phosphatidylinositol-linked HSPG of Schwann cells binds to laminin and fibronectin, apparently through heparan sulfate, localizes to points of cell contact with the basement membrane (Carey and Stahl, 1990), and appears to be involved in process extension (Carey *et al.,* 1990). In addition, a melanoma phosphatidylinositol-linked HSPG binds to a peptide from the heparin-binding domain of fibronectin and can mediate cell attachment to this peptide (Drake *et al.,* 1992). Interestingly, the binding appears to be through heparan sulfate chains, but adhesion is also disrupted by antibodies against the core protein that do not interfere with the binding of isolated proteoglycan to the peptide, indicating an additional role for the core protein in cell attachment (Drake *et al.,* 1992).

What role cell surface proteoglycans play in cell adhesion is still unclear. It does, however, appear that they can cooperate with other more specific matrix receptors (such as integrins) to augment normally low-affinity interactions and thus giving interactions high-avidity status. Early studies, for example, showed that the interaction of fibronectin and collagen was augmented by the addition of heparin (Johansson and Höök, 1980), as is the case with thrombin–antithrombin III binding (Kjellén and Lindahl, 1991). Later studies have shown that heparin or cell surface HSPG is needed for binding of growth factors to cells (Yayon *et al.,* 1991), and thus, these molecules may play a regulatory role. Transformed cells often have reduced cell surface HSPG or reduced sulfation of the HSPG (Woods and Couchman, 1988), which may allow for increased motility by prevention of focal adhesion formation. Alteration of sulfation level, and hence decreased interactions with heparin-binding matrix molecules, may be one way of regulating cell adhesion phenotype. In addition, differential binding of extracellular matrix molecules could destabilize adhesion. For example, thrombospondin and fibronectin both bind HSPG, and the heparin-binding domain of thrombospondin destabilizes focal adhesions in cells on fibronectin (Murphy-Ullrich and Höök, 1989). If cell surface proteoglycans play a role in regulating cell adhesion, it might be expected that they could generate some form of transmembrane signaling. As discussed above, transmembrane proteoglycans have consensus motifs for tyrosine, threonine, and serine phosphorylation. In addition, the phosphatidylinositol-linked proteoglycans could generate signals through phospholipid turnover. Recent studies (Woods and Couchman, 1992a) have indicated that focal adhesion and stress fiber formation in response to fibronectin may be under the control of protein kinase C (PKC). Inhibitors of PKC prevent focal adhesion formation in cells on fibronectin, and brief treatment with phorbol esters promotes their formation in cells prespread on the "cell-binding" fragment of fibronectin (Woods and Couchman, 1992a). In addition, treatment of cells with PKC inhibitors prevents the promotion of focal adhesion formation by heparin-binding fibronectin fragments (Woods and Couchman, 1992b). This indirectly indicates that cell surface HSPGs, which bind the heparin-binding fragments of fibronectin, may generate a signal to activate PKC. This signal appears to cause a redistribution of cytoskeletal components, including talin and vinculin, and of membrane integrin $\beta 1$ subunits, into focal adhesions. This is similar to the situation in lymphocytes, where activation of PKC appears to augment low-affinity interactions of integrins with their ligands and promote the formation of stable high-avidity multimolecular complexes (O'Rourke and Mescher, 1990).

III. Basement Membranes

Basement membranes are thin sheets of highly specialized extracellular matrix that underlie most epithelial and endothelial tissues and that surround muscle fibers, peripheral nerves, and fat cells. Knowledge of these structures has greatly increased over the past few years, especially since molecular and cell biological techniques have determined the properties of many basement membrane components. Once thought to be merely structural scaffolds, it is now apparent that many of the components of basement membranes have potent biological activity. Many aspects of growth, differentiation, and cell migration, as well as tissue development and repair, are strongly influenced by basement membranes and their components (Beck *et al.*, 1990; Kleinman and Weeks, 1989; LeBlond and Inoue, 1989; Martin *et al.*, 1988; Paulsson, 1987; Timpl, 1989; Yamada, 1991; Yurchenco and Schittny, 1990). More recently, it has also become apparent that, while basement membranes in different locations share morphological similarity, several isoforms of the major components exist (Ehrig *et al.*, 1990; Gunwar *et al.*, 1990; Hostikka *et al.*, 1990; Hunter *et al.*, 1989). This is also true with respect to the proteoglycans. Some basement membranes are structurally specialized, such as those composing the Reichert's membrane of rodent embryos (Inoue *et al.*, 1983), the fused basement membrane of the kidney glomerulus (Farquhar, 1991), and those underlying stratified squamous epithelia, where anchoring complexes of hemidesmosomes and anchoring fibrils provide important structural integrity and adhesion for the overlying epithelium (Briggaman, 1983; Holbrook and Smith, 1981; Tidman and Eady, 1984). Furthermore, there is tissue specificity of basement membranes, in terms of composition and organization, together with additional temporal differences, since these structures undergo a maturation process through embryonic and fetal life. This is reflected in both subtle and overt changes in the distribution of some basement membrane components, including proteoglycans, when viewed by light and electron microscopic immunohistochemistry (Abrahamson and St. John, 1990; Couchman *et al.*, 1992; Gibson *et al.*, 1983; Horiguchi *et al.*, 1991; Klein, G. *et al.*, 1990; Smith *et al.*, 1988; Wan *et al.*, 1984).

Basement membranes are biochemically complex, with several collagenous and noncollagenous proteins as well as proteoglycans. One major impetus for basement membrane research has been the analysis of the Engelbreth–Holm–Swarm (EHS) tumor matrix. Several basement membrane components have been purified from this source, including type IV collagen, laminin, entactin/nidogen, and proteoglycans (Hassell *et al.*, 1985; Timpl, 1989; Timpl and Dziadek, 1986). Type IV collagen, which is usually restricted to basement membranes, forms a structural scaffold to which the other components may be bound. There are at least five chains of type IV collagen (Butkowski *et al.*, 1989; Gunwar *et al.*, 1990; Hostikka *et al.*, 1990), but only two of these, α_1 (IV) and α_2 (IV), appear to be represented in the EHS tumor matrix. Laminin is a large heterotrimeric complex of approximately 10^6 kDa, which forms an asymmetrical cross shape (Beck *et al.*, 1990; Kleinman and Weeks, 1989; Martin *et al.*, 1988; Timpl, 1989). There are many isoforms of laminin that appear to have a tissue-specific distribution. At least two isotypes of the A chain and three short B chain variants exist (Ehrig *et al.*, 1990; Hunter *et al.*, 1989; Sanes *et al.*, 1990). As with virtually all extracellular matrix molecules, laminin is a multidomain molecule, which interacts with a variety of integrin and nonintegrin receptors, as well as other basement membrane components (Beck *et al.*, 1990; Begovac and Shur, 1990; Hynes, 1992; Kleinman *et al.*, 1991; Ruoslahti, 1991; Timpl, 1989). One of these is a 150-kDa glycoprotein, entactin (Durkin *et al.*, 1988; Mann *et al.*, 1989), which binds tightly to laminin and may bridge this molecule and the type IV collagen meshwork (Aumailley *et al.*, 1989; Fox *et al.*, 1991).

Unlike type IV collagen, laminin, and entactin, other components may not be ubiquitously distributed in all basement membranes. Type VII collagen, for example, appears to be a major component of the anchoring fibrils of basement membranes underlying stratified squamous epithelia (Burgeson *et al.*, 1990). Fibronectin is also present in some basement membranes, especially those undergoing morphogenesis or repair (reviewed in Hynes, 1990). In contrast, proteoglycans appear to be present in all basement membranes. They are now receiving considerable attention, especially as some forms of basement membrane HSPG may act as sinks or reservoirs of heparin-binding growth factors, such as basic fibroblast growth factor.

A. *Perlecan, a Large Basement Membrane HSPG*

1. Structure

The best described basement membrane proteoglycan is the molecule known as perlecan (reviewed in Noonan and Hassell, 1993). This appears to be a ubiquitous basement membrane proteoglycan, since it has been located with immunochemical reagents in all basement membranes examined (Fig. 1). The molecule was first isolated from the EHS tumor matrix and was described as having a core protein of approximately 400 kDa, with three large heparan sulfate chains, each of approximately 60 kDa. Rotary shadowing revealed that the three glycosaminoglycan chains of the EHS tumor proteoglycan were present at one end of the 80-nm-long core protein, with the remainder comprising six globular regions interlinked by rodlike domains. The name arises from the appearance of a string of pearls (reviewed in Noonan and Hassell, 1993). In addition to being present in basement membranes (Couchman and Ljubimov, 1989; Hassell *et al.*, 1980; Heremans *et al.*, 1989; Isemura *et al.*, 1987; Kato *et al*, 1988; Laurie *et al.*, 1982; Murdoch *et al.*, 1992; Paulsson *et al.*, 1986), it is present in some stroma, including those surrounding tumors (Heremans *et al.*, 1989; Horiguchi *et al.*, 1989; Ljubimov *et al.*, 1992; Murdoch *et al.*, 1992), and has been isolated from some fibroblast cultures (Heremans *et al.*, 1989).

The glycosaminoglycan content of perlecan can vary. When isolated from the EHS tumor, the core protein bears only heparan sulfate chains (Hassell *et al.*, 1980; Kato *et al.*, 1988; Paulsson *et al.*, 1987), but it can be substituted with both heparan and chondroitin/dermatan sulfate chains. A hybrid molecule has been purified from human placenta

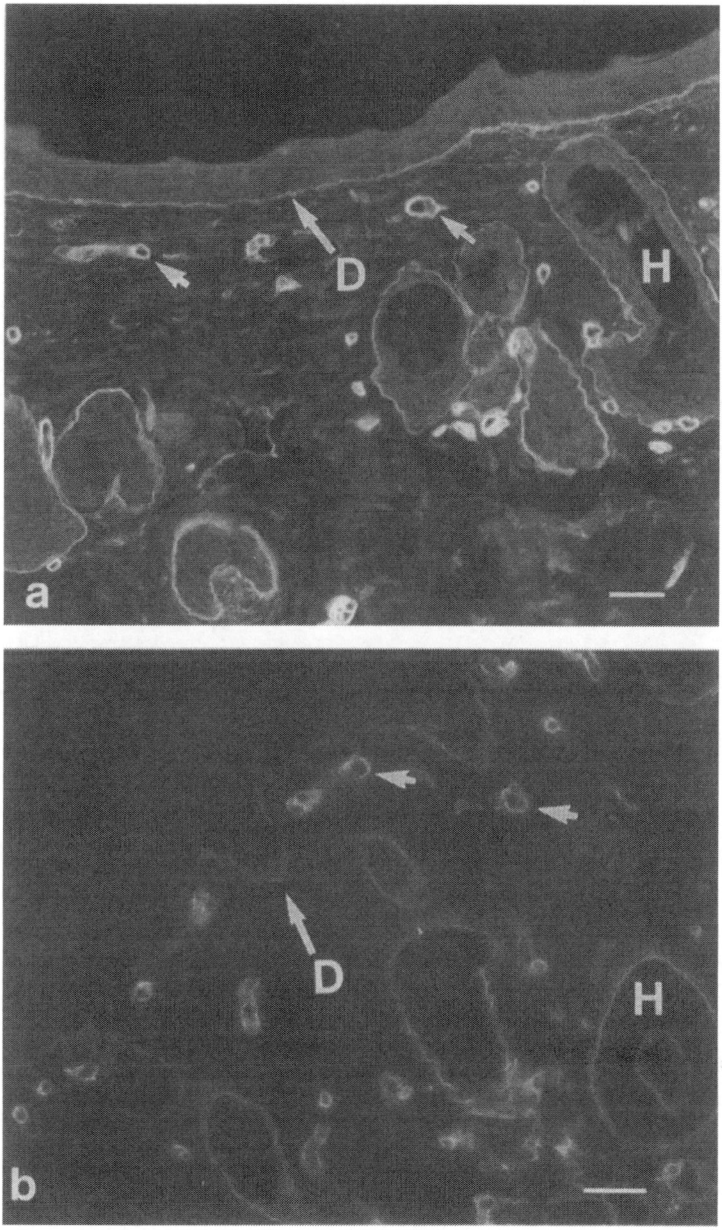

FIG. 1. Basement membrane localization of perlecan (a) and BM-CSPG (b) in rodent skin, detected by core protein-specific monoclonal antibodies. All basement membranes are stained, including those of the dermal–epidermal junction (D), hair follicles (H), and capillaries (arrows). Scale bars = 25 μm.

(Isemura *et al.*, 1987), and, interestingly, a cell line developed from the EHS tumor secretes hybrid perlecan (Danielson *et al.*, 1992). In addition, perlecan may be expressed solely as a chondroitin sulfate proteoglycan in the cornea (Hassell *et al.*, 1992). The entire core protein sequences of murine and human perlecan have now been reported (Kallunki and Tryggvason, 1992; Murdoch *et al.*, 1992; Noonan *et al.*, 1991) and show that perlecan core protein is a highly complex multidomain molecule. In fact, the core protein of perlecan is among the most complex and diverse of all extracellular matrix molecules for which sequence data are available. Most of the important sequence features are conserved between mouse and human, although there are some interesting differences between these two molecules. Mouse perlecan has a core protein of 396 kDa compared with the human homolog of 467 kDa. Both homologs appear to consist of five domains (Fig. 2). The

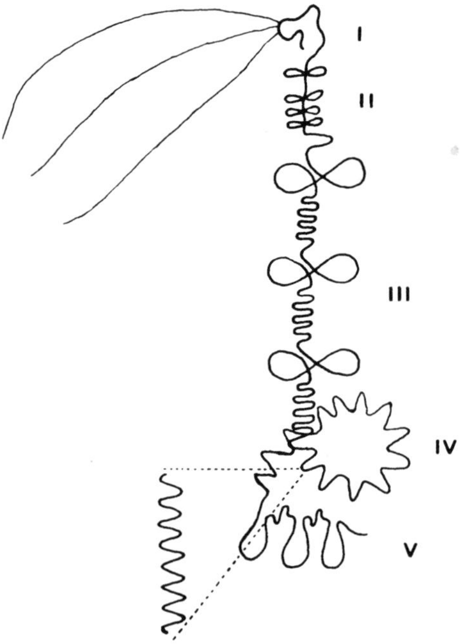

FIG. 2. Model of the perlecan molecule with its five domains (I–V, see text). Three heparan sulfate chains are shown in domain I, as proposed for the molecule isolated from the EHS tumor. The arrangement of domains is based on that proposed (Noonan *et al.*, 1991; Noonan and Hassell, 1993) and followed by Kallunki and Tryggvason (1992) and Murdoch *et al.* (1992). The extra seven Ig-like repeats reported for the human perlecan core protein are shown as a potential alternately spliced segment in Ig repeat 2 of domain IV (Kallunki and Tryggvason, 1992).

first of these, at the N terminus, consists of approximately 170 amino acids bearing virtually no homology to previously described extracellular matrix components. The mouse and human homologs are closely similar, particularly in a 14-amino-acid sequence containing three SGD repeats that is near the N terminus (Kallunki and Tryggvason, 1992; Murdoch *et al.,* 1992; Noonan *et al.,* 1991). These repeats are similar to sites in syndecan, glypican, NG2 proteoglycan, and the large aggregating CSPG versican. These three SGD sequences may, therefore, serve as heparan sulfate substitution sites, consistent with rotary shadowing and other data. Domain II comprises approximately 210 amino acids with homology to the low-density lipoprotein receptor. The function of this part of perlecan is conjectural, and whether the molecule can indeed bind LDL is currently unknown. It is also possible, as has been suggested by Noonan and Hassell (1993), that this region may direct the proteoglycan to the basolateral surfaces of epithelial cells for vectorial deposition in basement membranes. Domain III is large, with 1172 amino acids, and bears homology to domains III–V of the short arms of laminin A and B2 chains (Kallunki and Tryggvason, 1992; Murdoch *et al.,* 1992; Noonan *et al.,* 1991; Noonan and Hassell, 1993). It is therefore quite possible that this domain of the perlecan molecule interacts with cell surfaces and/or with other basement membrane components. In this context, it is interesting that murine perlecan has an RGD motif in domain III, consistent with recent hypotheses that this molecule interacts with integrin, as well as nonintegrin, receptors (Clément *et al.,* 1989; Hayashi *et al.,* 1990). However, the RGD sequence is not conserved in the human homolog (Kallunki and Tryggvason, 1992; Murdoch *et al.,* 1992). This parallels the situation in murine and human laminin A chains (Haarparanta *et al.,* 1991; Sasaki *et al.,* 1988).

Interposed between domain II and domain III is a single 100-amino-acid immunoglobulin (Ig)-like motif. Domain IV consists of multiple Ig repeats, each of approximately 100 amino acids, which most closely resemble those found in the cell adhesion molecule N-CAM. Domain IV shows the most difference between murine and human molecules. There are 14 Ig repeats in the murine perlecan, but both reports of the human sequence demonstrate 21 (Kallunki and Tryggvason, 1992; Murdoch *et al.,* 1992; Noonan *et al.,* 1991). Thus, 7 additional repeats exist in human perlecan, inserted in the middle of the second repeat in the mouse sequence (see Fig. 1) (Kallunki and Tryggvason, 1992). Whether this is a very recent duplication event in evolution or a result of alternate splicing remains to be seen. However, Noonan and Hassell (1993) report that they have identified an 856-bp sequence in the mouse that could not be assimilated into the published murine sequence. Since there is only one gene for perlecan, alternate splicing may occur in the mouse under certain circumstances. It has also been noted that the conserved position of the cysteine residues in domain IV may cause this region of the molecule to adopt a globular shape, so that the overall dimensions of the human and mouse molecules may not be substantially different despite extra Ig-like repeats (see Fig. 1) (Noonan and Hassell, 1993). N-CAM undergoes homotypic interactions and is a heparin-binding molecule involved in cell–cell adhesion (Rutishauser and Jessell, 1988). Therefore it is possible that this region in perlecan may assist in dimerization (Inoue *et al.*, 1989) or intramolecular self-association to form a globular domain. In addition, both reported sequences of domain IV of the human perlecan core protein indicate the presence of an SGXG motif, which has been proposed to be a site of potential galactosaminoglycan substitution (Bourdon *et al.*, 1987). However, differences in the cDNA sequences in these two reports place the SGXG motif in either the 2nd (Kallunki and Tryggvason, 1992) or the 14th (Murdoch *et al.*, 1992) Ig repeat of domain IV.

The C-terminal domain of perlecan, domain V, consists of approximately 700 amino acids that have homology with the C-terminal globular domains of the laminin A chain and merosin. However, there are some clear differences, including the fact that there are only three globular domains in perlecan compared with five in laminin A chain and merosin (Ehrig *et al.*, 1990; Sasaki *et al.*, 1988). In addition, subdomains consisting of two EGF-like repeats are interposed between the first and second, and the second and third globular regions of domain V. Once again, there are SGXG motifs, which may be substituted with glycosaminoglycan chains. These are present on the second and fourth EGF repeats, and are common to both the reported human sequences, but are not represented in the murine sequence (Kallunki and Tryggvason, 1992; Murdoch *et al.*, 1992; Noonan *et al.*, 1991). Evidence from rotary shadowing indicates that this domain of perlecan may undergo oligomerization, contributing to the overall structure of the basement membrane matrix (Yurchenco *et al.*, 1987). However, if some forms of perlecan are substituted with glycosaminoglycan chains in this region, this may preclude self-association and may alter the structural organization of perlecan in the basement membrane matrix.

The complexity of the perlecan molecule is increased by the great potential for post-translational modification. As mentioned earlier, there is the capacity for glycosaminoglycan chain substitution in domains I and V with additional possibilities in domain IV. In fact there are 42 SG pairs in the murine molecule and 52 or 53 in the human. However, it would seem that rather few of these are sites of glycanation.

In addition, there is considerable potential for N- and O-glycosylation as well as fatty acylation (Iozzo *et al.*, 1990). Domain IV also contains a hydrophobic stretch of approximately 28 amino acids that may indicate the potential for association with cell membranes (Murdoch *et al.*, 1992).

2. Biology of Perlecan

It is currently unclear what functions perlecan may have. As with other multidomain extracellular matrix molecules, it may undergo multiple interactions with other basement membrane components, including laminin and type IV collagen (Fujiwara *et al.*, 1984; Grant *et al.*, 1989). In addition, the molecule has the potential for self-association (Yurchenco *et al.*, 1987). Interactions through the heparan sulfate chains with laminin and fibronectin may be important for the integration of perlecan in the basement membrane matrix. Interactions with fibronectin may be further enhanced by an association of the core protein with a type III repeat in the fibronectin molecule (Heremans *et al.*, 1990). This generates the possibility that HSPGs such as perlecan may participate in cell adhesion, a finding supported by evidence that perlecan can be located at or near focal adhesion sites in cultured fibroblasts (Singer *et al.*, 1987). Perlecan can interact with several cell types, through multiple receptors, and this may, in turn, influence differentiation and gene expression (Li *et al.*, 1987). In this regard, the molecule resembles laminin, which can modulate cell behavior. The homology between perlecan domains III and V and portions of the laminin A chain, for instance, may partly explain this analogous behavior. A variety of evidence, reviewed by Noonan and Hassell (1993), indicates that perlecan may in some circumstances be specifically oriented within the basement membrane. Using cationic dyes, heparan sulfate chains have been shown to be located principally in the laminae rarae of basement membranes. Antibody studies, in contrast, indicate that core protein domains are present within the lamina densa. Therefore, the N-terminal domain of perlecan with the heparan sulfate chains may be oriented toward the cell surface, with the remainder of the core protein being embedded in the lamina densa. If so, this is an orientation somewhat different to that of laminin, which in some cases appears to have the C terminus of the A chain (the long arm of the laminin trimer) oriented toward the cell membrane (Abrahamson *et al.*, 1989). This part of the laminin molecule does interact with several cell surface receptors (reviewed in Hynes, 1992; Ruoslahti, 1991). These orientations augment two hypotheses. One is that the heparin-binding G domain on the long arm of the laminin A chain may interact with adjacent heparan sulfate on perlecan. The second is that bFGF residing in the basement membrane by association with heparan sulfate chains is adjacent to the cell surface, where it then may interact with high-affinity receptors.

B. *Small Basement Membrane HSPGs*

In addition to the large HSPG perlecan, there are one or more small HSPGs present in the EHS tumor matrix (Dziadek *et al.*, 1985; Hassell *et al.*, 1985; Kato *et al.*, 1987). Additionally, small HSPGs have been identified in various cell cultures and purified from liver sinusoids (Couchman, 1987; Oohira *et al.*, 1982; Soroka and Farquhar, 1991), as well as kidney (discussed later). None of these small HSPGs have been sequenced to date, so any homology with perlecan has yet to be determined. There has been some debate whether small HSPGs isolated from the EHS tumor are derived from perlecan by proteolytic processing or are unique (Dziadek *et al.*, 1985; Kato *et al.*, 1987; Ledbetter *et al.*, 1985). The HSPG isolated from liver, however, shares no known immunological similarity to perlecan and thus may be an entirely different proteoglycan resident within basement membranes (Soroka and Farquhar, 1991). Antibodies against this small HSPG stain many basement membranes, including those of the kidney glomerulus. Other studies have shown that small HSPGs purified from kidney glomerulus, lens, and murine parietal endoderm cells share some epitopes with perlecan, but also have additional unique epitopes (Couchman, 1987; Mohan and Spiro, 1991). The latter may be the result of post-translational modifications or alternate splicing, or may represent other members of a basement membrane proteoglycan family. One of the more interesting prospects for basement membrane research is to establish the relationship among the many proteoglycans that have been isolated thus far. Little is known about the potential function of these small proteoglycans, although presumably they share the same interactive capabilities as perlecan with respect to their heparan sulfate component. However, since evidence indicates that the fine structure of heparan sulfate chains may vary from site to site and may be rigorously controlled by the cells that synthesize them (Gallagher *et al.*, 1992), it may be that there are distinct populations of heparan sulfate chains on the various basement membrane HSPG core proteins. This could engender specificity in terms of interactions with other matrix molecules, growth factors, proteases and inhibitors, and cell surfaces.

C. Basement Membrane CSPGs

The presence of chondroitin sulfate in basement membrane has been suspected for a very long time. It was first shown by Cohn *et al.* (1977) that submandibular gland epithelium may secrete chondroitin sulfate into the basement membrane. Later immunohistochemical studies using monoclonal antibodies showed that many basement membranes contain chondroitin 6-sulfate (Couchman *et al.*, 1984). A candidate CSPG of core protein of 27–34 kDa was later identified in the culture supernatants of parietal endoderm cells (Couchman *et al.*, 1985). Since that time, it has been determined that perlecan may bear dermatan or chondroitin sulfate chains when isolated from human placenta or cornea (Hassell *et al.*, 1992; Isemura *et al.*, 1987). In 1986, Iozzo and Clark purified a CSPG from rat Reichert's membranes. This had a core protein of 140–150 kDa substituted with as many as 35 chondroitin 6-sulfate chains. Subsequently, monoclonal antibodies against the core protein of this proteoglycan have been shown to stain nearly all basement membranes in the adult rat (Fig. 1B). These monoclonal antibodies appear to have no immunological cross-reactivity to perlecan or related molecules, and this CSPG therefore appears to be a distinct basement membrane component (BM-CSPG; McCarthy and Couchman, 1990; McCarthy *et al.*, 1989, 1990). Evidence from skin, hair follicle, and kidney development indicates that this proteoglycan may be developmentally regulated in its expression and suggests that BM–CSPG may have a role in basement membrane stability (Couchman *et al.*, 1990, 1992). In agreement with this hypothesis, tissues undergoing morphogenesis appear to lack or have reduced amounts of BM–CSPG (discussed later). These studies support the hypothesis put forward by Bernfield and co-workers, which suggested that the presence of CSPG in basement membranes stabilizes those structures. For example, during branching morphogenesis of the salivary gland, basement membrane turnover at points where this proteoglycan was lacking could facilitate further branching (Bernfield *et al.*, 1984). Areas of branching were, on the other hand, enriched in HSPG. How basement membrane CSPGs may impart stability is presently unclear, and further studies on the interactions of this proteoglycan with other basement membrane components and core protein structural analysis are needed to elucidate potential function.

A large CSPG, PG-1000, has been isolated from the electric organs of elasmobranchs. This basement membrane proteoglycan is disulfide-bonded to other matrix components, including a second, smaller proteoglycan (Iwata and Carlson, 1991). Rather little is known of this proteoglycan at the present time, and the presence of an avian or mammalian homolog has yet to be demonstrated. From its distribution, it may be presumed to have a role in the neuromuscular junction.

D. Basement Membrane Proteoglycans in the Kidney Glomerulus

The glomerular basement membrane has long been a source of interest in terms of its proteoglycan structure and function (reviewed in Farquhar, 1991). While antibodies against the core protein of perlecan stain the structure (Grant and LeBlond, 1988; Laurie *et al.*, 1984), a molecule with the same physical properties as the EHS tumor perlecan has not been purified directly from this matrix. Numerous reports have indicated the presence of small HSPGs in this specialized basement membrane. Core protein sizes range from 18 to 200 kDa (van den Heuvel *et al.*, 1989; Kanwar *et al.*, 1984; Parthasarathy and Spiro, 1984). Perlecan core proteins of approximately 400 kDa have been identified in whole glomerulus, which may indicate that the HSPGs of the glomerular basement membrane are derived from perlecan by specific cleavage of the core protein (Kato *et al.*, 1988; Klein *et al.*, 1988). However, antibodies against the small liver-derived basement membrane HSPG, which is immunologically distinct from perlecan, also stain the glomerular basement membrane (Soroka and Farquhar, 1991). This structure is formed by the fusion of two basement membranes, one derived from the vascular endothelium and the other from the glomerular epithelium (podocytes). The major function of this basement membrane is selective filtration, a process that appears to be largely dependent on the presence of intact heparan sulfate chains. Staining of the glomerular basement membrane with cationic dyes reveals the presence of heparan sulfate chains in the laminae rarae. Selective removal by heparitinase destroyed the permeability characteristics of the glomeruli, leading to proteinuria (Kanwar *et al.*, 1980). Therefore, one specialized function of glomerular basement membrane HSPG is to contribute to the permselective properties of the glomerulus. What is not clear at the present time is which proteoglycan(s) in this tissue is responsible for this function, since there appear to be at least two basement membrane HSPGs, in addition to cell surface proteoglycans. This interesting area awaits further investigation with the use of selective immunological and molecular techniques.

One other interesting characteristic of the glomerular basement membrane is that it is the only basement membrane that lacks BM–CSPG in the adult rat (McCarthy and Couchman, 1990). This, along with other evidence for selective expression of BM-CSPG in development, indicates that the integrity of basement membranes per se is not dependent on the presence of BM-CSPG. Evidence on the turnover of glomerular basement membrane HSPGs further supports the hypothesis that the presence of BM-CSPG is correlated with stability. Studies by Beavan *et al.* (1989) have shown that the half-life of glomerular basement membrane HSPGs is remarkably short, in the order of 5–20 hr. This time scale is more reminiscent of the turnover of membrane-associated HSPGs (Yanagishita and Hascall, 1992). However, rapid turnover may be an important feature of continued renewal to maintain the vital function of selective permeability. Although absent in mature glomerular basement membranes, BM-CSPG is present early in development and is totally removed only when the newborn rat has reached 5 to 6 weeks of age (Couchman *et al.*, 1992). It would be interesting to know whether the turnover characteristics of HSPG in the newborn rat kidney glomerulus are different to those in the adult. It is clear, however, that, as with the EHS tumor matrix, the glomerular basement membrane is highly specialized and cannot be taken as a model for all basement membranes. This is underscored by the recent findings that distinct isotypes of laminin and type IV collagen are found in the glomerular basement membrane (Butkowski *et al.*, 1989; Sanes *et al.*, 1990). Therefore, while an important function of HSPGs in the glomerular basement membrane may be to contribute to the filtration function of the glomerulus, this is probably not so for the proteoglycans found in other tissue basement membranes. Indeed, the physiology of many tissues would preclude such a barrier function.

E. Basement Membrane Proteoglycans and Disease

The possibility that basement membrane proteoglycan metabolism is altered in some diseases is receiving considerable attention. Not surprisingly, considerable emphasis has been placed on a possible disruption of proteoglycan synthesis and/or stability, associated with proteinuria and the thickened glomerular basement membranes characteristic of diabetes. However, there seems to be a lack of consensus at the present time. While many studies indicate that the loss of anionic sites may result from decreased synthesis of perlecan and/or other HSPGs in the diabetic glomerulus, others indicate that amounts of HSPG per glomerulus may be unchanged (Brown *et al.*, 1982; Cohen and Surma, 1981; Ledbetter *et al.*, 1990; Nerlich and Schleicher, 1991; Rohrbach *et al.*, 1983; Shimomura and Spiro, 1987; Templeton 1989). This picture may be complicated by the presence of at least two distinct HSPGs in the normal kidney glomerular basement membrane, and these may have distinct ultrastructural distributions, as discussed later. In addition, changes in the extent of sulfation of heparan sulfate chains (Spiro, 1987) or other changes in post-translational modifications may alter the properties and functions of HSPGs, without necessarily influencing the rate of core protein synthesis. A variety of evidence indicates that an important facet of pathogenic changes in the glomerular basement membrane is a fundamental alteration in the stability and organization of the matrix (Klein *et al.*, 1986, 1989; Wu *et al.*, 1987). Alterations in turnover of glomerular HSPG may be involved in the pathogenesis of the diabetic state. Injection of labeled HSPG antibodies in nondiabetic rats, presumably affecting HSPG clearance, led to a long residence time for the label (60 days, cf. Beavan *et al.*, 1989) in the glomerular basement membrane, which underwent concomitant thickening (Miettinen *et al.*, 1986).

Very recently, it has been reported that, in a streptozotocin model of rat diabetes, thickened glomerular basement membranes are associated with the *de novo* expression of BM-CSPG, possibly by the capillary endothelium (McCarthy *et al.*, 1991) or mesangial interposition. As mentioned earlier, this proteoglycan, while present in early stages of development, is not normally a detectable component of the adult rat glomerular basement membrane. It is, however, abundant at all stages in the mesangial matrix. An abnormal expression of this proteoglycan in the diabetic state may contribute to changes in the permeability status, and turnover or organization of basement membrane components. This will clearly be a fruitful area for future research.

A second disease in which proteoglycans have been implicated is polycystic kidney disease. This can occur in autosomal dominant or recessive forms and in some cases is acquired (e.g., by patients on long-term dialysis). Studies using an antibody against a small glomerular basement membrane proteoglycan indicated a decreased presence of HSPGs in the basement membranes of the dilated distal tubules and collecting ducts (Carone *et al.*, 1988). Other data indicate that the synthesis of perlecan core protein is unaffected in human tubulocystic epithelial cells (Wilson and Sherwood, 1991). The heparan sulfate chains synthesized by affected tissues may, in addition, be of smaller size or undersulfated (Beavan *et al.*, 1991; Lelongt *et al.*, 1988). It remains to be established if proteoglycan defects are causal, but a variety of evidence points again to alterations in basement membrane stability (Carone *et al.*, 1989; Ebihara *et al.*, 1988; Wilson and Sherwood, 1991). Studies have mapped the locus for human autosomal dominant polycystic kidney disease to chromosome 16p13.3 (Germino *et al.*, 1992; Somlo *et al.*, 1992). This precludes a primary involvement of the perlecan core protein, mapped to chromosome 1p34 (Dodge *et al.*, 1991).

Third, puromycin aminonucleoside (PAN) nephrosis, a model of minimal change nephrotic syndrome in humans, involves proteinuria, some loss of anionic sites in the glomerular basement membrane, and podocyte foot process effacement. Evidence suggests that two different basement membrane HSPGs may be independently regulated in this disorder. Rada and Carlson (1991) have shown by immunoelectron microscopy that a loss of perlecan antibody binding sites in PAN-nephrotic rats was not matched by a loss of binding sites for an antibody against a 130-kDa glomerular HSPG core protein. This data resolved discrepancies seen previously in which only one population of HSPGs had been analyzed (Lelongt *et al.*, 1987; Mynderse *et al.*, 1983). Rada and Carlson (1991) also demonstrated differences in the ultrastructural localization of the two HSPG core proteins in normal glomerulus, the small species being resident in the two laminae rarae, while epitopes for perlecan core protein were present throughout the glomerular basement membrane, including the lamina densa. Similar results have been obtained previously, but again only one antibody, against perlecan or small glomerular basement membrane HSPG, was used on each occasion (reviewed in Farquhar, 1991).

An interesting finding in human brain, which has been reproduced in animal models, is that amyloid plaques of Alzheimer's disease and other amyloidoses have deposits of perlecan associated with them (Snow *et al.*, 1988, 1990, 1991; Young *et al.*, 1989). The reasons for this are unclear, although strong interactions between perlecan core protein and amyloid are suspected ($K_D = 9 \times 10^{-9}$–$10^{-10}M$) (Narindrasorasak *et al.*, 1991). It has also been shown in human senile plaques and a murine model of amyloidosis that perlecan is not alone, but is codeposited with other basement membrane components such as laminin and type IV collagen (Lyon *et al.*, 1991; Perlmutter *et al.*, 1991). It may well be, therefore, that one component of Alzheimer's disease is a basement membrane defect.

Some severe, and often lethal, diseases are associated with defects in the basement membranes underlying stratified squamous epithelia, notably the skin. Among these are dominant and recessive dystrophic forms of epidermolysis bullosa (Fine, 1987). There is now clear evidence linking defects in type VII collagen fibril assembly with this disease, which is characterized by a decrease in the number of anchoring fibrils present beneath the basement membrane of affected individuals (Tidman and Eady, 1985). In addition, however, there may be a proteoglycan involvement. It has been shown with antibodies against chondroitin 6-sulfate that the dermo–epidermal junction of patients with both dominant and recessive forms of dystrophic epidermolysis bullosa lacks this component (Fine and Couchman, 1989). In contrast, patients with junctional epidermolysis bullosa, in which lesions in basement membrane are also present, have apparently normal amounts of chondroitin 6-sulfate in skin basement membranes. Whether this defect in dystrophic epidermolysis bullosa involves the core protein of a basement membrane proteoglycan or results from abnormal post-translational modification remains to be resolved. However, consistent with the idea that BM-CSPG may be in some way involved with basement membrane stability, dystrophic epidermolysis bullosa patients are characterized by extreme fragility of the basement membrane between the epidermis and the underlying dermis.

Additionally, there is a similar report of lack of chondroitin 6-sulfate in the skin basement membranes associated with some squamous carcinoma lesions (Oguro *et al.*, 1991). This applied both to poorly differentiated and to poorly keratinized tumors in addition to more well-differentiated tumors where laminin, type IV collagen and perlecan were present in the basement membrane. Loss of basement membrane chondroitin 6-sulfate may therefore be a useful marker of carcinoma tissue, a feature deserving of more attention. Which core protein is involved in this observation is unknown.

IV. Biglycan

Five members of the small interstitial proteoglycan family have been sequenced (Table III). One of these is biglycan, which is unlike the other three in having a predominantly cell surface localization. All members of the family have a common leucine-rich repeating motif that has the consensus sequence LXXLXLXXNXL/I. From 8 to 12 of these repeats are found in these proteoglycans. This family can be divided into subgroups, based both on sequence homology and on glycosaminoglycan substitution patterns. Thus, lumican is most similar to fibromodulin and both are substituted with a small number of *N*-linked keratan sulfate chains (Blochberger *et al.*, 1992). While bovine fibro-

TABLE III

Small Leucine-Rich Proteoglycans

	Core protein (M_r)	GAG chain type (number)	Distribution	References (sequence data)
Decorin	36×10^3	CS/DS (1)	Fibrillar, collagenous matrices	Krusius and Ruoslahti (1986); Day *et al.* (1987)
Biglycan	38×10^3	CS/DS (1–2)	Selected matrices, cell surfaces	Fisher *et al.* (1989); Neame *et al.* (1989)
Fibromodulin	42×10^3	KS (4)	Fibrillar, collagenous matrices	Oldberg *et al.* (1989)
Lumican	38×10^3	KS (2–3)	Cornea, muscle, intestine, liver	Blochberger *et al.* (1992)
Lb	36×10^3	DS (1–2)	Cartilage	Shinomura and Kimata (1992)

modulin has no SG sequences, chick lumican does have one, although this is apparently not substituted with O-linked glycosaminoglycan. The avian Lb proteoglycan is more distantly related to the others and has homology with human osteoinductive factor (Shinomura and Kimata, 1992). Biglycan is most similar to decorin. In fact, a comparison of human biglycan and decorin sequences shows high homology. While this might imply a recent evolutionary divergence of the two molecules, human and bovine decorin core proteins, and human and bovine biglycan core proteins have diverged less than 10%, implying that these molecules are evolving slowly and represent a sustained divergence (Fisher *et al.*, 1989). This is supported by evidence of disparate distributions and interactions of the two proteoglycans. In sharp contrast, the extracellular domains of the syndecans show wide interspecies variation in sequence, implying a rapid evolution (Bernfield *et al.*, 1992). The highly conserved biglycan core protein also indicates that its structure is essential for function. On the other hand, since syndecans interact with ligands through their glycosaminoglycan chains, selection pressure may be maintaining the disposition of substitution sites, but not other core protein features.

Biglycan, also previously known as PG-1, and PG-S1 (reviewed in Wight *et al.*, 1991) has a quite different distribution compared with that of decorin and appears to be under different control with respect to the effect of TGF-β. Biglycan and decorin can be purified from bone as CSPGs, but from other sources are purified as dermatan sulfate proteoglycans (Neame *et al.*, 1989). Decorin is commonly substituted with one glycosaminoglycan chain close to the N terminus. In contrast, biglycan normally bears two glycosaminoglycan chains, both substituted close to the N terminus, giving rise to its name. Human biglycan has two more SG sequences located centrally in the core protein, but these do not appear to be substituted with glycosaminoglycan chains.

Studies on the distributions of biglycan and decorin have shown that these are substantially different and very frequently mutually exclusive (reviewed in Bianco *et al.*, 1990). While biglycan has been identified in some connective tissues, (e.g., the outer cap of prospective articular cartilage and articular soft tissues, including synovium, in embryonic human joints), it is most frequently associated with the surfaces of differentiating cells. Biglycan is restricted to the territorial matrix of chondrocyte isogenous groups in the upper proliferative zone of growth plates. Decorin, in contrast, is present in the interterritorial matrix. Decorin is abundant in such tissues as dermis, tendon, and cornea, but biglycan is either a minor component or totally absent. Biglycan is also a pericellular proteoglycan in capillary endothelial cells and in the differentiating layers of the epidermis, where it is associated with the surfaces of stratum spinosum and stratum granulosum keratinocytes. Interestingly, the proteoglycan appears to be absent from developing hair follicles. Biglycan is also present on the apical surfaces of epithelial cells lining the kidney collecting tubules. Decorin is not present in any of these locations (Bianco *et al.*, 1990).

In major blood vessels there is some overlap in distribution; both biglycan and decorin are present in the media and adventitia of the aorta, but biglycan is additionally found in the intima (Bianco *et al.*, 1990). This is consistent with tissue culture studies, which show that decorin and biglycan are both synthesized by vascular smooth muscle cells, but biglycan alone is synthesized by vascular endothelial cells (Järveläinen *et al.*, 1991). In the latter case, biglycan may be substituted with either one or two galactosaminoglycan chains. The cell surface distribution of biglycan, quite unlike that of decorin, fibromodulin, and lumican, which are all enriched in fibrillar collagenous matrices, is consistent with a lack of interaction with fibrillar collagens (reviewed in Wight *et al.*, 1991). Biglycan interactions with other pericellular molecules are not yet clear, but homology of biglycan with platelet glycoprotein Ib may indicate interaction with matrix glycoproteins such as von Willebrand factor (Fisher *et al.*, 1989). Biglycan is also homologous to morphogenesis regulating factors in *Drosophila* (Fisher *et al.*, 1989; Neame *et al.*, 1989), consistent with its association with differentiating cells. The leucine-rich repeats of decorin can sequester TGF-β and render it inactive (Yamaguchi *et al.*, 1990) and an analogous function for biglycan may allow this molecule to regulate growth and differentiation.

Another striking contrast between biglycan and decorin is that their synthesis is differentially regulated by TGF-β. Several studies have indicated that TGF-β upregulates biglycan mRNA levels and translated product in cultured fibroblasts (Breuer *et al.*, 1990; Kähäri *et al.*, 1991; Westergren-Thorsson *et al.*, 1991). In contrast decorin synthesis either is unaffected or decreases in response to TGF-β, and fibromodulin synthesis is not influenced. This diversity in response may be related to the divergent distribution of these two molecules *in vivo*. However, at the present time it is clear that very little of the interactions and function of the biglycan molecule at cell surfaces is understood.

V. Concluding Remarks

The pericellular region of many cells and tissues is populated by a variety of proteoglycans. It has become increasingly apparent that they are involved in the modulation of many cellular functions. Among these are regulation of growth factor activities, and adhesion responses of anchorage-dependent cells. In concert with increasing knowledge of proteoglycan activities, a great deal of molecular information concerning the structure of pericellular proteoglycans has emerged. It seems likely that there are more members of this group of molecules awaiting description, and there is much more to learn about the manner in which they exert their influence on cellular behavior. One of the more intriguing aspects of this group of molecules is also common to the cell surface group of adhesion proteins, the integrins. There appears to be both degeneracy (one protein species interacting with many ligands, e.g., syndecan 1) and redundancy (many different pericellular proteoglycans interacting with the same ligands, e.g., FGF binding to the cell surface and basement membrane proteoglycans). These phenomena are almost certainly more apparent than real and reflect the embryonic state of our knowledge. However, it is clear that cell surface proteoglycans can modulate the behavior of cells *in vitro* and presumably *in vivo*. Cell surface proteoglycans, through interactions with matrix molecules, alter cellular adhesive responses and cytoskeletal organization and thereby can determine shape and subsequent behavior, e.g., migration or anchorage, and differentiation. The mechanisms underlying the signaling are still unclear, but the example of the syndecans, in which the cytoplasmic domains are highly homologous between species and between syndecan types, is indicative of important signaling functions.

Molecular studies in the years ahead will likely reveal the mechanisms by which core proteins function. An added dimension for the proteoglycans is the importance of their glycosaminoglycan chains, whose structural requirements for ligand binding are still largely not understood. Neither is it clear how ligand binding to carbohydrate may induce core protein-mediated transmembrane signaling events. The next decade of proteoglycan biology should be exciting and provide some answers to these questions. Many of the most interesting challenges still lie ahead.

ACKNOWLEDGMENTS

We thank our colleagues at U.A.B. and elsewhere for helpful suggestions and discussions. We are particularly grateful to those who provided manuscripts not yet published. Some of the work reviewed here was supported by NIH Grants AR36457 and AR39741 to J.R.C. and AR20614 to the Multipurpose Arthritis Center, U.A.B. J.R.C. is an Established Investigator of the American Heart Association. Finally, we are deeply indebted to Ms. Cynthia Webster for excellent secretarial assistance.

References

Abrahamson, D. R., and St. John, P. L. (1990). Temporal changes in laminin distribution during glomerular development. *J. Am. Soc. Nephrol.* **1,** 49P.

Abrahamson, D. R., Irwin, M. H., St. John, P. L., Perry, E. W., Accavitti, M. A., Heck, L. W., and Couchman, J. R. (1989). Selective immunoreactivities of kidney basement membranes to monoclonal antibodies against laminin: Localization of the end of the long arm and the short arms to discrete microdomains. *J. Cell Biol.* **109,** 3477–3491.

Andres, J. L., DeFalcis, D., Noda, M., and Massagué, J. (1992). Binding of two growth factor families to separate domains of the proteoglycan betaglycan. *J. Biol. Chem.* **267,** 5927–5930.

Andres, J. L., Stanley, K., Cheifetz, S., and Massagué, J. (1989). Membrane-anchored and soluble forms of betaglycan, a polymorphic proteoglycan that binds transforming growth factor-β. *J. Cell Biol.* **109,** 3137–3145.

Aruffo, A., Stamenkovic, I., Melnick, M., Underhill, C. B., and Seed, B. (1990). CD44 is the principal cell surface receptor for hyaluronate. *Cell* **61,** 1303–1313.

Aumailley, M., Wiedemann, H., Mann, K., and Timpl, R. (1989). Binding of nidogen and the laminin–nidogen complex to basement membrane collagen type IV. *Eur. J. Biochem.* **184,** 241–248.

Beavan, L. A., Davies, M., Couchman, J. R., Williams, M. A., and Mason, R. M. (1989). *In vivo* turnover of the basement membrane and other heparan sulfate proteoglycans of rat glomerulus. *Arch. Biochem. Biophys.* **269,** 576–585.

Beavan, L. A., Carone, F. A., Nakamura, S., Jones, J. K., Reindel, J. F., and Price, R. G. (1991). Comparison of proteoglycans synthesized by porcine normal and polycystic renal tubular epithelial cells *in vitro*. *Arch. Biochem. Biophys.* **284,** 392–399.

Beck, K., Hunter, I., and Engel, J. (1990). Structure and function of laminin: Anatomy of a multidomain glycoprotein. *FASEB J.* **4,** 148–160.

Begovac, P. C., and Shur, B. D. (1990). Cell surface galactosyltransferase mediates the initiation of neurite outgrowth from PC12 cells on laminin. *J. Cell Biol.* **110,** 461–470.

Berg, E. L., Goldstein, L. A., Jutila, M. M., Nakache, M., Picker, L. J., Streeter, P. R., Wu, N. W., Zhou, D., and Butcher, E. C. (1989). Homing receptors and vascular addressins: Cell adhesion molecules that direct lymphocyte traffic. *Immunol. Rev.* **105,** 5–18.

Bernfield, M., and Sanderson, R. D. (1990). Syndecan, a developmentally regulated cell surface proteoglycan that binds extracellular matrix and growth factors. *Philos. Trans. R. Soc. London Ser. B* **327,** 171–186.

Bernfield, M., Banerjee, S. D., Koda, J. E., and Rapraeger, A. C. (1984). Remodelling of the basement membrane: Morphogenesis and maturation. *Ciba Found. Symp.* **108,** 179–196.

Bernfield, M., Kokenyesi, R., Kato, M., Hinkes, M. T., Spring, J., Gallo, R. L., and Lose, E. J. (1992). Biology of the syndecans. *Annu. Rev. Cell Biol.* **8,** 365–393.

Bianco, P., Fisher, L. W., Young, M. F., Termine, J. D., and Robey, P. G. (1990). Expression and localization of the two small proteoglycans biglycan and decorin in developing human skeletal and non-skeletal tissues. *J. Histochem. Cytochem.* **38,** 1549–1563.

Blochberger, T. C., Vergnes, J.-P., Hempel, J., and Hassell, J. R. (1992). cDNA to chick lumican (corneal keratan sulfate proteoglycan) reveals homology to the small interstitial proteoglycan gene family and expression in muscle and intestine. *J. Biol. Chem.* **267,** 347–352.

Boudreau, N., Turley, E., and Rabinovitch, M. (1991). Fibronectin, hyaluronan, and a hyaluronan binding protein contribute to increased ductus arteriosus smooth muscle cell migration. *Dev. Biol.* **143,** 235–247.

Bourdon, M. A., Krusius, T., Campbell, S., Schwartz, N. B., and Ruoslahti, E. (1987). Identification and synthesis of a recognition signal for the attachment of glycosaminoglycans to proteins. *Proc. Natl. Acad. Sci. U.S.A.* **84,** 3194–3198.

Breuer, B., Schmidt, G., and Kresse, H. (1990). Non-uniform influence of transforming growth factor-β on the biosynthesis of different forms of small chondroitin sulphate/dermatan sulphate proteoglycan. *Biochem. J.* **269,** 551–554.

Briggaman, R. A. (1983). The epidermal–dermal junction and genetic disorders of this area. *In* "Biochemistry and Physiology of the Skin" (L. A. Goldsmith, ed.), pp. 1001–1024. Oxford Univ. Press, New York.

Brown, D. M., Klein, D. J., Michael, A. F., and Oegema, T. R. (1982). ^{35}S-Glycosaminoglycan and ^{35}S-glycopeptide metabolism by diabetic glomeruli and aorta. *Diabetes* **31,** 418–425.

Brown, T. A., Bouchard, T., St. John, T., Wayner, E., and Carter, W. G. (1991). Human keratinocytes express a new CD44 core protein (CD44E) as a heparan-sulfate intrinsic membrane proteoglycan with additional exons. *J. Cell Biol.* **113,** 207–221.

Brunner, G., Gabrilove, J., Rifkin, D. B., and Wilson, E. L. (1991). Phospholipase C release of basic fibroblast growth factor from human bone marrow cultures as a biologically active complex with a phosphatidylinositol-anchored heparan sulfate proteoglycan. *J. Cell Biol.* **114,** 1275–1283.

Burgeson, R. E., Lunstrum, G. P., Rokosova, B., Rimberg, C. S., Rosenbaum, L. M., and Keene, D. R. (1990). The structure and function of type VII collagen. *Annu. N. Y. Acad. Sci.* **580,** 32–43.

Burridge, K., Fath, K., Kelley, T., Nuckolls, G., and Turner, C. (1988). Focal adhesions: Transmembrane junctions between the extracellular matrix and the cytoskeleton. *Annu. Rev. Cell Biol.* **4,** 487–525.

Butkowski, R. J., Wieslander, J., Kleppel, M., Michael, A. F., and Fish, A. J. (1989). Basement membrane collagen in the kidney: Regional localization of novel chains related to collagen IV. *Kidney Int.* **35,** 1195–1202.

Carey, D. J., and Evans, D. M. (1989). Membrane anchoring of heparan sulfate proteoglycans by phosphatidylinositol and kinetics of synthesis of peripheral and detergent-solubilized proteoglycans in Schwann cells. *J. Cell Biol.* **108,** 1891–1897.

Carey, D. J., Crumbling, D. M., Stahl, R. C., and Evans, D. M. (1990). Association of cell surface heparan sulfate proteoglycans of Schwann cells with extracellular matrix proteins. *J. Biol. Chem.* **265,** 20627–20633.

Carey, D. J., and Stahl, R. C. (1990). Identification of a lipid-anchored heparan sulfate proteoglycan in Schwann cells. *J. Cell Biol.* **111,** 2053–2062.

Carey, D. J., Evans, D. M., Stahl, R. C., Asundi, V. K., Conner, K. J., Garbes, P., and Cizmeci-Smith, G. (1992). Molecular cloning and characterization of *N*-syndecan, a novel transmembrane heparan sulfate proteoglycan. *J. Cell Biol.* **117,** 191–201.

Carone, F. A., Makino, H., and Kanwar, Y. S. (1988). Basement membrane antigens in renal polycystic disease. *Am. J. Pathol.* **130,** 466–471.

Carone, F. A., Hollenberg, P. F., Nakamura, S., Punyarit, P., Glagowski, W., and Flouret, G. (1989). Tubular basement membrane change occurs pari passu with development of cyst formation. *Kidney Int.* **35,** 1034–1040.

Chajek-Shaul, T., Halima, O., Ben-Naim, M., Stein, O., and Stein, Y. (1989). Phosphatidylinositol-specific phospholipase C releases lipoprotein lipase from the heparin releasable pool in rat heart cell cultures. *Biochim. Biophys. Acta.* **1014,** 178–183.

Cheifetz, S., and Massagué, J. (1989). The TGF-β receptor proteoglycan: Cell surface expression and ligand binding in the absence of glycosaminoglycan chains. *J. Biol. Chem.* **264,** 12025–12028.

Claesson, L., Larhammar, D., Rask, L., and Peterson, P. A. (1983). cDNA clone for the human invariant gamma chain of class II histocompatibility antigens and its implications for the protein structure. *Proc. Natl. Acad. Sci. USA* **80,** 7395–7399.

Clément, B., Segui-Real, B., Hassell, J. R., Martin, G. R., and Yamada, Y. (1989). Identification of a cell surface-binding protein for the core protein of the basement membrane proteoglycan. *J. Biol. Chem.* **264,** 12467–12471.

Cohen, M. P., and Surma, M. L. (1981). [^{35}S]Sulfate incorporation in glomerular basement membrane glycosaminoglycans is decreased in experimental diabetes. *J. Lab. Clin. Med.* **98,** 715–722.

Cohn, R. H., Banerjee, S. D., and Bernfield, M. R. (1977). Basal lamina of embryonic salivary epithelia. *J. Cell Biol.* **73,** 464–478.

Couchman, J. R. (1987). Heterogeneous distribution of a basement membrane heparan sulfate proteoglycan in rat tissues. *J. Cell Biol.* **105,** 1901–1916.

Couchman, J. R., and Ljubimov, A. V. (1989). Mammalian tissue distribution of a large heparan sulfate proteoglycan detected by monoclonal antibodies. *Matrix* **9,** 311–321.

Couchman, J. R., Caterson, B., Christner, J. E., and Baker, J. R. (1984). Mapping by monoclonal antibody detection of glycosaminoglycans in connective tissues. *Nature (London)* **307,** 650–652.

Couchman, J. R., Woods, A., Höök, M., and Christner, J. E. (1985). Characterization of a dermatan sulfate proteoglycan synthesized by murine parietal yolk sac (PYS-2) cells. *J. Biol. Chem.* **260,** 13755–13762.

Couchman, J. R., Austria, R., Woods, A., and Hughes, R. C. (1988). Adhesion defective BHK cell mutant has cell surface heparan sulfate proteoglycan of altered properties. *J. Cell Physiol.* **136,** 226–236.

Couchman, J. R., King, J. L., and McCarthy, K. J. (1990). Distribution of two basement membrane proteoglycans through hair follicle development and hair growth cycle in the rat. *J. Invest. Dermatol.* **94,** 65–70.

Couchman, J. R., Abrahamson, D. R., and McCarthy, K. J. (1993). Basement membrane proteoglycans and development. *Kidney Int.* **43,** 79–84.

Culp, L. A., Laterra, J., Lark, M. W., Beyth, R. J., and Tobey, S. L. (1986). Heparan sulphate proteoglycan as mediator of some adhesive responses and cytoskeletal reorganization of cells on fibronectin matrices: Independent versus cooperative functions. *Ciba Found. Symp.* **124,** 158–183.

Danielson, K. G., Martinez-Hernandez, A., Hassell, J. R., and Iozzo, R. (1992). Establishment of a cell line from the EHS tumor: Biosynthesis of basement membrane constituents and characterization of a hybrid proteoglycan containing heparan and chondroitin sulfate chains. *Matrix* **11,** 22–35.

David, G., Lories, V., Decock, B., Marynen, P., Cassiman, J.-J., and Van den Berghe, H. (1990). Molecular cloning of a phosphatidylinositol-anchored membrane heparan sulfate proteoglycan from human lung fibroblasts. *J. Cell Biol.* **111,** 3165–3176.

David, G., van der Schueren, B., Marynen, P., Cassiman, J.-J., and Van den Berghe, H. (1992). Molecular cloning of amphiglycan, a novel integral membrane heparan sulfate proteoglycan expressed by epithelial and fibroblastic cells. *J. Cell Biol.* **118,** 961–969.

Davis, C. G., Lehrman, M. A., Russell, D. W., Anderson, R. G. W., Brown, M. S., and Goldstein, J. L. (1986). The J. D. mutation in familial hypercholesterolemia: Amino acid substitution in cytoplasmic domain impedes internalization of LDL receptors. *Cell* **45,** 15–24.

Day, A. A., McQuillan, C. I., Termine, J. D., and Young, M. R. (1987). Molecular cloning and sequence analysis of the cDNA for small proteoglycan II of bovine bone. *Biochem. J.* **248,** 801–805.

Dodge, G. R., Kovalszky, I., Chu, M.-L., Hassell, J. R., McBride, O. W., Yi, H. F., and Iozzo, R. V. (1991). Heparan sulfate proteoglycan of human colon: Partial molecular cloning, cellular expression, and mapping of the gene (HSPG2) to the short arm of human chromosome 1. *Genomics* **10,** 673–680.

Drake, S., Klein, D. J., Mickelson, D. J., Oegema, T. R., Furcht, L. T., and McCarthy, J. B. (1992). Cell surface phosphatidylinositol-anchored heparan sulfate proteoglycan initiates mouse melanoma cell adhesion to a fibronectin-derived heparin binding synthetic peptide. *J. Cell Biol.* **117,** 1331–1342.

Durkin, M. E., Chakravarti, S., Bartos, B. B., Liu, S. H., Friedman, R. L., and Chung, A. E. (1988). Amino acid sequence and domain structure of entactin: Homology with epidermal growth factor precursor and low density lipoprotein receptor. *J. Cell Biol.* **107,** 2749–2756.

Dziadek, M., Fujiwara, S., Paulsson, M., and Timpl, R. (1985). Immunological characterization of basement membrane types of heparan sulfate proteoglycans. *EMBO J.* **4,** 905–912.

Ebihara, I., Killen, P. D., Laurie, G. W., Huang, T., Yamada, Y., Martin, G. R., and Brown, K. S. (1988). Altered mRNA expression of basement membrane components in a murine model of polycystic kidney disease. *Lab. Invest.* **58,** 262–269.

Ehrig, K., Leivo, I., Argraves, W. S., Ruoslahti, E., and Engvall, E. (1990). Merosin, a tissue-specific basement membrane protein, is a laminin-like protein. *Proc. Natl. Acad. Sci. USA* **87,** 3264–3268.

Elenius, K., Salmivirta, M., Inta P., Mali, M., and Jalkanen, M. (1990). Binding of human syndecan to extracellular matrix proteins. *J. Biol. Chem.* **265,** 17837–17843.

Elenius, K., Vaino, S., Laato, M., Salmivirta, M., Thesleff, I., and Jalkanen, M. (1991). Induced expression of syndecan in healing wounds. *J. Cell Biol.* **114,** 585–595.

Elenius, K., Määttä, A., Salmivirta, M., and Jalkanen, M. (1992). Growth factors induce 3T3 cells to express bFGF-binding syndecan. *J. Biol. Chem.* **267,** 6435–6441.

Esko, J. D. (1991). Genetic analysis of proteoglycan structure, function and metabolism. *Curr. Opin. Cell Biol.* **3,** 805–816.

Faassen, A. E., Schrager, J. A., Klein, D. J., Oegema, T. R., Couchman, J. R., and McCarthy, J. B. (1992). A cell surface chondroitin sulfate proteoglycan, immunologically related to CD44, is involved in Type I collagen-mediated melanoma cell motility and invasion. *J. Cell Biol.* **116,** 521–531.

Farquhar, M. G. (1991). The glomerular basement membrane: A selective macromolecular filter. *In* "Cell Biology of Extracellular Matrix" (E. D. Hay, ed.), 2nd Ed., pp. 365–418. Plenum, New York.

Fedarko, N. S., Ishihara, M., and Conrad, H. E. (1989). Control of cell division in hepatoma cells by exogenous heparan sulfate proteoglycan. *J. Cell Physiol.* **139,** 287–294.

Fine, J.-D. (1987). Altered skin basement membrane antigen defect in dystrophic forms of epidermolysis bullosa. *Curr. Prob. Dermatol.* **17,** 111–126.

Fine, J.-D., and Couchman, J. R. (1989). Chondroitin 6-sulfate proteoglycan but not heparan sulfate proteoglycan is abnormally expressed in skin basement membrane from patients with dominant and recessive dystrophic epidermolysis bullosa. *J. Invest. Dermatol.* **92,** 611–616.

Fisher, L. W., Termine, J. D., and Young, M. F. (1989). Deduced protein sequence of bone small proteoglycan I (biglycan) shows homology with proteoglycan II (decorin) and several nonconnective tissue proteins in a variety of species. *J. Biol. Chem.* **264,** 4571–4576.

Fox, J. W., Mayer, U., Nischt, R., Aumailley, M., Reinhardt, D., Wiedemann, H., Mann, K., Timpl, R., Krieg, T., Engel, J., and Chu, M.-L. (1991). Recombinant nidogen consists of three globular domains and mediates binding of laminin to collagen type IV. *EMBO J.* **10,** 3137–3146.

Fujiwara, S., Wiedemann, H., Timpl, R., Lustig, A., and Engel, J. (1984). Structure and interactions of heparan sulfate proteoglycans from a mouse tumor basement membrane. *Eur. J. Biochem.* **143,** 145–157.

Funderburg, F. M., and Markwald, R. R. (1986). Conditioning of native substrates by chondroitin sulfate proteoglycans during cardiac mesenchymal migration. *J. Cell Biol.* **103,** 2475–2487.

Gallagher, J. T. (1989). The extended family of proteoglycans: Social residents of the pericellular zone. *Curr. Opin. Cell Biol.* **1,** 1201–1218.

Gallagher, J. T., Turnbull, J. E., and Lyon, M. (1992). Patterns of sulphation in heparan sulphate: Polymorphism based on a common structural theme. *Int. J. Biochem.* **24,** 553–560.

Germino, G. G., Weinstat-Saslow, D., Himmelbauer, H., Gillespie, G. A. J., Somlo, S., Wirth, B., Barton, N., Harris, K. L., Frischauf, A.-M., and Reeders, S. T. (1992). The gene for autosomal dominant polycystic kidney disease lies in a 750-kb CpG-rich region. *Genomics* **13,** 144–151.

Gibson, W. T., Couchman, J. R., and Weaver, A. C. (1983). Fibronectin distribution during the development of fetal rat skin. *J. Invest. Dermatol.* **81,** 480–485.

Goldstein, L. A., Zhou, D. F. H., Picker, L. J., Minty, C. N., Bargatze, R. F., Ding, J. F., and Butcher, E. C. (1989). A human lymphocyte homing receptor, the hermes antigen, is related to cartilage proteoglycan core and link proteins. *Cell* **56,** 1063–1072.

Gougos, A., and Letarte, M. (1990). Primary structure of endoglin, an RGD-containing glycoprotein of human endothelial cells. *J. Biol. Chem.* **265,** 8361–8364.

Gould, S. E., Upholt, W. B., and Kosher, R. A. (1992). Syndecan 3: A member of the syndecan family of membrane-intercalated proteoglycans that is expressed in high amounts at the onset of chicken limb cartilage differentiation. *Proc. Natl. Acad. Sci. USA* **89,** 3271–3275.

Grant, D. S., and LeBlond, C. P. (1988). Immunogold quantitation of laminin, type IV collagen, and heparan sulfate proteoglycan in a variety of basement membranes. *J. Histochem. Cytochem.* **36,** 271–283.

Grant, D. S., LeBlond, C. P., Kleinman, H. K., Inoue, S., and Hassell, J. R. (1989). The incubation of laminin, collagen IV, and heparan sulfate proteoglycan at 35°C yields basement membrane-like structures. *J. Cell Biol.* **198,** 1567–1574.

Günthert, U., Hofmann, M., Rudy, W., Reber, S., Zöller, M., Haußman, I., Matzku, S., Wenzel, A., Ponta, H., and Herrlich, P. (1991). A new variant of glycoprotein CD44 confers metastatic potential to rat carcinoma cells. *Cell* **65,** 13–24.

Gunwar, S., Saus, J., Noelken, M. E., and Hudson, B. G. (1990). Glomerular basement membrane: Identification of a fourth chain, $\alpha 4$, of type IV collagen. *J. Biol. Chem.* **265,** 5466–5469.

Haaparanta, T., Uitto, J., Ruoslahti, E., and Engvall, E. (1991). Molecular cloning of the cDNA encoding human laminin A chain. *Matrix* **11,** 151–160.

Harper, J. R., and Reisfeld, R. A. (1987). Cell-associated proteoglycans in human malignant melanoma. *In* "Biology of Proteoglycans" (T. N. Wight and R. P. Mecham, eds.), pp. 345–366. Academic Press, Orlando.

Hassell, J. R., Robey, P. G., Barrach, H.-J., Wilczek, J., Rennard, S. I., and Martin, G. R. (1980). Isolation of a heparan sulfate-containing proteoglycan from basement membrane. *Proc. Natl. Acad. Sci. USA* **77,** 4494–4498.

Hassell, J. R., Leyshon, W. C., Ledbetter, S. R., Tyree, B., Suzuki, S., Kato, M., Kimata, K., and Kleinman, H. K. (1985). Isolation of two forms of basement membrane proteoglycans. *J. Biol. Chem.* **260,** 8098–8105.

Hassell, J. R., Schrecengost, P. K., Rada, J. A., SundarRaj, N., Sossi, G., and Thoft, R. A. (1992). Biosynthesis of stromal matrix proteoglycans and basement membrane components by human corneal fibroblasts. *Invest. Ophthalmol. Visual Sci.* **33,** 547–557.

Hayashi, K., Madri, J., Trelstad, R., and Yurchenco, P. (1990). Heparan sulfate proteoglycan–vascular cell interactions. *J. Cell Biol.* **111,** 11a.

Haynes, B. F., Telen, M. J., Hale, L. P., and Denning, S. M. (1989). CD44: A molecular involved in leukocyte adherence and T-cell activation. *Immunol. Today.* **10,** 423–428.

Heremans, A., Van Der Schueren, B., De Cock, B., Paulsson, M., Cassiman, J.-J., Van Den Berghe, H., and David, G. (1989). Matrix-associated heparan sulfate proteoglycan: Core protein-specific monoclonal antibodies decorate the pericellular matrix of connective tissue cells and the stromal side of basement membranes. *J. Cell Biol.* **109,** 3199–3211.

Heremans, A., De Cock, B., Cassiman, J.-J., Van Den Berghe, H., and David, G. (1990). The core protein of matrix-associated heparan sulfate proteoglycan binds to fibronectin. *J. Biol. Chem.* **265,** 8716–8724.

Holbrook, K. A., and Smith, L. T. (1981). Ultrastructural aspects of human skin during embryonic, fetal, neonatal and adult periods of life; *In* "Morphogenesis and Malformations of the Skin" (R. J. Blandau, ed.), Vol. 17, No. 2, pp. 9–38. Alan R. Liss, New York.

Höök, M., Woods, A., Johansson, S., Kjellén, L., and Couchman, J. R. (1986). Functions of proteoglycans at the cell surface. *Ciba Found. Symp.* **124,** 143–157.

Horiguchi, Y., Couchman, J. R., Ljubimov, A. V., Yamasaki, H., and Fine, J.-D. (1989). Distribution, ultrastructural localization, and ontogeny of the core protein of a heparan sulfate proteoglycan in human skin and other basement membranes. *J. Histochem. Cytochem.* **37,** 961–970.

Horiguchi, Y., Fine, J.-D., and Couchman, J. R. (1991). Human skin basement membrane-associated heparan sulphate proteoglycan: Distinctive differences in ultrastructural localization as a function of developmental age. *Br. J. Dermatol.* **124,** 410–414.

Hostikka, S. L., Eddy, R. L. Byers, M. G., Höyhtyä, M., Shows, T. B., and Tryggvason, K. (1990). Identification of a distinct type IV collagen α chain with restricted kidney distribution and assignment of its gene to the locus of X chromosome-linked Alport syndrome. *Proc. Natl. Acad. Sci. USA* **87,** 1606–1610.

Hunter, D. D., Shah, V., Merlie, J. P., and Sanes, J. R. (1989). A laminin-like adhesive protein concentrated in the synaptic cleft of the neuromuscular junction. *Nature (London)* **338,** 229–234.

Hynes, R. O. (1990). "Fibronectins," p. 546. Springer-Verlag, New York.

Hynes, R. O. (1992). Integrins: Versatility, modulation, and signaling in cell adhesion. *Cell* **69,** 11–25.

Inoue, S., LeBlond, C. P., and Laurie, G. W. (1983). Ultrastructure of Reichert's membrane, a multilayered basement membrane in the parietal wall of the rat yolk sac. *J. Cell Biol.* **97,** 1524–1537.

Inoue, S., Grant, D., and LeBlond, C. P. (1989). Heparan sulfate proteoglycan is present in basement membrane as a double-tracked structure. *J. Histochem. Cytochem.* **37,** 597–602.

Iozzo, R. V., and Clark, C. C. (1986). Biosynthesis of proteoglycans by rat embryo parietal yolk sac cells in organ culture. *J. Biol. Chem.* **261,** 6658–6669.

Iozzo, R. V., Kovalszky, I., Hacobian, N., Schick, P. K., Ellingson, J. S., and Dodge, G. R. (1990). Fatty acylation of heparan sulfate proteoglycan from human colon carcinoma cells. *J. Biol. Chem.* **265,** 19980–19989.

Ishihara, M., and Conrad, H. E. (1989). Correlations between heparan sulfate metabolism and hepatoma cell growth. *J. Cell Physiol.* **138,** 467–476.

Ishihara, M., Fedarko, N. S., and Conrad, H. E. (1987). Involvement of phosphatidylinositol and insulin in the coordinate regulation of proteoheparan sulfate metabolism and hepatocyte growth. *J. Biol. Chem.* **262,** 4708–4716.

Isemura, M., Sato, N., Yamaguchi, Y., Aikawa, J., Munakata, H., Hayashi, N., Yosizawa, Z., Nakamura, T., Kubota, A., Arakawa, M., and Hsu, C. (1987). Isolation and characterization of a fibronectin-binding proteoglycan carrying both heparan sulfate and dermatan sulfate chains from human placenta. *J. Biol. Chem.* **262,** 8926–8933.

Iwata, M., and Carlson, S. S. (1991). A large chondroitin sulfate basement membrane-associated proteoglycan exists as a disulfide-stabilized complex of several proteins. *J. Biol. Chem.* **266,** 323–333.

Izzard, C. S., Radinsky, R., and Culp, L. A. (1986). Substratum contacts and cytoskeletal reorganization of Balb/3T3 cells on a cell-binding fragment and heparin-binding fragments of plasma fibronectin. *Exp. Cell Res.* **165,** 320–336.

Jackman, R. W., Beeler, D. L., van der Water, L., and Rosenberg, R. D. (1986). Characterization of a thrombomodulin cDNA reveals structural similarity to the low density lipoprotein receptor. *Proc. Natl. Acad. Sci. USA* **83,** 8834–8838.

Jackson, R. L., Busch, S. J., and Cardin, A. D. (1991). Glycosaminoglycans: Molecular properties, protein interactions, and role in physiological processes. *Physiol Rev.* **2,** 481–485.

Jacobson, K., O'Dell, D., and August, T. J. (1984). Lateral diffusion of an 80,000-dalton glycoprotein in the plasma membrane of murine fibroblasts: Relationships to cell structure and function. *J. Cell Biol.* **99,** 1624–1633.

Jalkanen, M., Rapraeger, A., Saunders, S., and Bernfield, M. (1987). Cell surface proteoglycan of mouse mammary epithelial cells is shed by cleavage of its matrix-binding ectodomain from its membrane-associated domain. *J. Cell Biol.* **105,** 3087–3096.

Jalkanen, M., Jalkanen, S., and Bernfield, M. (1991). Binding of extracellular effector molecules by cell surface proteoglycans. *In* "Receptors for Extracellular Matrix" (J. A. McDonald and R. P. Mecham, eds.), pp. 1–37. Academic Press, San Diego.

Jalkanen, S., and Jalkanen, M. (1992). Lymphocyte CD44 binds the COOH-terminal heparin-binding domain of fibronectin. *J. Cell Biol.* **116,** 817–825.

Jalkanen, S., Bargatze, R., Herron, L. R., and Butcher, E. C. (1986). A lymphoid cell surface glycoprotein involved in endothelial cell recognition and lymphocyte homing in man. *Eur. J. Immunol.* **16,** 1195–1202.

Jalkanen, S., Jalkanen, M., Bargatze, R., Tammi, M., and Butcher, E. C. (1988). Biochemical properties of glycoproteins involved in lymphocyte recognition of high endothelial venules in man. *J. Immunol.* **141,** 1615–1623.

Järveläninen, H. T., Kinsella, M. G., Wight, T. N., and Sandell, L. J. (1991). Differential expression of small chondroitin/dermatan sulfate proteoglycans, PG-I/biglycan and PG-II/decorin, by vascular smooth muscle and endothelial cells in culture. *J. Biol. Chem.* **266,** 23274–23281.

Johansson, S., and Höök, M. (1980). Heparin enhances the rate of binding of fibronectin to collagen. *Biochem. J.* **187,** 521–524.

Kähäri, V.-M., Larjava, H., and Uitto, J. (1991). Differential regulation of extracellular matrix proteoglycan (PG) gene expression: Transforming growth factor-β1 up-regulates biglycan (PGI), and versican (large fibroblast PG) but down-regulates decorin (PGII) mRNA levels in human fibroblasts in culture. *J. Biol. Chem.* **266,** 10608–10615.

Kallunki, P., and Tryggvason, K. (1992). Human basement membrane heparan sulfate proteoglycan core protein: A 467-kD protein containing multiple domains resembling elements of the low density lipoprotein receptor, laminin, neural cell adhesion molecules, and epidermal growth factor. *J. Cell Biol.* **116,** 559–571.

Kanwar, Y. S., Linker, A., and Farquhar, M. G. (1980). Increased permeability of glomerular basement membrane to ferritin after removal of glycosaminoglycans (heparan sulfate) by enzyme digestion. *J. Cell Biol.* **86,** 688–693.

Kanwar, Y. S., Veis, A., Kimura, J. H., and Jakubowski, M. L. (1984). Characterization of heparan sulfate-proteoglycan of glomerular basement membranes. *Proc. Natl. Acad. Sci. USA* **81,** 762–766.

Kato, M., and Bernfield, M. (1990). Epithelial cells made deficient in syndecan lose cell surface expression of E-cadherin and β1-integrins. *J. Cell Biol.* **111,** 263a.

Kato, M., Koike, Y., Ito, Y., Suzuki, S., and Kimata, K. (1987). Multiple forms of heparan sulfate proteoglycans in the Engelbreth–Holm–Swarm tumor: The occurrence of high density forms bearing heparan sulfate and chondroitin sulfate side chains. *J. Biol. Chem.* **262,** 7180–7188.

Kato, M., Koike, Y., Suzuki, S., and Kimata, K. (1988). Basement membrane proteoglycan in various tissues: Characterization using monoclonal antibodies to the Englebreth–Holm–Swarm mouse tumor low density heparan sulfate proteoglycan. *J. Cell Biol.* **106,** 2203–2210.

Kiefer, M. C., Stephans, J. C., Crawford, K., Okino, K., and Barr, P. J. (1990). Ligand-affinity cloning and structure of a cell surface heparan sulfate proteoglycan that binds basic fibroblast growth factor. *Proc. Natl. Acad. Sci. USA* **87,** 6985–6989.

Kinsella, M. G., and Wight, T. N. (1986). Modulation of sulfated proteoglycan synthesis by bovine aortic endothelial cells during migration. *J. Cell Biol.* **102,** 679–687.

Kjellén, L., and Lindahl, U. (1991). Proteoglycans: Structures and interactions. *Annu. Rev. Biochem.* **60,** 443–475.

Klein, D. J., Brown, D. M., and Oegema, T. R. (1986). Glomerular proteoglycans in diabetes. *Diabetes* **36,** 374–381.

Klein, D. J., Brown, D. M., Oegema, T. R., Brenchley, P. E., Anderson, J. C., Dickinson, M. A. J., Horigan, E. A., and Hassell, J. R. (1988). Glomerular basement membrane proteoglycans are derived from a large precursor. *J. Cell Biol.* **106,** 963–970.

Klein, D. J., Oegema, T. R., and Brown, D. M. (1989). Release of glomerular heparan-$^{35}SO_4$ proteoglycan by heparin from glomeruli of streptozocin-induced diabetic rats. *Diabetes* **38,** 130–139.

Klein, G., Ekblom, M., Fecker, L., Timpl, R., and Ekblom, P. (1990). Differential expression of laminin A and B chains during development of embryonic mouse organs. *Development* **110,** 823–837.

Kleinman, H. K., and Weeks, B. S. (1989). Laminin: Structure, functions and receptors. *Curr. Opin. Cell Biol.* **1,** 964–967.

Kleinman, H. K., Weeks, B. S., Cannon, F. B., Sweeney, T. M., Sephel, G. C., Clément, B., Zain, M., Olson, M. O., Jucker, M., and Burrous, B. A. (1991). Identification of a 110-kDa nonintegrin cell surface laminin-binding protein which recognizes an A chain neurite-promoting peptide. *Arch. Biochem. Biophys.* **290,** 320–325.

Koda, J. E., Rapraeger, A., and Bernfield, M. (1985). Heparan sulfate proteoglycans from mouse mammary epithelial cells: Cell surface proteoglycan as a receptor for interstitial collagens. *J. Biol. Chem.* **260,** 8157–8162.

Kojima, T., Leone, C. W., Marchildon, G. A., Marcum, J. A., and Rosenberg, R. D. (1992a). Isolation and characterization of heparan sufate proteoglycans produced by cloned rat microvascular endothelial cells. *J. Biol. Chem.* **267,** 4859–4869.

Kojima, T., Shworak, N. W., and Rosenberg, R. D. (1992b). Molecular cloning and expression of two distinct cDNA-encoding heparan sulfate proteoglycan core proteins from a rat endothelial cell line. *J. Biol. Chem.* **267,** 4870–4877.

Kolset, S. O., and Gallagher, J. T. (1990). Proteoglycans in haemopoietic cells. *Biochim. Biophys. Acta.* **1032,** 191–211.

Krusius, T., and Ruoslahti, E. (1986). Primary structure of an extracellular matrix proteoglycan core protein deduced from cloned cDNA. *Proc. Natl. Acad. Sci. USA* **83,** 7683–7687.

Ktistakis, N. T., Thomas, D., and Roth, M. G. (1990). Characteristics of the tyrosine recognition signal for internalization of transmembrane surface glycoprotein. *J. Cell Biol.* **111,** 1393–1407.

Kvist, S., Wiman, K., Claesson, L., Peterson, P. A., and Dobberstein, B. (1982). Membrane insertion and oligomeric assembly of HLA-DR histocompatibility antigens. *Cell* **29,** 61–69.

Lane, M. C., and Solursh, M. (1991). Primary mesenchyme cell migration requires a chondroitin sulfate/dermatan sulfate proteoglycan. *Dev. Biol.* **143,** 389–397.

Laurie, G. W., LeBlond, C. P., and Martin, G. R. (1982). Localization of type IV collagen, laminin, heparan sulfate proteoglycan, and fibronectin to the basal lamina of basement membranes. *J. Cell Biol.* **95,** 340–344.

Laurie, G. W., LeBlond, C. P., Inoue, S., Martin, G. R., and Chung, A. (1984). Fine structure of the glomerular basement membrane and immunolocalization of five basement membrane components to the lamina densa (basal lamina) and its extensions in both glomeruli and tubules of the rat kidney. *Am. J. Anat.* **169,** 463–481.

LeBaron, R. G., Esko, J. D., Woods, A., Johansson, S., and Höök, M. (1988). Adhesion of glycosaminoglycan-deficient Chinese Hamster ovary cell mutants to fibronectin substrata. *J. Cell Biol.* **106,** 945–952.

LeBlond, C. P., and Inoue, S. (1989). Structure, composition, and assembly of basement membrane. *Am. J. Anat.* **185,** 367–390.

Ledbetter, S. R., Tyree, B., Hassell, J. R., and Horigan, E. A. (1985). Identification of the precursor protein to basement membrane heparan sulfate proteoglycans. *J. Biol. Chem.* **260,** 8106–8113.

Ledbetter, S., Copeland, E. J., Noonan, D., Vogeli, G., and Hassell, J. R. (1990). Altered steady-state mRNA levels of basement membrane proteins in diabetic mouse kidneys and thromboxane synthase inhibition. *Diabetes* **39,** 196–203.

Lelongt, B., Makino, H., and Kanwar, Y. S. (1987). Status of glomerular proteoglycans in aminonucleoside nephrosis. *Kidney Int.* **31,** 1299–1310.

Lelongt, B., Carone, F. A., and Kanwar, Y. S. (1988). Decreased de novo synthesis of proteoglycans in drug-induced renal cystic disease. *Proc. Natl. Acad. Sci. USA* **85,** 9047–9051.

Leppä, S., Härkönen, P., and Jalkanen, M. (1991). Steroid-induced epithelial-fibroblastic conversion associated with syndecan suppression in S115 mouse mammary tumor cells. *Cell Regul.* **2,** 1–11.

Lesley, J., He, Q., Miyake, K., Hamann, A., Hyman, R., and Kincade, P. W. (1992). Requirements for hyaluronic acid binding by CD44: A role for the cytoplasmic domain and activation by antibody. *J. Exp. Med.* **175,** 257–266.

Lewandowska, K. Choi, H. U., Rosenberg, L. C., Zardi, L., and Culp, L. A. (1987). Fibronectin-mediated adhesion of fibroblasts: Inhibition by dermatan sulfate and evidence for a cryptic glycosaminoglycan-binding domain. *J. Cell Biol.* **105,** 1443–1454.

Li, M. L., Aggeler, J., Farson, D. A., Hatier, C., Hassell, J., and Bissell, M. J. (1987). Influence of a reconstituted basement membrane and its components on casein gene expression and secretion in mouse mammary epithelial cells. *Proc. Natl. Acad. Sci. USA* **84,** 136–140.

Ljubimov, A. V., Bartek, J., Couchman, J. R., Kapuller, L. L., Veselov, V. V., Kovarik, J., Peravoshchikov, A. G., and Krutovskikh, V. S. (1992). Distribution of individual components of basement membrane in human colon polyps and adenocarcinomas as revealed by monoclonal antibodies. *Int. J. Cancer* **50,** 562–566.

López-Casillas, F., Cheifetz, S., Doody, J., Andres, J. L., Lane, W. S., and Massagué, J. (1991). Structure and expression of the membrane proteoglycan betaglycan, a component of the TGF-β receptor system. *Cell* **67,** 785–795.

Lories, V., Cassiman, J.-J., Van den Berghe, H., and David, G. (1992). Differential expression of cell surface heparan sulfate proteoglycans in human mammary epithelial cells and lung fibroblasts. *J. Biol. Chem.* **267,** 1116–1122.

Lyon, A. W., Narindrasorasak, S., Young, I. D., Anastassiades, T., Couchman, J. R., McCarthy, K. J., and Kisilevsky, R. (1991). Co-deposition of basement membrane components during the induction of murine splenic AA amyloid. *Lab. Invest.* **64,** 785–790.

Mali, M., Jaakkola, P., Arvilommi, A.-M., and Jalkanen, M. (1990). Sequence of human syndecan indicates a novel gene family of integral membrane proteoglycans. *J. Biol. Chem.* **265,** 6884–6889.

Mann, K., Deutzmann, R., Aumailley, M., Timpl, R., Raimondi, L., Yamada, Y., Pan, T. C., Conway, D., and Chu, M. L. (1988). Amino acid sequence of mouse nidogen, a multidomain basement membrane protein with binding activity for laminin, collagen IV and cells. *EMBO J.* **8,** 65–72.

Martin, G. R., Timpl, R., and Kühn, K. (1988). Basement membrane proteins: Molecular structure and function. *Adv. Protein Chem.* **39,** 1–50.

Marynen, P., Zhang, J., Cassiman, J.-J., Van den Berghe, H., and David, G. (1989). Partial primary structure of the 48- and 90-kilodalton core proteins of cell surface-associated heparan sulfate proteoglycans of lung fibroblasts: Prediction of an integral membrane domain and evidence for multiple distinct core proteins at the cell surface of human lung fibroblasts. *J. Biol. Chem.* **264,** 7017–7024.

McCarthy, K. J., and Couchman, J. R. (1990). Basement membrane chondroitin sulfate proteoglycans: Localization in adult rat tissues. *J. Histochem. Cytochem.* **38,** 1479–1486.

McCarthy, K. J., Accavitti, M. A., and Couchman, J. R. (1989). Immunological characterization of a basement membrane-specific chondroitin sulfate proteoglycan. *J. Cell Biol.* **109,** 3187–3198.

McCarthy, K. J., Horiguchi, Y., Couchman, J. R., and Fine, J.-D. (1990). Ultrastructural localization of the core protein of a basement membrane-specific chondroitin sulfate proteoglycan in adult rat skin. *Arch. Dermatol.* **282,** 397–401.

McCarthy, K. J., Abrahamson, D. R., and Couchman, J. R. (1991). Basement membrane-specific chondroitin sulfate proteoglycan is abnormally expressed in the glomerular basement membrane of chemically-induced diabetic rats. *Anat. Rec.* **229,** 57.

Miettinen, A., Stow, J. L., Mentone, S., and Farquhar, M. G. (1986). Antibodies to basement membrane heparan sulfate proteoglycans bind to the laminae rarae of the glomerular basement membrane (GBM) and induce subepithelial GBM thickening. *J. Exp. Med.* **163,** 1064–1084.

Miyake, K., Medina, K. L., Hayashi, S.-I., Ono, S., Hamaoka, T., and Kincade, P. W. (1990). Monoclonal antibodies to Pgp-1/CD44 block lympho-hemopoiesis in long-term bone marrow cultures. *J. Exp. Med.* **171,** 477–488.

Mohan, P. S., and Spiro, R. G. (1991). Characterization of heparan sulfate proteoglycan from calf lens capsule and proteoglycans synthesized by cultured lens epithelial cells. *J. Biol. Chem.* **266,** 8567–8575.

Murakami, S., Miyake, K., June, C. H., Kincade, P. W., and Hodes, R. J. (1990). IL-5 induces a Pgp-1 (CD44) bright B cell subpopulation that is highly enriched in proliferative and Ig secretory activity and binds to hyaluronate. *J. Immunol.* **145,** 3618–3627.

Murdoch, A. D., Dodge, G. R., Cohen, I., Tuan, R. S., and Iozzo, R. V. (1992). Primary structure of the human heparan sulfate proteoglycan from basement membrane (HSPG2/perlecan): A chimeric molecule with multiple domains homologous to the low density lipoprotein receptor, laminin, neural cell adhesion molecules, and epidermal growth factor. *J. Biol. Chem.* **267,** 8544–8557.

Murphy-Ullrich, J. E., and Höök, M. (1989). Thrombospondin modulates focal adhesions in endothelial cells. *J. Cell Biol.* **109,** 1309–1319.

Mynderse, L. A., Hassell, J., Kleinman, H. K., Martin, G. R., and Martinez-Hernandez, A. (1983). Loss of heparan sulfate proteoglycan from glomerular basement membrane of nephrotic rats. *Lab. Invest.* **48,** 292–302.

Narindrasorasak, S., Lowery, D., Gonzalez-DeWhitt, P., Poorman, R. A., Greenberg, B., and Kisilevsky, R. (1991). High affinity interactions between the Alzheimer's β-amyloid precursor proteins and the basement membrane form of heparan sulfate proteoglycan. *J. Biol. Chem.* **266,** 12878–12883.

Neame, P. J., Choi, H. U., and Rosenberg, L. C. (1989). The primary structure of the core protein of the small, leucine-rich proteoglycan (PG 1) from bovine articular cartilage. *J. Biol. Chem.* **264,** 8653–8660.

Nerlich, A., and Schleicher, E. (1991). Immunohistochemical localization of extracellular matrix components in human diabetic glomerular lesions. *Am. J. Pathol.* **139,** 889–899.

Nishiyama, A., Dahlin, K. J., and Stallcup, W. B. (1991a). The expression of NG2 proteoglycan in the developing rat limb. *Development* **111,** 933–944.

Nishiyama, A., Dahlin, K. J., Prince J. T., Johnstone, S. R., and Stallcup, W. B. (1991b). The primary structure of NG2, a novel membrane-spanning proteoglycan. *J. Cell Biol.* **114,** 359–371.

Noonan, D. M., and Hassell, J. R. (1993). Proteoglycans of the basement membrane. *In* "Molecular and Cellular Aspects of Basement Membranes" (D. H. Rohrbach and R. Timpl, eds.). Academic Press, New York.

Noonan, D. M., Fulle, A., Valente, P., Cai, S., Horigan, E., Sasaki, M., Yamada, Y., and Hassell, J. R. (1991). The complete sequence of perlecan, a basement membrane heparan sulfate proteoglycan, reveals extensive similarity with laminin A chain, low density lipoprotein-receptor, and the neural cell adhesion molecule. *J. Biol. Chem.* **266,** 22939–22947.

Oguro, K., Kazama, T., Isemura, M., Nakamura, T., Akai, S., and Sato, Y. (1991). Immunohistochemical alterations in basement membrane components of squamous cell carcinoma. *J. Invest. Dermatol.* **96,** 250–254.

Oldberg, Å, Antonsson, P., Lindblom, K., and Heinegård, D. (1989). A collagen-binding 59-kd protein (fibromodulin) is structurally related to the small interstitial proteoglycans PG-S1 and PG-S2 (decorin). *EMBO J.* **8,** 2601–2604.

Oohira, A., Wight, T. N., McPherson, J., and Bornstein, P. (1982). Biochemical and ultrastructural studies of proteoheparan sulfate synthesized by PYS-2, a basement membrane-producing cell line. *J. Cell Biol.* **92,** 357–367.

O'Rourke, A. M., and Mescher, M. F. (1990). T-cell receptor activated adhesion systems. *Curr. Opin. Cell Biol.* **2,** 888–893.

Parthasarathy, N., and Spiro, R. G. (1984). Isolation and characterization of the heparan sulfate proteoglycan of the bovine glomerular basement membrane. *J. Biol. Chem.* **259,** 12749–12755.

Paulsson, M. (1987). Noncollagenous proteins of basement membranes. *Collagen Rel. Res.* **7,** 443–461.

Paulsson, M., Fujiwara, S., Dziadek, M., Timpl, R., Pejler, G., Backstrom, G., Lindahl, U., and Engel, J. (1986). Structure and function of basement membrane proteoglycans. *Ciba Found. Symp.* **124,** 189–203.

Paulsson, M., Yurchenco, P. D., Ruben, G. C., Engel, J., and Timpl, R. (1987). Structure of low density heparan sulfate proteoglycan isolated from a mouse tumor basement membrane. *J. Mol. Biol.* **197,** 297–313.

Perris, R., and Johansson, S., (1990). Inhibition of neural crest cell migration by aggregating chondroitin sulfate proteoglycans is mediated by their hyaluronan-binding region. *Dev. Biol.* **137,** 1–12.

Perlmutter, L. S., Barron, E., Saperia, D., and Chui, H. C. (1991). Association between vascular basement membrane components and the lesions of Alzheimer's disease. *J. Neurosci. Res.* **30,** 673–681.

Pierce, A., Lyon, M., Hampson, I. N., Cowling, G. J., and Gallagher, J. T. (1992). Molecular cloning of the major cell surface heparan sulfate proteoglycan from rat liver. *J. Biol. Chem.* **267,** 3894–3900.

Rada, J. A., and Carlson, E. C. (1991). Anionic site and immunogold quantitation of heparan sulfate proteoglycans in glomerular basement membranes of puromycin aminonucleoside nephrotic rats. *Anat. Rec.* **231,** 35–47.

Rapraeger, A. (1989). Transforming growth factor (type beta) promotes the addition of chrondroitin sulfate chains to the cell surface proteoglycan (syndecan) of mouse mammary epithelia. *J. Cell Biol.* **109,** 2509–2518.

Rapraeger, A., Jalkanen, M., Endo, E., Koda, J., and Bernfield, M. (1985). The cell surface proteoglycan from mouse mammary epithelial cells bears chondroitin sulfate and heparan sulfate glycosaminoglycans. *J. Biol. Chem.* **260,** 11046–11052.

Rapraeger, A., Jalkanen, M., and Bernfield, M. (1986). Cell surface proteoglycan associates with the cytoskeleton at the basolateral cell surface of mouse mammary epithelial cells. *J. Cell Biol.* **103,** 2683–2696.

Rohrbach, D. H., Wagner, C. W., Star, V. L., Martin, G. R., Brown, A. S., and Yoon, J. W. (1983). Reduced synthesis of basement membrane heparan sulfate proteoglycan in streptozotocin-induced diabetic mice. *J. Biol. Chem.* **258,** 11672–11677.

Rosenberg, L. C., Choi, H. U., Poole, A. R., Lewandowska, K., and Culp, L. A. (1986). Biological roles of dermatan sulphate proteoglycans. *Ciba Found. Sym.* **124,** 47–68.

Ruoslahti, E. (1991). Integrins as receptors for extracellular matrix. *In* "Cell Biology of Extracellular Matrix" (E. D. Hay, ed.), 2nd Ed., pp. 343–363. Plenum, New York.

Ruoslahti, E., and Yamaguchi, Y. (1991). Proteoglycans as modulators of growth factor activities. *Cell* **64,** 867–869.

Rutishauser, U., and Jessel, T. M. (1988). Cell adhesion molecules in vertebrate neural development. *Physiol. Rev.* **68,** 819–857.

Salmivirta, M., Elenius, K., Vaino, S., Hofer, U., Chiquet-Ehrismann, R., Thesleff, I., and Jalkanen, M. (1991). Syndecan from embryonic tooth mesenchyme binds tenascin. *J. Biol. Chem.* **266,** 7733–7739.

Salmivirta, M., Rauvala, H., Elenius, K., and Jalkanen, M. (1992). Neurite growth-promoting protein (amphoterin, p30) binds syndecan. *Exp. Cell Res.* **200,** 444–451.

Sanderson, R. D., and Bernfield, M. (1988). Molecular polymorphism of a cell surface proteoglycan: Distinct structure on simple and stratified epithelia. *Proc. Natl. Acad. Sci. USA* **85,** 9562–9566.

Sanderson, R. D., Lalor, P., and Bernfield, M. (1989). B lymphocytes express and lose syndecan at specific stages of differentiation. *Cell Regul.* **1,** 27–35.

Sanes, J. R., Engvall, E., Butkowski, R., and Hunter, D. D. (1990). Molecular heterogeneity of basal laminae: Isoforms of laminin and collagen IV at the neuromuscular junction and elsewhere. *J. Cell Biol.* **111,** 1685–1700.

Sasaki, M., Kleinman, H. K., Huber, H., Deutzmann, R., and Yamada, Y. (1988). Laminin, a multidomain protein: The A chain has a unique globular domain and homology with the basement membrane proteoglycan and laminin B chains. *J. Biol. Chem.* **263,** 16536–16544.

Saunders, S., Jalkanen, M., and Bernfield, M. (1989a). Molecular cloning of syndecan, an integral membrane proteoglycan. *J. Cell Biol.* **108,** 1547–1556.

Saunders, S., Nguyen, H., and Bernfield, M. (1989b). Mammary epithelial cells transfected with antisense cDNA reduce cell surface syndecan and become fibroblastic in morphology. *J. Cell Biol.* **109,** 5a.

Schmidtchen, A., Sundler, R., and Fransson, L.-Å. (1990). A fibroblast heparan sulfate proteoglycan with a 70 kDa core protein is linked to membrane phophatidylinositol. *Glycoconjugate J.* **7,** 563–572.

Shieh, M.-T., WuDunn, D., Montgomery, R. I., Esko, J. D., and Spear, P. G. (1992). Cell surface receptors for herpes simplex virus are heparan sulfate proteoglycans. *J. Cell Biol.* **116,** 1273–1281.

Shimizu, Y., and Shaw, S. (1991). Lymphocyte interactions with extracellular matrix. *FASEB J.* **5,** 2292–2299.

Shimizu, Y., van Seventer, G. A., Siraganian, R., Wahl, L., and Shaw, S. (1989). Dual role of the CD44 molecule in T cell adhesion and activation. *J. Immunol.* **143,** 2457–2463.

Shimomura, H., and Spiro, R. G. (1987). Studies on macromolecular components of human glomerular basement membrane and alterations in diabetes. *Diabetes* **36,** 374–381.

Shinomura, T., and Kimata, K. (1992). Proteoglycan-Lb, a small dermatan sulfate proteoglycan expressed in embryonic chick epiphyseal cartilage, is structurally related to osteoinductive factor. *J. Biol. Chem.* **267,** 1265–1270.

Singer, I. I., Solomon, S., Kawka, D. W., and Hassell, J. R. (1987). Extracellular matrix fibers containing fibronectin and basement membrane heparan sulfate proteoglycan coalign with focal contacts and microfilament bundles in stationary fibroblasts. *Exp. Cell Res.* **173,** 558–571.

Smith, L. T., Sakai, L. Y., Burgeson, R. E., and Holbrook, K. A. (1988). Ontogeny of structural components at the dermal–epidermal junction in human embryonic and fetal skin: The appearance of anchoring fibrils and type VII collagen. *J. Invest. Dermatol.* **81,** 480–485.

Snow, A. D., Mar, H., Notchlin, D., Kimata, K., Sato, M., Suzuki, S., Hassell, J., and Wight, T. N. (1988). The presence of heparan sulfate proteoglycans in the neuritic plaques and congophilic angiopathy in Alzheimer's disease. *Am. J. Pathol.* **133,** 456–463.

Snow, A. D., Mar, H., Notchlin, D., Sekiguchi, R. T., Kimata, K., Koike, Y., and Wight, T. N. (1990). Early accumulation of heparan sulfate in neurons and in the beta amyloid protein containing lesions of Alzheimer's disease and Down's syndrome. *Am. J. Pathol.* **137,** 1253–1270.

Snow, A. D., Bramson, R., Henderson, M., Wight, T. N., and Kisilevsky, R. (1991). A temporal and ultrastructural relationship between heparan sulfate proteoglycans and AA amyloid in experimental amyloidosis. *J. Histochem. Cytochem.* **39,** 1321–1330.

Somlo, S., Wirth, B., Germino, G. G., Weinstat-Saslow, D., Gillespie, G. A. J., Himmelbauer, H., Stevens, L., Coucke, P., Willems, P., Bachner, L., Coto, E., López-Larrea, C., Peral, B., San Millán, J. L., Saris, J. J., Breuning, M. H., Frischauf, A.-M., and Reeders, S. T. (1992). Fine genetic localization of the gene for autosomal dominant polycystic kidney disease (PKD1) with respect to physically mapped markers. *Genomics* **13,** 152–158.

Solursh, M., Reiter, R. S., Jensen, K. L., Kato, M., and Bernfield, M. (1990). Transient expression of a cell surface heparan sulfate proteoglycan (syndecan) during limb development. *Dev. Biol.* **140,** 83–92.

Soroka, C. J., and Farquhar, M. G. (1991). Characterization of a novel heparan sulfate proteoglycan found in the extracellular matrix of liver sinusoids and basement membranes. *J. Cell Biol.* **113,** 1231–1241.

Spiro, M. J. (1987). Sulfate metabolism in the alloxan-diabetic rat: Relationship of altered sulfate pools to proteoglycan sulfation in heart and other tissues. *Diabetologia* **30,** 259–267.

Stallcup, W. B. (1981). The NG2 antigen, a putative lineage marker: Immunofluorescent localization in primary cultures of rat brain. *Dev. Biol.* **83,** 154–165.

Stallcup, W. B., Dahlin, K., and Healy, P. (1990). Interaction of the NG2 chondroitin sulfate proteoglycan with type VI collagen. *J. Cell Biol.* **111,** 3177–3188.

Stamenkovic, I., Aruffo, A., Amiot, M., and Seed, B. (1991). The hematopoietic and epithelial forms of CD44 are distinct polypeptides with different adhesion potentials for hyaluronate-bearing cells. *EMBO J.* **10,** 343–348.

Sung, E., and Jones, P. P. (1982). The invariant chain of murine Ia antigens: Its glycosylation, abundance and subcellular localization. *Mol. Immunol.* **18,** 899–913.

Sutherland, A. E., Sanderson, R. D., Mayes, M., Siebert, M., Calarco, P. G., Bernfield, M., and Damsky, C. (1991). Expression of syndecan, a putative low affinity growth factor receptor, in the early mouse embryo. *Development* **113,** 339–351.

Templeton, D. M. (1989). Retention of glomerular basement membrane proteoglycans accompanying loss of anionic site staining in experimental diabetes. *Lab. Invest.* **61,** 202–211.

Tidman, M. J., and Eady, R. A. J. (1984). Ultrastructural morphometry of normal human dermal–epidermal junction: The influence of age, sex, and body region on laminar and nonlaminar components. *J. Invest. Dermatol.* **83,** 448–453.

Tidman, M. J., and Eady, R. A. J. (1985). Evaluation of anchoring fibrils and other components of the dermal–epidermal junction in dystrophic epidermolysis by a quantitative ultrastructural technique. *J. Invest. Dermatol.* **84,** 374–377.

Timpl, R. (1989). Structure and biological activity of basement membrane proteins. *Eur. J. Biochem.* **180,** 487–502.

Timpl, R., and Dziadek, M. (1986). Structure, development, and molecular pathology of basement membranes. *Int. Rev. Exp. Pathol.* **29,** 1–112.

Toole, B. P. (1990). Hyaluronan and its binding proteins, the hyaladherins. *Curr. Opin. Cell Biol.* **2,** 839–844.

Tsiang, M., Lentz, S. R., and Sadler, J. E. (1992). Functional domains of membrane-bound human thrombomodulin: EGF-like domains four to six and the serine/threonine-rich domain are required for cofactor activity. *J. Biol. Chem.* **267,** 6164–6170.

Trautman, M. S., Kimelman, J., and Bernfield, M. (1991). Developmental expression of syndecan, an integral membrane proteoglycan, correlates with cell differentiation. *Development* **111,** 213–220.

Van Den Heuvel, L. P. W. J., Van Den Born, J., Van De Velden, T. J. A. M., Veerkamp, J. H., Monnens, L. A. H., Schroder, C. H., and Berden, J. H. M. (1989). Isolation and partial characterization of heparan sulfate proteoglycan from the human glomerular basement membrane. *Biochem. J.* **264,** 457–465.

Wan, Y.-J., Wu, T.-C., Chung, A. E., and Damjarov, I. (1984). Monoclonal antibodies to laminin reveal the heterogeneity of basement membranes in the developing and adult mouse tissues. *J. Cell Biol.* **98,** 971–979.

Wang., X.-F., Lin, H., Ng-Eaton, E., Downward, J., Lodish, H. F., and Weinberg, R. A. (1991). Expression cloning and characterization of the TGF-β type III receptor. *Cell* **67,** 797–805.

Wen, D., Dittman, W. A., Ye, R. D., Deaven, L. L., Majerus, P. W., and Sadler, J. E. (1987). Human thrombomodulin: Complete cDNA sequence and chromosome localization of the gene. *Biochemistry* **26,** 4350–4357.

Westergren-Thorsson, G., Antonsson, P., Malmström, A., Heinegård, D., and Oldberg, Å. (1991). The synthesis of a family of structurally related proteoglycans in fibroblasts is differently regulated by TGF-β. *Matrix* **11,** 177–183.

Wight, T. N., Heinegård, D. K., and Hascall, V. C. (1991). Proteoglycans: Structure and Function. *In* "Biology of Extracellular Matrix" (E. D. Hay, ed.), 2nd Ed., pp. 45–78. Plenum, New York.

Wilson, P. D., and Sherwood, A. C. (1991). Tubulocystic epithelium. *Kidney Int.* **39,** 450–463.

Woods, A., and Couchman, J. R. (1988). Focal adhesions and cell–matrix interactions. *Collagen Rel. Res.* **8,** 155–182.

Woods, A., and Couchman, J. R. (1992a). Protein kinase C involvement in focal adhesion formation. *J. Cell Sci.* **101,** 277–290.

Woods, A., and Couchman, J. R. (1992b). Heparan sulfate proteoglycans and signalling in cell adhesion. *Adv. Exp. Med. Biol.* **313,** 87–96.

Woods, A., Höök, M., Kjellén, L., Smith, C. G., and Rees, D. A. (1984). Relationship of heparan sulfate proteoglycans to the cytoskeleton and extracellular matrix of cultured fibroblasts. *J. Cell Biol.* **99,** 1743–1753.

Woods, A., Couchman, J. R., and Höök, M. (1985). Heparan sulfate proteoglycans of rat fibroblasts: A hydrophobic form may link cytoskeleton and matrix components. *J. Biol. Chem.* **260,** 10872–10879.

Woods, A., Couchman, J. R., Johansson, S., and Höök, M. (1986). Adhesion and cytoskeletal organisation of fibroblasts in response to fibronectin fragments. *EMBO J.* **5,** 665–670.

Wu, V.-Y., Wilson, B., and Cohen, M. P. (1987). Disturbances in glomerular basement membrane glycosaminoglycans in experimental diabetes. *Diabetes* **36,** 679–683.

Yamada, K. M. (1991). Fibronectin and other cell interactive glycoproteins. *In* "Biology of Extracellular Matrix" (E. D. Hay, ed.), 2nd Ed., pp. 111–146. Plenum, New York.

Yamaguchi, Y., Mann, D. M., and Ruoslahti, E. (1990). Negative regulation of transforming growth factor-β by the proteoglycan decorin. *Nature (London)* **336,** 244–246.

Yanagashita, M., and Hascall, V. C. (1992). Cell surface heparan sulfate proteoglycans. *J. Biol. Chem.* **267,** 9451–9454.

Yanagashita, M., and McQuillan, D. J. (1989). Two forms of plasma membrane-intercalated heparan sulfate proteoglycans in rat ovarian granulosa cells. *J. Biol. Chem.* **264,** 17551–17558.

Yayon, A. M., Klagsbrun, M., Esko, J. D., Leder, P., and Ornitz, D. M. (1991). Cell surface heparin-like molecules are required for binding of basic fibroblast growth factor. *Cell* **64,** 841–848.

Young, I. D., Willmer, J. P., and Kisilevsky, R. (1989). The ultrastructural localization of sulfated proteoglycans is identical in the amyloids of Alzheimer's disease and AA, AL senile cardiac and medullary carcinoma-associated amyloidosis. *Acta Neuropathol.* **78,** 202–209.

Yurchenco, P. D., and Schittny, J. C. (1990). Molecular architecture of basement membranes. *FASEB J.* **4,** 1577–1590.

Yurchenco, P. D., Cheng, Y. S., and Ruben, G. C. (1987). Self-assembly of a high molecular weight basement membrane heparan sulfate proteoglycan into dimers and oligomers. *J. Biol. Chem.* **262,** 17668–17676.

Glycosyl Phosphatidylinositol-Linked Membrane Proteins: Structure, Biosynthesis, and Function

Tamara L. Doering

Department of Biological Chemistry, Johns Hopkins Medical School, Baltimore, Maryland 21205

I. Background

Glycosyl phosphatidylinositols (GPIs) only recently entered the arena of biological glycoconjugates (reviewed in Ferguson and Williams, 1988; Low and Saltiel, 1988; Thomas *et al.*, 1990; Doering *et al.*, 1990a; Cross, 1990; Lisanti *et al.*, 1990; Mensa-Wilmot *et al.*, 1991; Field and Menon, 1991; Ferguson, 1991). This exciting new field has grown rapidly over the last few years, and the pace of discovery continues to accelerate.

While initial studies focused on identifying GPIs in nature and examining the details of their structure, other fascinating questions under investigation concern how these compounds are assembled and what functions they perform.

GPI structures have been detected in organisms ranging from yeast to man, but are most prevalent in lower eukaryotes such as protozoa. They have been widely studied in their role as anchors for cell surface proteins, the topic of this chapter. In addition to proteins, GPIs anchor complex glycoconjugates, such as the lipophosphoglycan (LPG) of the protozoan parasite *Leishmania*. LPG has a tripartite structure consisting of a GPI anchor, an extensive series of phosphorylated disaccharide repeats, and a neutral oligosaccharide cap (Turco, 1988; McConville, 1991). This macromolecule plays a crucial role in the biology of *Leishmania* and is instrumental in allowing that parasite to survive within the hostile environment of its host.

Other GPIs exist in cell membranes, unencumbered by additional glycoconjugate moieties or by protein. These low molecular weight species have been extensively studied in *Leishmania* and are termed glycoinositolphospholipids (GIPLs). Some GIPLs are structurally homologous to GPI protein anchors, possibly representing intermediates in anchor biosynthesis; others exhibit broad variation on the GPI structural theme and are apparently metabolic end products of unknown function (McConville, 1991). It has also been suggested that free GPIs serve as mediators of insulin action (reviewed in Saltiel, 1991; Jacquemin, 1991; Romero, 1991).

Proteins anchored by GPIs exhibit incredible diversity, ranging from hydrolases, adhesion molecules, and parasitic or mammalian surface antigens to members of the immunoglobulin superfamily. Cell types bearing these proteins include yeast, protozoa, lymphocytes, epithelial cells, and numerous others. In this chapter structures, biosynthesis, and analysis of GPI protein anchors will be described, followed by a discussion of potential functions of GPI anchorage and major current research questions.[1]

A. *Glycosyl Phosphatidylinositol Structure*

The inital investigations of GPI structure concentrated on a GPI-anchored parasite surface antigen, the variant surface glycoprotein (VSG) of the bloodstream-form African trypanosome (Cross, 1975, 1984; Turner *et al.*, 1985; Hajduk *et al.*, 1989). VSG is extremely abundant (10^7 molecules per cell, composing at least 10% of total cell protein)

[1] Where possible, citations refer to review articles; these should be consulted for more detailed listings of primary research papers.

and may be readily purified, making bloodstream trypanosomes an ideal system for study of GPIs. The VSG anchor structure was first completely described in 1988 (Ferguson *et al.*). Since then, other GPI structures have also been determined (Homans *et al.*, 1988; Roberts *et al.*, 1988a,b; Schneider *et al.*, 1990; Mayor *et al.*, 1990) and found to share a linear glycan core, shown in boldface type (not shaded) in Fig. 1. This shared region consists of ethanolamine in amide linkage to the C-terminal residue of the anchored protein, followed by phosphate, three mannose residues, and finally a nonacetylated glucosamine linked to the *myo*-inositol of a phospholipid. Interestingly, this core glycan is precisely conserved between species, down to the level of

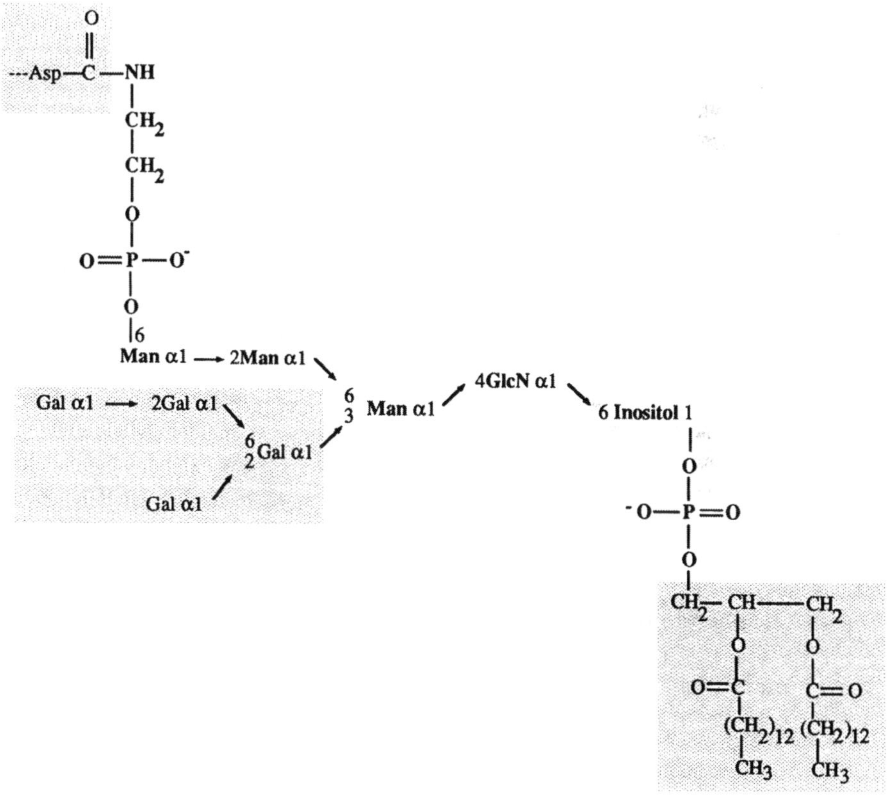

FIG. 1. Structure of the GPI anchor of VSG (Ferguson *et al.*, 1988). Shaded areas are specific to VSG; the remaining (in bold face type, unshaded) structure represents the core glycan common to all GPIs. See text and Ferguson (1991) for possible modifications of this core. Asp, C-terminal amino acid of this variant of VSG; Man, mannose; GlcN, glucosamine; Gal, galactose. Arrows denote glycosidic linkages.

individual glycosidic linkages. Additional groups may modify the core, introducing variation dependent on species, cell type, or protein (summarized in Ferguson, 1991). For example, the nonreducing terminal mannose may be decorated with an additional α-linked mannose residue (Homans *et al.*, 1988; unpublished data cited in Ferguson, 1991), the inositol ring is frequently acylated (Roberts *et al.*, 1988a; Mayor *et al.*, 1990; Walter *et al.*, 1990), and the reducing terminal mannose may bear a variably branched structure of α-linked galactose (Ferguson *et al.*, 1988), GalNAc in β-linkage (Homans *et al.*, 1988), or additional ethanolamine phosphate groups (Homans *et al.*, 1988; Roberts *et al.*, 1988b; Walter *et al.*, 1990).

Notwithstanding the variety of potential side chain modifications, GPI glycans from evolutionarily disparate organisms are closely related by their common core. Studies of the lipid portions of these anchors present a more heterogeneous picture. The lipid moiety of the anchoring phospholipid may consist of diacyl glycerol, alkyl-acyl glycerol, *lyso*-acyl compounds, or ceramide (in slime mold and yeast; Stadler *et al.*, 1989; Conzelmann *et al.*, 1992). The hydrocarbon chains vary in length and degree of saturation and are frequently present as mixtures at any one position. Many GPIs, especially in higher eukaryotes, are acylated (generally with palmitate) on inositol, adding to the structural diversity. The lipid species that are present in GPIs do not neccessarily reflect the endogenous population of the parent compound (e.g., phosphatidylinositol). The most extreme example of this is the trypanosome VSG, which bears exclusively dimyristoylglycerol on its anchor (Ferguson *et al.*, 1985; Ferguson, 1991) although dimyristoylphosphatidylinositol constitutes only 1.5% of cellular PI (D. S. Raben, T. L. Doering, G. W. Hart, and P. T. Englund, unpublished data). Similar observations of compositional discrepancies have been made on acetylcholine esterase and other proteins (Roberts *et al.*, 1988a; Butikofer *et al.*, 1991).

It is unlikely that the full spectrum of possible GPI modifications, either of glycan or lipid, has been described to date. The first studied anchor, that of trypanosome VSG, now appears exceptional in terms of its lipid composition, its galactose side chain, and its single ethanolamine phosphate. New anchor modifications continue to be revealed, and others undoubtedly remain to be discovered.

B. *Glycosyl Phosphatidylinositol Attachment to Protein*

Proteins destined for GPI anchorage are polymerized on membrane-bound ribosomes and directed into the endoplasmic reticulum lumen by means of an N-terminal signal sequence. A C-terminal sequence then directs addition of the GPI anchor, concomitant with removal of a C-terminal hydrophobic peptide (reviewed in Ferguson and Williams, 1988). In the case of VSG, the newly synthesized polypeptide is linked to its GPI anchor within 30 sec of synthesis (Bangs *et al.*, 1988). This rapid attachment suggested that the anchor exists as a preformed precursor in the endoplasmic reticulum (Bangs *et al.*, 1985; Ferguson *et al.*, 1986; Conzelmann *et al.*, 1987). GPI precursor compounds have been isolated in several laboratories (Krakow *et al.*, 1986; Menon *et al.*, 1988) and have been successfully transfered to protein *in vitro* (Mayor *et al.*, 1991).

The precursor for the trypanosome VSG is termed glycolipid A (in this chapter; Krakow *et al.*, 1986) or P2 (Menon *et al.*, 1988). This GPI consists of the core glycan (Fig. 1), bereft of protein consistent with its precursor status and also missing the galactose side chain found on the mature VSG protein. Its lipid moiety, like that of the mature anchored protein, is dimyristoyl phosphatidylinositol. A related GPI precursor termed PP1 (Field *et al.*, 1991a) corresponds to the anchor of a major surface protein of insect form trypanosomes (procyclic acidic repetitive protein, Roditi *et al.*, 1989). PP1 consists of the same core glycan, this time anchored by *lyso*-phosphatidylinositol and modified by a fatty acyl chain on the inositol ring (Field *et al.*, 1991b). The conservation of structure between GPI protein anchors suggests that other precursors may also consist of the basic core region, perhaps obtaining side chain modifications later in biosynthesis (Bangs *et al.*, 1988).[2]

II. Experimental Approaches

A. *Is a Protein Glycosyl Phosphatidylinositol-Anchored?*

There are several ways to ascertain whether a protein bears a GPI anchor (methods reviewed in Mayor and Menon, 1990; Doering *et al.*, 1993). GPI-linked proteins are released from cell surfaces by phosphatidylinositol-specific enzymes such as bacterial phosphatidylinositol-specific phospholipase C (PI-PLC; see Fig. 2).[3] PI-PLC treatment offers

[2] In trypanosome cell-free systems, some anchor precursors bearing galactose residues are produced. These may be nonfunctional products of biosynthetic reactions (Mayor *et al.*, 1992).

[3] Early studies with PI-PLC demonstrated specific release of a set of cell surface proteins. The released proteins were water soluble, indicating alteration of some fundamental property of the polypeptide that had mediated membrane attachment (reviewed in Low and Saltiel, 1988).

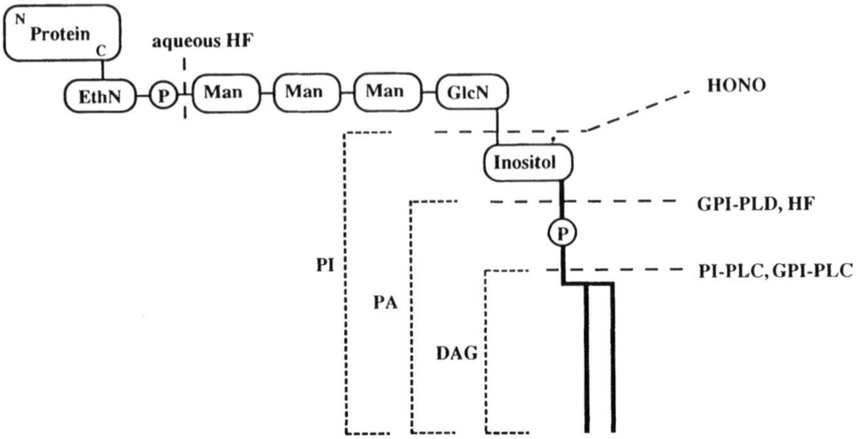

FIG. 2. Schematic representation of a GPI with sites of cleavage for analysis indicated. See text for details. N, N terminus of anchored protein; C, C terminus of anchored protein; EthN, ethanolamine; P, phosphate; HF, hydrogen fluoride; HONO; nitrous acid; PI, phosphatidylinositol; GPI-PLD, GPI-specific phospholipase D; PA, phosphatidic acid; PI-PLC, PI-specific phospholipase C; GPI-PLC, GPI-specific phospholipase C; DAG, diacylglycerol. Other abbreviations are as in Fig. 1.

a simple and specific test of GPI anchorage, as proteins released by this enzyme have all been found to be GPI-anchored (reviewed in Ikezawa, 1991). Other lipases useful in analysis of GPI-anchored proteins include a trypanosomal GPI-specific PLC (Hereld *et al.*, 1986; Bulow and Overath, 1986; Fox *et al.*, 1986) and a GPI-specific phospholipase D (PLD) found in mammalian serum (Metz, 1991).

One useful and straightforward approach to identification and analysis of GPI-anchored proteins involves differential extraction of cells or tissues with Triton X-114 detergent. Extraction provides a profile of amphiphilic proteins, including integral membrane proteins as well as species bearing GPI anchors (Hooper and Bashir, 1991). Triton X-114 forms a clear micellar solution at low temperatures, which separates into two phases when warmed above 20°C due to aggregation of detergent micelles. GPI-anchored proteins partition with the detergent phase unless they are released from their anchors; release allows them to enter the aqueous phase (Bordier, 1981).

Alteration of detergent partitioning behavior by phospholipase digestion of anchors is a convenient tool for assessing the presence of GPIs. PI-specific or GPI-specific PLCs generally release anchored proteins in a soluble form. Exceptions include proteins with anchors containing acylated inositols; these GPIs are resistant to PLC action although they are cleaved by the GPI-specific PLD (Fig. 2; reviewed in Rosenberry, 1991). Another useful method for anchor cleavage is nitrous acid deamination. As mentioned above, all GPIs contain a nonacetylated glucosamine. This allows nitrous acid to cleave the core glycan between glucosamine and inositol (Fig. 2), releasing PI and a glycan that terminates in anhydromannose (Ferguson and Williams, 1988).

Digestion of GPIs with phospholipase C releases diacylglycerol and reveals a cryptic epitope termed the cross-reacting determinant (CRD; Holder and Cross, 1981). Polyclonal antibodies to this determinant have been prepared and analyzed. The major moiety contributing to the CRD is the inositol 1,2-(cyclic)monophosphate formed during phospholipase C cleavage, although other structural features may also contribute (Shak *et al.,* 1988; Zamze *et al.,* 1988). Reactivity of a protein with anti-CRD antibody after cleavage by phospholipase C strongly indicates the presence of a GPI anchor.

Ability to radiolabel a protein with GPI precursors suggests GPI anchorage and provides a useful means of tagging the species for other studies. Precursors may include lipids, *myo*-inositol, ethanolamine, or monosaccharide precursors (Doering *et al.,* 1990b). GPI species commonly represent a small fraction of glycoconjugates within the eukaryotic cell, so metabolic radiolabeling may be inefficient. In the case of monosaccharide precursors, tunicamycin may be used to inhibit N-glycosylation and increase specific incorporation into GPIs (Masterson *et al.,* 1989).

B. Structural Analysis of Glycosyl Phosphatidylinositols

The methods described above indicate whether molecules are GPI-anchored and provide useful information about anchor structures. Rigorous identification requires detailed structural analysis, as has been performed for the trypanosome VSG anchor and precursor (Ferguson *et al.,* 1988; Mayor *et al.,* 1990), the rat brain Thy-1 antigen (Homans *et al.,* 1988), human erythrocyte acetylcholinesterase (Roberts *et al.,* 1988a,b), and a surface protease from *Leishmania major* (Schneider *et al.,* 1990). An excellent guide to chemical and enzymatic analysis of GPI fine structure has been compiled by Ferguson (1992). His chapter describes a general strategy for such analysis, including organized protocols and helpful tables. The suggested approach relies on the use of gas chromatography–mass spectrometry (GC–MS) and liquid chromatography.[4] First, the glycoprotein to be studied is purified, and the GPI-containing C-terminal peptide is released by protease digestion

[4] Published studies frequently utilize two-dimensional ^{1}H-NMR studies (Ferguson *et al.,* 1988; Homans *et al.,* 1988; Schneider *et al.,* 1990); this requires more material.

and purified by octyl–Sepharose chromatography. Compositional analyses (utilizing GC–MS, GC, or HPLC) are then performed in concert with the following methods: methanolysis and trimethylsilyl derivitization to detect monosaccharides and lipids; strong acid hydrolysis and trimethylsilyl derivitization for inositol analysis; hydroxyester fatty acid analysis for lipids; and acid hydrolysis and phenyl isothiocyanate derivitization for amino acid and ethanolamine detection. Peptide sequence data may also be obtained at this point.

Following compositional analysis, the GPI–peptide is delipidated using base hydrolysis and/or phosphatidylinositol specific phospholipases (Fig. 3). Aqueous HF dephosphorylation of the GPI and nitrous acid deamination with sodium borohydride reduction produce a neutral glycan. This species is subjected to methylation analysis and exoglycosidase sequencing. Exoglycosidases used include jack bean α-mannosidase (to remove nonsubstituted terminal α-D-mannopyranose residues), *Aspergillus phoenicus* α-mannosidase (for α-D-Man(1$\rightarrow$2)D-Man linkages), coffee bean α-galactosidase (for nonsubstituted terminal α-D-galactopyranose residues), and jack bean β-hexosaminidase (for nonsubstituted terminal β-D-GlcNAc or GalNAc). Product analysis is performed by either Bio-Gel P4 gel filtration or high-performance anion-exchange chromatography. Integration of the results from all these techniques allows determination of the GPI structure.

C. Investigations of Glycosyl Phosphatidylinositol Biosynthesis

From the core glycan structure shown in Fig. 1, it is apparent that GPIs demonstrate several novel carbohydrate structures. The trypanosome system, with its massive production of GPI components, has been instrumental in allowing definition of the activities involved in producing these structures. Initial studies in bloodstream trypanosomes used a cell-free system for synthesis of the trypanosome anchor precursor glycolipid A (Masterson *et al.*, 1989). Hypotonic lysates of cells were washed and incubated with radiolabeled biosynthetic precursors such as sugar nucleotides and acyl CoA (Masterson *et al.*, 1989). The radiolabel was incorporated into various glycolipid intermediates; these were analyzed by phospholipase and glycosidase digestion, incorporation of various radiolabels, nitrous acid deamination, and appropriate chromatography. Kinetic studies of their interdependence permitted elucidation of the biosynthetic pathway of glycolipid A (Fig. 4; Masterson *et al.*, 1989, 1990; Doering *et al*, 1989). Similar studies have been accomplished using crude microsomal fractions from bloodstream trypano-

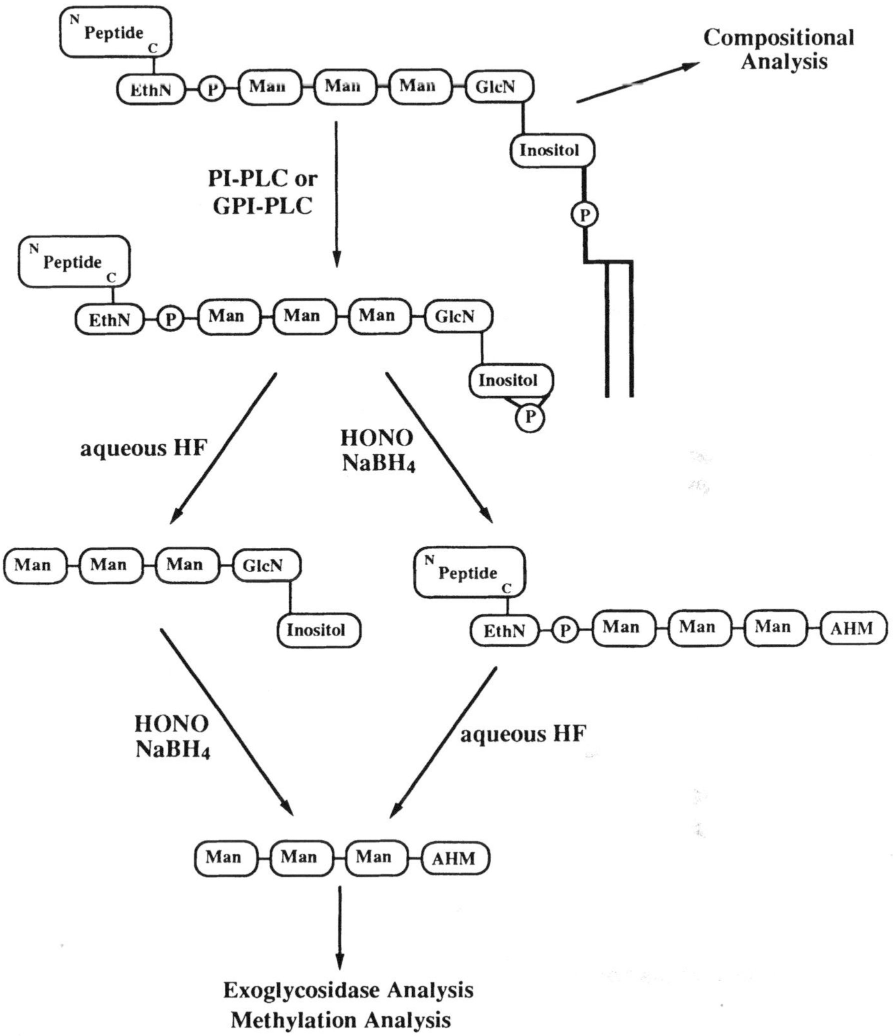

FIG. 3. Simplified scheme for GPI analysis. The glycan depicted is unmodified. Note that PLC digestion yields a cyclic phosphate on inositol. In anchors where the inositol is acylated, the fatty acid must be removed (e. g., by mild base treatment) to allow PLC digestion to proceed. See text for explanation and Ferguson (1992) for detailed methods. $NaBH_4$, sodium borohydride; AHM, anhydromannitol. Other abbreviations are as in previous figure legends.

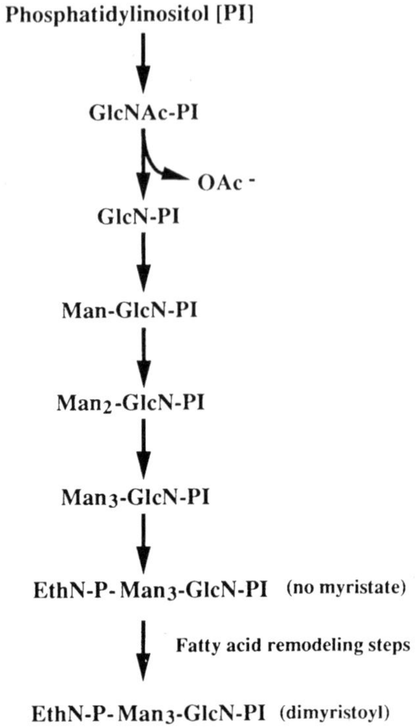

FIG. 4. Pathway for biosynthesis of the VSG anchor from bloodstream form trypanosomes. The final species, EthN-P-Man_3-GlcN-PI with two myristates in the PI portion, is termed glycolipid A. See text for explanation and details. Abbreviations are as in previous figure legends.

somes (Menon *et al.*, 1990a,b) and using hypotonic lysates of insect-form trypanosomes (Field *et al.*, 1992). Useful references for the methodologies employed in these studies are Doering *et al.* (1990b), Mayor and Menon (1990), and Field and Menon (1992).

GPI biosynthesis in trypanosomes begins with the addition of *N*-acetylglucosamine to endogenous phosphatidylinositol (PI) from a UDP–*N*-acetylglucosamine donor (Doering *et al.*, 1989). The acetylated intermediate is efficiently deacetylated to form glucosaminyl-PI, the first species with the characteristic properties of a GPI. These properties include susceptibility to PI-specific phospholipase C, to GPI-specific phospholipases, and to nitrous acid deamination.

The next steps in GPI biosynthesis add a linear chain of three mannose residues to glucosaminyl-PI (Masterson *et al.*, 1989; Menon *et al.*, 1990b). Although kinetic data suggest sequential appearance of glycolipid intermediates containing one, two, and three mannose residues (Fig. 4; Masterson *et al.*, 1989), it has also been proposed that mannose incorporation occurs via a combination of addition and trimming, analogous to processing of N-linked oligosaccharides (Menon *et al.*, 1990a; Field and Menon, 1991). Several cell-free experiments indicate that the mannose source is mannosyl phosphoryl dolichol: kinetic studies indicate that GDP–mannose is not the immediate mannose donor; amphomycin inhibits GPI precursor biosynthesis; and dolichol phosphate stimulates mannose incorporation into GPI species. More compelling, GPI species may be directly radiolabeled via [^{3}H]mannosyl phosphoryl dolichol (Menon *et al.*, 1990a). There are indications that the same mannose donor is used for GPI biosynthesis in several eukaryotic systems. Yeast mutants with temperature-sensitive defects in either phosphomannomutase or dolicholphosphomannose synthase do not perform GPI biosynthesis at the nonpermissive temperature (Conzelmann *et al.*, 1990; Orlean, 1990); these may be rescued by expression of the cloned gene (Orlean *et al.*, 1988). Also, T cell hybridoma mutants deficient in mannosyl phosphoryl dolichol synthesis do not produce GPI anchors, but may similarly be rescued by transfection with the yeast dolicholphosphomannose synthase gene (DeGasperi *et al.*, 1990).

The GPI core glycan is completed by the addition of ethanolamine phosphate. CDP–ethanolamine is not involved with this pathway (Masterson *et al.*, 1989; Menon, 1991). Rather, radiolabeling experiments in trypanosomes (unpublished data, cited in Menon, 1991) and studies of yeast mutants defective in ethanolaminephosphotransferase suggest phosphatidylethanolamine as the donor (Menon, 1991).

In the case of the trypanosome VSG anchor, glycan assembly is followed by a process of fatty acid remodeling, whereby myristate is incorporated into both positions of the anchor PI (Masterson *et al.*, 1990; Doering *et al.*, 1990a). Fatty acid remodeling may be unique to the VSG and is responsible for the unusual presence of exclusively myristate in its GPI anchor. A similar remodeling process may occur in yeast, where some form of anchor evolution occurs as GPI-linked proteins are processed (Conzelmann *et al.*, 1992). Pulse–chase experiments indicate that while the lipid portions of yeast anchor precursors and newly attached anchors are predominantly sensitive to alkali treatment, these structures become base resistant during maturation (Conzelmann *et al.*, 1992).

The yeast system, with numerous GPI-linked proteins, is potentially valuable for studying GPI biosynthesis, especially because it is so amenable to genetic manipulation. Studies in yeast have provided information on GPI attachment (Conzelmann *et al.*, 1988; Nuoffer *et al.*, 1991), on the involvement of mannosyl phosphoryl dolichol in GPI biosynthesis, and on interesting GPI lipid modifications. To date, however, no specific mutants in GPI biosynthesis have been described, despite attempts by several laboratories. It is possible that defects in this pathway are lethal, necessitating new screening approaches to advance yeast as a model system for GPI investigation.

Biosynthetic studies in mammalian systems (reviewed in Fasel *et al.*, 1991) provide an interesting complement to those in protozoa. One fruitful approach has been examination of a series of lymphoma mutants deficient in surface expression of Thy-1, a GPI-anchored antigen (Hyman, 1988). Analysis of GPI production in various complementation groups of these mutants has yielded information about the mammalian biosynthetic pathway (Stevens and Raetz, 1991; Sugiyama *et al.*, 1991; Puoti *et al.*, 1991). For example, three complementation groups are unable to synthesize *N*-acetylglucosaminyl PI, the first biosynthetic intermediate (Stevens and Raetz, 1991; Sugiyama *et al.*, 1991), suggesting that the initial step in GPI biosynthesis requires at least three gene products.[5] Other mutants exhibit defects later in the biosynthetic pathway, e.g., in addition of the final mannose or completion of the anchor precursor (Puoti *et al.*, 1991; Singh *et al.*, 1991).

Studies in mammalian systems also highlight possible variations in GPI biosynthetic pathways (Fasel *et al.*, 1991). A number of GPI anchors bear acylated inositol residues. Several partially mannosylated intermediates in the trypanosome biosynthetic pathway also exhibit this modification, although the VSG anchor itself does not (Ferguson *et al.*, 1988). It is at present unclear whether acylated species represent a distinct but parallel synthetic pathway or whether there is some more complex relationship between biosynthetic intermediates with and those without this modification. In mammalian systems acylated species are observed earlier in the GPI pathway than in bloodstream trypanosomes. In some cases these are en route to an acylated precursor, but in other cases their specific role is uncertain. Mammalian intermediates may also bear additional ethanolamine phosphate residues, reflecting the prevalence of this modification among anchors of higher eukaryotes (Ferguson and Williams, 1988).

D. Study of Glycosyl Phosphatidylinositol Attachment

A final area of interest in the biosynthesis of GPI-anchored proteins relates to the attachment of the polypeptide C-terminus to a preformed anchor (reviewed in Caras, 1991a; Udenfriend *et al.*, 1991). The requirements for this C-terminal processing have been the focus of intensive

[5] Interestingly, the defect in one of these complementation groups (class H) may be partially alleviated in a reversible fashion by treatment of cells with sodium butyrate, which causes accumulation of unique RNA transcripts (Tisdale *et al.*, 1991).

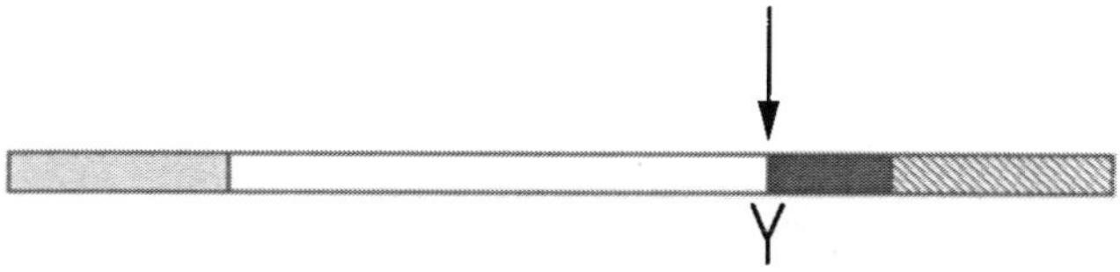

FIG. 5. Schematic representation of the requirements for GPI attachment. Light stippled region, N-terminal signal sequence; dark stippled region, spacer sequence; striped region, C-terminal hydrophobic sequence. See text and references (Udenfriend *et al.*, 1991); Caras, 1991a) for details.

effort by several groups. Experiments typically use constructs of non-GPI-anchored proteins combined with C-terminal sequences appropriate for anchorage. Such chimeric genes, when transfected into cells, direct the synthesis of hybrid polypeptide products bearing GPI anchors. By altering GPI-specifying sequences and analyzing the products for lipid attachment, the necessary elements for GPI addition have been established (Caras, 1991a; Udenfriend *et al.*, 1991).

In order to become GPI-linked, a protein must contain an N-terminal signal sequence as well as a signal for GPI addition. The latter consists of a 15–20 residue hydrophobic region at the C terminus, preceded by a 10- to 12-amino-acid spacer region (Fig. 5).[6] The actual cleavage/attachment site N-terminal to the spacer consists of a pair of small amino acids with monosubstituted β carbon atoms (Micanovic *et al.*, 1990b; Moran *et al.*, 1991; Moran and Caras, 1991). In nature, only certain amino acids become linked to the ethanolamine of an anchor precursor. This restriction has been investigated using mutagenesis techniques, with good agreement between endogenous and engineered acceptor sites. Amino acids that allow GPI addition are Asp, Gly, Ala, Cys, Asn, and Ser; others do not (Micanovic *et al.*, 1990b; Moran *et al.*, 1991).[7]

Little is known about the enzyme(s) involved in catalyzing GPI addition to newly synthesized proteins. Addition is believed to occur by a transamidation reaction, which removes the C-terminal hydrophobic peptide from the protein with simultaneous addition of the GPI precursor. This reaction has been demonstrated in the trypanosome cell-free system using endogenous protein and exogenously added anchor precursors (Mayor *et al.*, 1991). The transfer reaction has also been

[6] Although natural anchor sequences are found at the C termini of proteins, this location is not an absolute requirement. A construct containing an appropriate GPI addition signal placed internal to a protein sequence still specified GPI addition (Caras, 1991b).

[7] Val or Glu allow low levels of GPI attachment in some systems (Moran *et al.*, 1991).

studied using rough microsomal membranes from CHO cells as a source of activity able to modify exogenously added proteins or peptides (Kodukula *et al.*, 1991, 1992; Udenfriend *et al.*, 1992). The peptides used are engineered portions of normally GPI-modified proteins; their smaller size enables subtle changes in biochemical character to be detected. In the peptide/microsome system specific cleavage of a polypeptide precursor has been demonstrated (Kodukula *et al.*, 1991), and the cleaved product was shown to be GPI-modified (based on phase partitioning in TX-114 detergent; Kodukula *et al.*, 1992).

III. Functions of Glycosyl Phosphatidylinositol Anchors

The main function of GPI anchors is to attach proteins to cell surfaces, a role they perform admirably. Simpler modes of linkage are available, however, raising the question of why these complex yet highly conserved structures have evolved. The question is particularly puzzling because no common theme among anchored proteins has been observed, although GPI anchoring appears to be ubiquitous in eukaryotes. Does it have any special functions?

A. Protein Mobility

Proteins anchored by GPI structures interact only with the extracellular space and the outer leaflet of the plasma membrane. In contrast, cell surface proteins containing transmembrane domains extend through the bilayer and may communicate with intracellular structures. Such communication might serve to localize transmembrane proteins, limiting their lateral mobility. The inability of GPI-linked polypeptides to directly contact cytoskeletal structures suggested that one function of these anchors could be to allow increased lateral freedom of proteins within the bilayer.

Several studies have assessed the diffusion of GPI-linked proteins, many of which do have higher lateral mobility than most transmembrane proteins (Ishihara *et al.*, 1987; Noda *et al.*, 1987). However, this varies with the individual protein; both VSG and a sperm antigen (PH-20) diffuse slowly (Bulow *et al.*, 1988; Phelps *et al.*, 1988). To investigate specific effects of the GPI portion upon mobility, the behavior of chimeric proteins has been addressed. These studies (Edidin and Stroynowski, 1991; Zhang *et al.*, 1991) conclude that replacement of transmembrane with GPI anchors contributes only marginally (Zhang *et al.*, 1991) or not at all (Edidin and Stroynowski, 1991) to increased diffusion of proteins. Effects of protein concentration (crowding), involvement of extracellular matrix, or interactions between protein ectodomains modulate diffusion rates in both transmembrane and GPI-anchored species; interactions mediated by cytoplasmic protein domains played a less important role in the chimeras (Zhang *et al.*, 1991).

One observed difference between GPI and transmembrane anchoring concerns membrane microdomains. These are micrometer-scale membrane regions whose composition, physical properties, and function differ from the average for a whole cell surface (Edidin and Stroynowski, 1991). Transmembrane-anchored proteins travel between cell surface membrane domains less readily than do their GPI-linked homologs. Although this does not alter measured lateral diffusion coefficients, it could be important in constraining or organizing proteins on cell surfaces. GPI-anchored proteins might not be subject to these domain restrictions, which are probably mediated by interactions with cytoplasmic elements (Edidin and Stroynowski, 1991; Edidin *et al.*, 1991).

B. *Potocytosis and Endocytosis*

Studies of the folate receptor suggest that GPI-anchored proteins play a fascinating role in sequestration and transport of small molecules (Anderson *et al.*, 1992). The folate receptor is a GPI-anchored protein (Luhrs and Slomiany, 1989) densely clustered on plasma membranes in association with caveolae (Rothberg *et al.*, 1990b). Caveolae are invaginated pits or vesicles on the cell surface with a diameter of about 50 nm (Rothberg *et al.*, 1990b; Anderson *et al.*, 1992). They are believed to reversibly alter their shape (Rothberg *et al.*, 1992), transiently invaginating to form closed compartments that trap resident proteins within their boundaries (Rothberg *et al.*, 1990b; Anderson *et al.*, 1992). In the case of folate uptake the resulting compartment may sequester the receptor with bound ligand and then acidify, causing the folate to dissociate from the receptor and allowing it to be transported into the cell (Rothberg *et al.*, 1990b). Cavelolae are decorated with a coat protein, caveolin (Rothberg *et al.*, 1992), and require a high concentration of cholesterol to maintain their structure.

Uptake via caveolae (termed potocytosis) may not require GPI-linked proteins. However, several GPI-anchored proteins are specifically associated with these specialized membrane structures (Anderson *et al.*, 1992). Also, both the folate receptor arrays and the caveolar structures themselves depend on cholesterol for their integrity (Rothberg *et al.*, 1990b). It is possible that interactions between certain GPI fatty acids and cholesterol lead to the formation of specific lipid clusters (Rothberg *et al.*, 1990a). The caveolae might work in concert with GPI-anchored receptors or enzymes to transiently control local concentrations or delivery of ligands, enzyme substrates, or reaction products (Anderson *et al.*, 1992).

GPI-linked proteins are internalized by cells, but the mechanism is poorly understood. Most cell surface receptors are directed to enter cells via clathrin-coated pits; the signals for this localization reside in their cytoplasmic domains (Brodsky *et al.*, 1991). Once present in these specialized plasma membrane sites such proteins are efficiently internalized (Lazarovitz and Roth, 1988). This suggests that GPI-anchored proteins, with no cytoplasmic domains, could not be endocytosed by the same mechanism. Consistent with this proposal, GPI-anchored proteins are poorly represented in coated pit areas of membrane (Lemansky *et al.*, 1990). Several (TAP and Thy-1) undergo extremely slow rates of turnover, remaining primarily at the cell surface (Lemansky *et al.*, 1990; Bamezai *et al.*, 1992). Their pathway of endocytosis appears to be distinct from that of proteins internalized via coated pits: TAP and Thy-1 are not observed in association with coated pits or vesicles; their internalization is not inhibited by agents that block endocytosis of proteins known to enter via coated pits; and their rate of endocytosis may be modulated independently of coated-pit-mediated uptake (Bamezai *et al.*, 1992; Bamezai and Rock, 1991). Hybrid proteins containing GPI anchors are internalized extremely slowly, with the rate of endocytosis up to 50 times slower than constitutive internalization of plasma membrane (Lisanti *et al.*, 1990). GPI anchors may serve to exclude attached proteins from coated pits, thereby preventing their efficient endocytosis and increasing their residence time on the cell surface.

C. Glycosyl Phosphatidylinositol Anchors and Protein Sorting

Epithelial cells are polarized, with distinct apical and basolateral domains of their plasma membranes. These domains are distinguished by their lipid, protein, and morphological profiles; these confer specific functions upon each region and thus polarity upon the cell. To create and maintain these specialized regions cells must sort both protein and lipid membrane components, directing their differential placement on the surface. Similar mechanisms may be involved on a less dramatic scale in cell types that are not polarized. Investigation of the role of GPI anchors in protein sorting was prompted by the observation that several proteins used as markers for apical domains were anchored by these structures (Lisanti and Rodriguez-Boulan, 1991). Six GPI–proteins were present in a polarized renal epithelial cell line, all restricted to the apical surface; this suggested that GPIs participate in targeting (Lisanti *et al.*, 1988). This hypothesis was tested with chimeras between proteins anchored by transmembrane domains and those usually anchored by GPIs (Brown *et al.*, 1989; Lisanti *et al.*, 1989). Proteins that were usually basolaterally localized were redirected to the apical surface by the substitution of a GPI-anchoring signal for their original transmembrane domain (Brown *et al.*, 1989; Lisanti *et al.*, 1989). Similarly, when GPI-linked proteins foreign to polarized cell types were introduced into such cells by transfection they were correctly expressed, anchored, and directed to the apical surface (Wilson *et al.*, 1990). Neural cell adhesion molecule (N-CAM), which occurs in both GPI- and transmembrane-anchored forms via differential splicing (Cunningham *et al.*, 1987; He *et al.*, 1987), provides a striking example. Upon transfection into polarized epithelial cells the GPI-anchored form was targeted to the apical surface and the transmembrane-anchored form to the basolateral membrane domain (Powell *et al.*, 1991a). This localization may mirror a natural situation in neurons, where in several cases GPI-linked proteins are asymetrically distributed, with localization to neuronal outgrowths (Lisanti *et al.*, 1990).

Although GPI anchorage appears to direct apical expression, the possibility of redundant sorting signals was also explored (Powell *et al.*, 1991b; Lisanti *et al.*, 1991). Chimeras were engineered to eliminate nearly all sequences originating from a GPI-anchored protein; this showed that the apical sorting information was related to the signal for anchor attachment, rather than to any residual sequence from the original GPI-linked protein (Lisanti *et al.*, 1991). Surprisingly, in constructs where GPI-linked proteins were truncated to abolish anchor addition, proteins were secreted apically (Brown *et al.*, 1989; Powell *et al.*, 1991b); this suggested that additional targeting information resides in the polypeptide sequence or that an apical default pathway exists (Mostov *et al.*, 1992). Overall, GPI-directed sorting may be controlled by a heirarchy of signals, with anchor attachment acting as a positive apical signal (Lisanti and Rodriguez-Boulan, 1991).

The pathway utilized by GPI-anchored proteins to reach the apical surface of cells may be different from that used by proteins localized by various polypeptide determinants. In Con A-resistant MDCK (Madin–Darby canine kidney) cells several GPI-linked proteins failed to exhibit apical polarity, while non-GPI-linked proteins were sorted properly (Lisanti and Rodriguez-Boulan, 1991). Similar patterns were observed in a poorly differentiated thyroid epithelial cell line (Nitsch *et al.*, 1985); this disjunction between proteins anchored by different structures implies independent pathways or requirements. Although most of the studies about polarized distribution of GPI-linked proteins have been performed in MDCK cells, specific localization of such proteins has also been noted in nonepithelial cells such as sperm, lymphocytes, and neurons (Dotti *et al.*, 1991; reviewed in Lisanti and Rodriguez-Boulan, 1990).

Apical membrane domains have an unusually high content of sphingolipid, mainly glycosphingolipids. The sorting of these compounds, like that of GPI-anchored and other proteins, occurs in the Golgi complex (Simons and van Meer, 1988). Apical transport of newly synthesized proteins and lipids may be intimately related, through sorting into common carrier vesicles for appropriate delivery (Simons and van Meer, 1988; Simons and Wandinger-Ness, 1990). Such sorting might be mediated by the ability of glycosphingolipids to self-associate into aggregated patches that demonstrate restricted mobility in membranes (probably mediated by interlipid hydrogen bonding; Thompson and Tilack, 1985). GPI-linked proteins fit well into this model, as they could presumably associate with lipid aggregates.

Recent data suggest that complexes of GPI-anchored proteins and lipids do occur in cells. GPI-linked species were found in a detergent-insoluble complex enriched in glycosphingolipids and other GPI-linked proteins, but not containing a basolateral marker protein (Brown and Rose, 1992). The GPI-linked protein became less soluble over time during residence in the Golgi complex, and this was not affected by agents known to disrupt cytoskeletal organization or even cell polarization. The proposed model is that GPI-linked proteins and glycosphingolipids first aggregate in the Golgi complex. Mediated by apical sorting signals contained in some of the polypeptides, these clusters are then packaged into apical transport vesicles and proceed to the cell surface (Brown and Rose, 1992). In this manner GPI-linked proteins, even without additional apical transport signals, could be properly localized.

The proposed model may bear on previous observations about the behavior of GPI-anchored proteins. Without anchor addition, these proteins sometimes remain trapped within cells (Micanovic *et al.*, 1990a; Moran *et al.*, 1991; Takami et al., 1992). This may be because without a domain for lipid association they lack all transport signals. Also, a fraction of GPI-anchored proteins demonstrates restricted mobility on cell surfaces (Ishihara *et al.*, 1987), perhaps because of maintained association in tightly bound aggregates. Last, the behavior of GPI-anchored proteins associated with caveolae is known to be drastically modified by agents that perturb cholesterol distribution (Rothberg *et al.*, 1990a). This may be yet another example of close association between GPI-linked proteins and their lipid environment.

D. *Glycosyl Phosphatidylinositol Anchors in the Immune and Nervous Systems*

GPI-anchored proteins have been suggested to participate in signaling pathways, and their mode of action appears to depend on proper anchorage. These proteins include numerous molecules of the immune system on the cell surfaces of T lymphocytes, B lymphocytes, monocytes, and granulocytes. Many of these have demonstrated signaling capability, leading to T lymphocyte proliferation, interleukin production, and other signs of cell activation (reviewed in Robinson, 1991a,b). A number of GPI-anchored proteins become expressed as lymphocytes mature, become activated, or differentiate, also suggesting their importance in immune regulation (Feldbush and Lafrenz, 1991).

To determine whether the GPI anchor itself participates in cell activation, transgenic mice expressing various forms of a GPI-anchored histocompatibility antigen, Qa-2 (Robinson *et al.*, 1989), were produced. Antibodies specific for Qa-2 are known to mediate T lymphocyte activation.[8] Constructs combined segments of Qa-2 with those of the normally transmembrane-anchored protein H-2; antibodies against the latter do not stimulate T-cell proliferation. Only molecules bearing GPI anchors were found to retain mitogenic activity, independent of the specific protein domain anchored (Robinson *et al.*, 1989). This result strongly indicated that the GPI anchor was crucial for this activation. Similar results have been obtained from experiments based on transfection of hybrid constructs (Su *et al.*, 1991).

The immunoglobulin F_c receptor ($F_c\gamma R$) plays an important role in triggering neutrophil effector functions. This receptor is present in both GPI- and transmembrane-anchored forms, of slightly different antigenic nature. The two forms of the receptor are differentially expressed in various cell types (Scallon *et al.*, 1989) and appear to have distinct functions in immune modulation. While the GPI-anchored form can mediate neutrophil stimulation, it is unable to autonomously trigger cell behavior as can the transmembrane-anchored form; some other intermediary protein is required (Salmon *et al.*, 1991).

Several models have been proposed to explain signal transduction by GPI-anchored proteins. Many immune functions stimulated by GPI-linked proteins require clustering of the antigens (e.g., by specific antibody binding). It has been suggested that this clustering stimulates antigen uptake and subsequent anchor degradation, allowing portions of the GPI to mediate further stimuli. No anchor degrading activity

[8] These antibodies stimulate T cells in the presence of anti-immunoglobulin antibodies or Fc receptor-bearing accessory cells (Robinson, 1991b).

has been identified in T lymphocytes, however (Robinson, 1991b), and rapid kinetics may preclude an internalization-based mechanism (Barboni *et al.*, 1991). Another possibility is that the GPI-anchored proteins act via other signal-transducing proteins, perhaps by direct physical interaction (Robinson, 1991a,b). Some intermediary does seem required, as these proteins interact only with the outer leaflet of the plasma membrane, out of reach of the cellular machinery. Indeed, when membrane immunoglobulin is altered to remove three cytoplasmic residues (resulting in GPI, in place of transmembrane, anchoring) it loses its functional capacity (Mitchell *et al.*, 1991).

Since many receptors involved with signal transduction are associated with kinase activities, this association was examined in the immune system. Immunoprecipitation of various GPI-anchored molecules resulted in specific coprecipitation of protein kinase activity; this association was dependent on intact GPI anchors (Stefanova *et al.*, 1991). Although specific interactions remain to be demonstrated, this does suggest one potential transducer for signals transmitted by GPI-anchored proteins. Based on characteristics of the initial stages of T cell activation, the GPI-induced signal transduction mechanism may involve a pathway distinct from that used by transmembrane proteins (Barboni *et al.*, 1991).

Many glycoproteins in the central nervous system, frequently members of the immunoglobulin gene superfamily, are attached to the cell membrane via GPI. These molecules have been implicated in critical neuronal processes such as axonal growth, guidance, and fasciculation (reviewed in Walsh and Doherty, 1991). The mechanism by which these proteins mediate signaling is unknown, analogous to the situation in the immune system. One prominant molecule, N-CAM, is present in isoforms that are either soluble, anchored by transmembrane domains, or GPI-linked (Cunningham *et al.*, 1987). Either membrane associated form is able to serve as a substrate for neurite extension, implying that the GPI-anchored form provides an effective recognition signal (Doherty *et al.*, 1990). This does not, however, address the mechanism(s) of signal transduction. The receptor for ciliary neurotrophic factor, a factor involved in development and function of neurons, is also GPI-anchored. This receptor is highly homologous to the transmembrane-anchored receptor for interleukin-6 (Davis *et al.*, 1991). The interleukin receptor is known to require a second protein in order to perform its signal transducing role; this indirectly suggests that the related GPI-linked receptor interacts with a second protein (Davis *et al.*, 1991).

These diverse examples all suggest that GPI anchors are important in cell signaling. While mechanisms and components remain to be determined, signal transduction may be an important function of GPIs.

E. Glycosyl Phosphatidylinositol Anchors as a Release Mechanism

Several phospholipases that cleave GPI structures, releasing anchored proteins from their hydrophobic glycolipid moiety, have been described. These include bacterial phosphatidylinositol-specific phospholipase C as well as more specific phospholipases that effectively degrade only GPI species (e.g., the trypanosome PLC and serum PLD mentioned earlier). The existence of GPI-specific enzymes has inspired speculation on their function. A role in secretion has been suggested in the case of the pancreatic granule membrane protein GP-2. This GPI-anchored protein is associated with granule and plasma membranes, but is also observed in a soluble form both within granules and secreted by the pancreas (Lebel and Beattie, 1988). A single gene encodes both forms, and the relationship has been suggested to be one of precursor and product (Fukuoka *et al.*, 1991). Initially, GP-2 may be synthesized and GPI-anchored within granules. Some of it could then be released by endogenous phospholipase C activity en route to the cell surface and secreted in an anchorless form (Paul *et al.*, 1991).

Phospholipases may serve to catalyze selective release of GPI-anchored proteins from cells, leaving other surface constituents unperturbed. This could either reduce the amount of a specific protein on the cell surface or increase the concentration of soluble protein in the extracellular fluid or serum. There are several difficulties in proposing cellular phospholipases as mediators of selective release of GPI-anchored proteins. First, the known phospholipase C enzymes capable of releasing GPI-linked proteins have been isolated from bacteria or parasites and have yet to be characterized in higher eukaryotes. Also, such enzymes would be expected to be intracellular, while the postulated substrates are confined to the outer leaflet of the plasma membrane; any hypothesis must address this. While serum enzymes (such as the GPI-specific phospholipase D) could access cell surface proteins, these enzymes are constitutively present and would require regulation. An additional problem is lack of selectivity: a variety of proteins are anchored to cell surfaces by GPI anchors, so some mode of discrimination would be required. This specificity could be provided by modifications that render anchors resistant to degradation, such as inositol acylation.

The trypanosome VSG is often cited as a potential example of selective release. At one point in its life cycle the parasite undergoes transformation from a form adapted to a mammalian host to a form suited to survival in an insect vector. At that time its GPI-anchored VSG coat, which is phospholipase C sensitive, is replaced by another surface protein (termed procyclin or procyclic acidic repetitive protein), also GPI-anchored but phospholipase C resistant (Roditi *et al.*, 1989; Clayton and Mowatt, 1989). One hypothesis is that the endogenous trypanosome GPI-specific phospholipase C is involved in VSG shedding; the procyclin replacement is immune to cleavage and remains on the cell membrane (Clayton and Mowatt, 1989). As the enzyme is intracellular, VSG cleavage could take place in endocytic vesicles, with exocytosis of the degradation products (reviewed in Ferguson and Williams, 1988). This hypothesis, while interesting, remains to be proved.

IV. Glycosyl Phosphatidylinositol Anchors in Disease

A. *Paroxysmal Nocturnal Hemoglobinuria*

The best studied example of disease caused by a defect in GPI anchoring is paroxysmal nocturnal hemoglobinuria (PNH; reviewed in Halperin and Nicholson-Weller, 1989; Davitz, 1987). Patients with PNH experience periodic hemolysis secondary to increased sensitivity of blood cells to autologous complement-mediated lysis. This condition is due to the absence of two GPI-anchored proteins, decay-accelerating factor and C8 binding protein, which normally serve to prevent host cell lysis during complement activation. Other GPI-anchored proteins are also reduced or absent from cell surfaces in PNH patients, suggesting a global defect in anchor synthesis or attachment, possibly resulting in improper localization or instability of the involved proteins. One study of anchor biosynthesis in cells from PNH patients concludes that the first few steps in biosynthesis are normal (Hirose *et al.*, 1991), but this work does not preclude other defects. It is likely that several defects contribute to the spectrum of this disease, in accordance with the occurrence of erythrocytes of varying sensitivity within the patient population. Consistent with this idea, neutrophils from different patients exhibit heterogeneity in levels and localization of a GPI-anchored protein (Jost *et al.*, 1991).

B. *Parasitic Infections*

GPI-anchored proteins constitute a prominent group of cell surface antigens in many pathogenic parasites, including trypanosomes, toxoplasma, malarial parasites, schistosomes, and others. Two of these organisms have been central in the study of GPIs. African trypanosomes bear a densely packed surface coat composed of VSG. This protein mediates the property of antigenic variation, whereby antigenically distinct variants are sequentially expressed by the trypanosome population, enabling the parasite to evade host immune defenses. VSG has also been shown to transfer from trypanosomes to erythrocyte membranes *in vitro*, suggesting a potential role for this protein in the red cell destruction and consequent anemia observed in infected individuals (Rifkin and Landsberger, 1990).[9] The trypanosome VSG is unique among GPI-anchored proteins in containing exclusively myristate in its PI portion. Analogs of myristate demonstrate selective toxicity against bloodstream trypanosomes and may provide a novel chemotherapeutic strategy (Doering *et al.*, 1991, and unpublished data).

Leishmania parasites, which cause visceral, cutaneous, and mucocutaneous leishmaniasis, express a broad range of GPIs and related glycolipids. These include free glycoinositol phospholipids and the complex lipophosphoglycan (described earlier); the latter has been implicated in parasite infectivity and defensive mechanisms (reviewed in Turco, 1988). In addition, leishmania express a GPI-anchored surface protease, gp63, which is important for parasite virulence. The abundance of gp63 is upregulated via differential expression of three distinct genes as this parasite evolves to a more infectious stage (Ramamoorthy *et al.*, 1992).

C. *Tumor Antigens and Tumorigenesis*

Carcinoembryonic antigen (CEA) is a well-characterized tumor antigen anchored by a GPI structure (Hefta *et al.*, 1988; Takami *et al.*, 1988). Although some CEA is found in normal tissues and fluids, elevated concentration of this antigen is considered a highly significant disease marker, indicating the presence of inflammation or malignancy in epithelial tissues. When tumors express CEA, relative serum concentrations correlate with the tumor burden (for review, see Lisanti *et al.*, 1990). This may be a direct consequence of GPI anchorage, as follows: CEA, as a GPI-anchored protein, is normally localized to the apical surface of epithelial cells and thus is isolated from the bloodstream. In poorly differentiated tumor cells normal surface distribution is lost, allowing CEA to appear on the basolateral region of the cell, and to be released into the serum (Lisanti *et al.*, 1990). At least one other tumor antigen, the CA125 marker of ovarian epithelial tumors, is probably GPI-anchored. CA125 from cancerous ovarian tissue is more susceptible to phosphatidylinositol-specific phospholipase C than the antigen derived from normal tissue. This could be due to an anchor modification, which presumably fails to occur in transformed cells (Nagata *et al.*, 1991).

[9] The possibility of this transfer occurring in serum, in the presence of resident phospholipases, was not addressed (Ferguson, 1991).

Cell lines have been developed from a radiation-induced T lymphoma, and Thy-1 negative subclones were examined. One cell variant has a defect in GPI biosynthesis (Teng *et al.*, 1989) and sheds Thy-1 into the medium. It is intriguing that these cells also have impaired ability to cause tumor formation (reviewed in Fasel *et al.*, 1991).

Another interesting observation is that the surface expression (on NIH3T3 cells) of Thy-1 and another GPI-anchored protein (TAP) decreases markedly upon the expression of the c-H-*ras* oncogene, while transmembrane markers are not affected. The decrease correlates with increased internalization of GPI-linked proteins and is modulated by *ras* expression (Bamezai and Rock, 1991).

D. Scrapie

Scrapie is one of several neurodegenerative diseases believed to be mediated by proteinaceous infectious particles, or prions (Stahl and Prusiner, 1991). The major component of the scrapie prion is PrP^{Sc}, derived from a cellular protein, PrP^{C}. Both forms of PrP are GPI-anchored sialoglycoproteins with identical primary sequences. The GPI anchor of PrP^{Sc} consists of a heterogeneous mixture of structures, several of which contain sialic acid. Interestingly, the cellular and prion forms of PrP exhibit differential susceptibility to proteases and to phosphatidylinositol-specific phospholipase C (Stahl *et al.*, 1990a,b). This disparity is observed when membranes containing these proteins are digested, but not once the proteins are denatured, suggesting it is due to distinct properties of conformation or aggregation that differ between the normal protein and its prion homolog (Stahl and Prusiner, 1991).

V. Current Research Priorities

There are many exciting areas of GPI research yet to be pursued. To date the basic core structure of glycosyl phosphatidylinositols has been established, with several anchors thoroughly characterized. Many of these compounds exhibit previously unknown modifications of lipid or glycan structure. The identification of these elements, and their correlation with cell type or specific protein should prove fascinating. Molecular modeling of the anchors and related structures may also elucidate their arrangement on cell surfaces.

Study of GPI biosynthesis, although burgeoning as a field in the last few years, is still in a descriptive stage. The basic reactions in one system, that of trypanosomes, have been defined, but many details of the pathway remain unclear. Specifics of several events have not been determined, including the sequence of mannose addition, the role of the ethanolamine donor, whether ethanolamine and phosphate are transferred together, and the role of acylated intermediates in the production of nonacylated anchor precursors. Beyond these issues, the reaction pathway in mammalian systems remains to be fully demonstrated. This is important because significant differences exist between mammalian and trypanosome anchors, which should be reflected in their biosynthesis. The characterization of additional anchor precursors from different systems may help to address these biosynthetic issues.

The enzymology of GPI biosynthesis is in its infancy. Novel enzymes must exist to build and degrade the unusual structures that compose anchors. Elucidation of GPI structure and biosynthesis has suggested the activities involved, but these have yet to be purified and studied. In addition to those enzymes that create the core glycan, others are required to modify the core structures, either before or after protein attachment (Doering *et al.*, 1990a). Once some of these proteins have been characterized their regulation may be addressed. Other questions may also become answerable, such as those involving the topology of biosynthetic reactions (Field and Menon, 1991) or the choice of lipid structure upon which anchors are constructed. Because of the abundance of GPIs in parasites such as trypanosomes, these organisms are valuable for studying the basic enzymology of GPI biosynthesis. However, because trypanosomes are not amenable to genetic manipulation, yeast and other eukaryotic systems may prove more tractable as research systems, especially to address topics such as regulation.

Another key enzyme to investigate is the putative transamidase responsible for transfer of polypeptide to GPI structures. Although the substrates of this reaction (anchor precursors and anchor attachment sequences) have now been thoroughly characterized, it is only recently that the reaction has been acheived *in vitro*. This accomplishment should lead to characterization of the enzyme involved, opening the way to studies of its regulation and specificity, and perhaps the biological effects of interfering with its activity.

As mentioned earlier, protozoan systems provide excellent model systems for investigation of GPI structure and biosynthesis. As similar studies in mammalian systems advance, many differences between mammalian systems and lower eukaryotes have surfaced. While the evolutionary conservation of many features of GPIs has been invaluable in expediting research in higher systems, the heterogeneity observed between species now promises interesting comparisons. Study of variation between GPI metabolism in mammals and that in pathogenic organisms may lead to development of selectively toxic agents.

Fascinating questions remain in the realm of GPI function. These structures are implicated in processes such as protein sorting, caveolar uptake, and control of lateral mobility and membrane microlocalization. Examination of sorting in nonepithelial cell types may address whether mechanisms of differential membrane localization are conserved among cell types. Exploration of the association between GPI anchors and cellular lipids is also a promising avenue of research to further understanding of cellular events.

Evidence is accumulating that GPI-anchored proteins participate in signal transduction, perhaps via association with other membrane proteins. Little is known about this postulated class of important secondary mediators, which presumably interact both with GPI-anchored proteins and with cellular machinery out of reach of GPI structures. Investigations of T-cell activation and neutrophil stimulation may illuminate this area.

Further studies of systems in which one protein is either transmembrane- or GPI-anchored may help illustrate specific functions of GPIs. At present there are examples in which such homologous proteins play indistinguishable roles, and others in which they have very different influences on cellular behavior. Particularly fascinating are cases in which the relative distribution of these proteins varies with development or differentiation of cells.

Finally, GPI-anchored proteins are important in the pathogenesis of several major diseases. Research in these areas may lead to improved understanding of these disease processes and potentially suggest new therapeutic approaches.

Acknowledgments

I have enjoyed many productive discussions with P. T. Englund, G. W. Hart, and T. A. B. Shapiro. I thank them and A. Varadhachary for helpful comments on this manuscript. Work performed in the laboratories of P. T. Englund and G. W. Hart was supported by National Institutes of Health Grant AI21334, Medical Scientist Training Grant 5T32GM07309, and a grant from the MacArthur Foundation.

References

Anderson, R. G. W., Kamen, B. A., Rothberg, K. G., and Lacey, S. W. (1992). Potocytosis: Sequestration and transport of small molecules by caveolae. *Science* **255,** 410–411.

Bamezai, A., and Rock, K. L. (1991). Effect of ras-activation on the expression of glycosylphosphatidylinositol-anchored proteins on the plasma membrane. *Oncogene* **6,** 1445–1451.

Bamezai, A., Goldmacher, V. S., and Rock, K. L. (1992). Internalization of glycosylphosphatidylinositol (GPI)-anchored lymphocyte proteins: GPI-anchored and transmembrane molecules internalize through distinct pathways. *Eur. J. Immunol.* **22,** 15–21.

Bangs, J. D., Hereld, D., Krakow, J. L., Hart, G. W., and Englund, P. T. (1985). Rapid processing of the carboxyl terminus of a trypanosome variant surface glycoprotein. *Proc. Natl. Acad. Sci. USA* **82,** 3207–3211.

Bangs, J. D., Doering, T. L., Englund, P. T., and Hart, G. W. (1988). Biosynthesis of a variant surface glycoprotein of *Trypanosoma brucei*. *J. Biol. Chem.* **263,** 17697–17705.

Barboni, E., Gormley, A. M., Pliego Rivero, F. B., Vidal, M., and Morris, R. J. (1991). Activation of T lymphocytes by cross-linking of glycophospholipid-anchored Thy-1 mobilizes separate pools of intracellular second-messengers to those induced by the antigen-receptor/CD3 complex. *Immunology* **72,** 457–463.

Bordier, C. (1981). Phase separation of integral membrane protein in Triton X-114 solution. *J. Biol. Chem.* **256,** 1604–1607.

Brodsky, F. M., Hill, B. L., Acton, S. L., Nathke, I., Wong, D. H., Ponnambalam, S., and Parham, P. (1991). Clathrin light chains: Arrays of protein motifs that regulate coated-vesicle dynamics. *TIBS* **16,** 208–213.

Brown, D. A., Crise, B., and Rose, J. K. (1989). Mechanism of membrane anchoring affects polarized expression of two proteins in MDCK cells. *Science* **245,** 1499–1501.

Brown, D. A., and Rose, J. K. (1992). Sorting of GPI-anchored proteins to glycolipid-enriched membrane subdomains during transport to the apical surface. *Cell* **68,** 533–544.

Bulow, R., and Overath, P. (1986). Purification and characterization of the membrane-form variant surface glycoprotein hydrolase of *Trypanosoma brucei*. *J. Biol. Chem.* **261,** 11918–11923.

Bulow, R., Overath, P., and Davoust, J. (1988). Rapid lateral diffusion of the variant surface glycoprotein in the coat of *Trypanosoma brucei*. *Biochemistry* **27,** 2384–2388.

Butikofer, P., Kuypers, F. A., Shackleton, C., Brodbeck, U., and Steiger, S. (1991). Molecular species analysis of the glycosyl phosphatidylinositol anchor of *Torpedo marmorata* acetylcholinesterase. *J. Biol. Chem.* **265,** 18983–18987.

Caras, I. W. (1991a). Probing the signal for glycophosphatidylinositol anchor attachment using decay accelerating factor as a model system. *Cell Biol. Int. Rep.* **15,** 815–826.

Caras, I. W. (1991b). An internally positioned signal can direct attachment of a glycophospholipid membrane anchor. *J. Cell Biol.* **113,** 77–85.

Clayton, C. E., and Mowatt, M. R. (1989). The procyclic acidic repetitive proteins of *Trypanosoma brucei*. *J. Biol. Chem.* **264,** 15088–15093.

Conzelmann, A., Spiazzi, A., and Bron, C. (1987). Glycolipid anchors are attached to Thy-1 glycoprotein rapidly after translation. *Biochem. J.* **246,** 605–610.

Conzelmann, A., Riezman, H., Desponds, C., and Bron, C. (1988). A major 125-kd membrane glycoprotein of *Saccharomyces cerevisiae* is attached to the lipid bilayer through an inositol-containing phospholipid. *EMBO J.* **7,** 2233–2240.

Conzelmann, A., Fankhauser, C., and Desponds, C. (1990). Myoinositol gets incorporated into numerous membrane glycoproteins of *Saccharomyces cerevisiae*: Incorporation is dependent on phosphomannomutase (SEC53). *EMBO J.* **9,** 653–661.

Conzelmann, A., Puoti, A., Lester, R. L., and Desponds, C. (1992). Two different types of lipid moieties are present in glycophosphoinositol-anchored membrane proteins of *Saccharomyces cerevisiae*. *EMBO J.* **11,** 457–466.

Cross, G. A. M. (1975). Identification, purification, and properties of clone-specific glycoprotein antigens constituting the surface coat of *Trypanosoma brucei*. *Parasitology* **71,** 393–417.

Cross, G. A. M. (1984). Release and purification of *Trypanosoma brucei* variant surface glycoprotein. *J. Cell Biochem.* **24,** 79–90.

Cross, G. A. M. (1990). Glycolipid anchoring of plasma membrane proteins. *Annu. Rev. Cell. Biol.* **6,** 1–39.

Cunningham, B. A., Hemperly, J. J., Murray, B. A., Prediger, E. A., Brackenbury, R., and Edelman, G. M. (1987). Neural cell adhesion molecule: Structure, immunoglobulin-like domains, cell surface modulation, and alternative RNA splicing. *Science* **236,** 799–806.

Davis, S., Aldrich, T. H., Valenzuela, D. M., Wong, V., Furth, M. E., Squinto, S. P., and Yancopoulos, G. D. (1991). The receptor for ciliary neurotrophic factor. *Science* **253,** 59–63.

Davitz, M. A. (1987). Decay-accelerating factor (DAF): A review of its function and structure. *Acta Med. Scand.* **715,** 111–121.

DeGasperi, R., Thomas, L. J., Sugiyama, E., Chang, H. M., Beck, P. J., Orlean, P., Albright, C., Waneck, G., Sambrook, J. F., Warren, C. D., and Yeh, E. T. H. (1990). Correction of a defect in mamalian GPI anchor biosynthesis by a transfected yeast gene. *Science* **250,** 988–991.

Doering, T. L., Masterson, W. J., Englund, P. T., and Hart, G. W. (1989). Biosynthesis of the glycosyl phosphatidylinositol membrane anchor of the trypanosome variant surface glycoprotein: Origin of the nonacetylated glucosamine. *J. Biol. Chem.* **264,** 11168–11173.

Doering, T. L., Masterson, W. J., Hart, G. W., and Englund, P. T. (1990a). Biosynthesis of glycosyl phosphatidylinositol membrane anchors. *J. Biol. Chem.* **265,** 611–614.

Doering, T. L., Raper, J., Buxbaum, L. U., Hart, G. W., and Englund, P. T. (1990b). Biosynthesis of glycosyl phosphatidylinositol protein anchors. *Methods* **1,** 288–296.

Doering, T. L., Raper, J., Buxbaum, L. U., Adams, S. P., Gordon, J. I., Hart, G. W., and Englund, P. T. (1991). An analog of myristic acid with selective toxicity for African trypanosomes. *Science* **252,** 1851–1854.

Doering, T. L., Englund, P. T., and Hart, G. W. (1993). Detection of glycophospholipid anchors on proteins. *In* "Current Protocols in Molecular Biology" (F. M. Ausubel, R. Brent, R. E. Kingston, D. D. Moore, J. G. Seidman, J. A. Smith, and K. Struhl, eds.), in press.

Doherty, P., Fruns, M., Seaton, P., Disckson, G., Barton, C. H., Sears, T. A., and Walsh, F. S. (1990). A threshold effect of the major isoforms of NCAM on neurite outgrowth. *Nature* **343,** 464–466.

Dotti, C. G., Parton, R. G., and Simons, K. (1991). Polarized sorting of glypiated proteins in hippocampal neurons. *Nature* **349,** 158–161.

Edidin, M., Kuo, S. C., and Sheetz, M. P. (1991). Lateral movements of membrane glycoproteins restricted by dynamic cytoplasmic barriers. *Science* **254,** 1379–1382.

Edidin, M., and Stroynowski, I. (1991). Differences between the lateral organization of conventional and inositol phospholipid-anchored membrane proteins: A further definition of micrometer scale membrane domains. *J. Cell Biol.* **112,** 1143–1150.

Fasel, N., Deglon, N., Rousseaux, M., Beghdadi-Rais, C., Bernard, M., and Bron, C. (1991). Evaluation of somatic cell variants deficient in glycosylphosphatidylinositol anchoring as candidates for genetic correction. *Cell Biol. Int. Rep.* **15,** 1051–1064.

Feldbush, T. L. and Lafrenz, D. (1991). Characterization of the expression of alkaline phosphatase on activated B cells. *J. Immunol.* **147,** 3690–3695.

Ferguson, M. A., Low, M. G., and Cross, G. A. M. (1985). Glycosyl-sn-1,2-dimyristylphosphatidylinositol is covalently linked to *Trypanosoma brucei* variant surface glycoprotein. *J. Biol. Chem.* **260,** 14547–14555.

Ferguson, M. A., Duszenko, M., Lamont, G. S., Overath, P., and Cross, G. A. M. (1986). Biosynthesis of *Trypanosoma brucei* variant surface glycoproteins *N*-glycosylation and addition of a phosphatidylinositol membrane anchor. *J. Biol. Chem.* **261,** 356–362.

Ferguson, M. A. J., Homans, S. W., Dwek, R. A., and Rademacher, T. W. (1988). Glycosyl-phosphatidylinositol moiety that anchors *Trypanosoma brucei* variant surface glycoprotein to the membrane. *Science* **239,** 753–759.

Ferguson, M. A. J. (1991). Lipid anchors on membrane proteins. *Curr. Opin. Struct. Biol.* **1,** 522–529.

Ferguson, M. A. J. (1992). The chemical and enzymic analysis of G-PI fine structure. *In* "Lipid Modifications of Proteins: A Practical Approach" (A. J. Turner and N. M. Hooper, eds.), pp. 191–231. IRL Press, Oxford.

Ferguson, M. A. J., and Williams, A. F. (1988). Cell-surface anchoring of proteins via glycosyl-phosphatidylinositol structures. *Annu. Rev. Biochem.* **57,** 285–320.

Field, M. C., and Menon, A. K. (1991). Biosynthesis of glycolipid anchors in *Trypanosoma brucei. Trends Glycosci. Glycotechnol.* **3,** 107–115.

Field, M. C., Menon, A. K., and Cross, G. A. M. (1991a). Developmental variation of glycosylphosphatidylinositol membrane anchors in *Trypanosoma brucei*: Identification of a candidate biosynthetic precursor of the GPI anchor of the major procyclic stage surface glycoprotein. *JBC* **266,** 8392–8400.

Field, M. C., Menon, A. K., and Cross, G. A. M. (1991b). A glycosylphosphatidylinositol protein anchor from procyclic stage *Trypanosoma brucei*: Lipid structure and biosynthesis. *EMBO J.* 2731–2739.

Field, M. C., and Menon, A. K. (1992). Biosynthesis of membrane protein glycolipid anchors: Methods for labeling, extraction, separation and analysis of glycoinositol phospholipids. *In* "Lipid Modifications of Proteins: A Practical Approach" (A. J. Turner and N. M. Hooper, eds.). IRL Press, Oxford.

Field, M. C., Menon, A. K., and Cross, G. A. M. (1992). Developmental variation of glycosylphosphatidylinositol membrane anchors in *Trypanosoma brucei*: In vitro biosynthesis of intermediates in the construction of the GPI-anchor of the major procyclic surface glycoprotein. *J. Biol. Chem.* **267,** 5324–5329.

Fox, J. A., Duszenko, M., Ferguson, M. A., Low, M. G., and Cross, G. A. M. (1986). Purification and characterization of a novel glycan-phosphatidylinositol-specific phospholipase C from *Trypanosoma brucei. J. Biol. Chem.* **261,** 15767–15771.

Fukuoka, S.-I., Freedman, S. D., and Scheele, G. A. (1991). A single gene encodes membrane bound and free forms of GP-2, the major glycoprotein in pancreatic secretory (zymogen) granule membranes. *Proc. Natl. Acad. Sci. USA* **88,** 2898–2902.

Hajduk, S. L., Englund, P. T., and Smith, D. H. (1989). African trypanosomiasis. *In* "Tropical and Geographic Medicine" (K. S. Warren and A. A. F. Mahmoud, eds.), pp. 268–281. McGraw–Hill, New York.

Halperin, J. A., and Nicholson-Weller, A. (1989). Paroxysmal nocturnal hemoglobinuria: A complement-mediated disease. *Complement Inflammation* **6,** 65–72.

He, H.-T., Finne, J., and Goridis, C. (1987). Biosynthesis, membrane association, and release of N-CAM-120, a phosphatidylinositol-linked form of the neural cell adhesion molecule. *J. Cell Biol.* **105,** 2489–2500.

Hefta, S. A., Hefta, L. J. F., Lee, T. D., Paxton, R. J., and Shively, J. E. (1988). Carcinoembryonic antigen is anchored to membranes by covalent attachment to a glycosyl phosphatidylinositol moiety: Identification of the ethanolamine linkage site. *Proc. Natl. Acad. Sci. USA* **85,** 4648–4652.

Hereld, D., Krakow, J. L., Bangs, J. D., Hart, G. W., and Englund, P. T. (1986). A phospholipase C from *Trypanosoma brucei* which selectively cleaves the glycolipid on the variant surface glycoprotein. *J. Biol. Chem.* **261,** 13813–13819.

Hirose, S., Ravi, L., Hazra, S. V., and Medof, M. E. (1991). Assembly and deacetylation of *N*-acetylglucosaminyl-plasmanylinositol in normal and affected paroxysmal nocturnal hemoglobinuria cells. *Proc. Natl. Acad. Sci. USA* **88,** 3762–3766.

Holder, A. A., and Cross, G. A. M. (1981). Glycopeptides from variant surface glycoproteins of *Trypanosoma brucei*. C-terminal location of antigenically cross-reacting carbohydrate moieties. *Mol. Biochem. Parasitol.* **2,** 135–150.

Homans, S. W., Ferguson, M. A. J., Dwek, R. A., Rademacher, T. W., Anand, R., and

Williams, A. F. (1988). Complete structure of the glycosyl phosphatidylinositol membrane of rat brain Thy-1 glycoprotein. *Nature* **333,** 269–272.

Hooper, N. M., and Bashir, A. (1991). Glycosyl-phosphatidylinositol-anchored membrane proteins can be distinguished from transmembrane polypeptide-anchored proteins by differential solubilization and temperature-induced phase separation in Triton X-114 . *Biochem. J.* **280,** 745–751.

Hyman, R. (1988). Somatic genetic analysis of the expression of cell surface molecules. *Trends Genet.* **4,** 5–8.

Ikezawa, H. (1991). Bacterial PIPLCs—Unique properties and usefulness in studies on GPI anchors. *Cell Biol. Int. Rep.* **15,** 1115–1131.

Ishihara, A., Hou, Y., and Jacobson, K. (1987). The Thy-1 antigen exhibits rapid lateral diffusion in the plasma membrane of rodent lymphoid cells and fibroblasts. *Proc. Natl. Acad. Sci. USA* **84,** 1290–1293.

Jacquemin, C. (1991). Glycosyl phosphatidylinositol in thyroid: Cell signalling or protein anchor? *Biochimie* **73,** 37–40.

Jost, C. R., Gaillard, M. L., Fransen, J. A. M., Daha, M. R., and Ginsel, L. A. (1991). Intracellular localization of glycosyl-phosphatidylinositol-anchored CD67 and FcRIII (CD16) in affected neutrophil granulocytes of patients with paroxysmal nocturnal hemoglobinuria. *Blood* **78,** 3030–3036.

Kodukula, K., Micanovic, R., Gerber, L., Tamburrini, M., Brink, L., and Udenfriend, S. (1991). Biosynthesis of phosphatidylinositol glycan-anchored membrane proteins. *J. Biol. Chem.* **266,** 4464–4470.

Kodukula, K., Cines, D., Amthauer, R., Gerber, L., and Udenfriend, S. (1992). Biosynthesis of glycosylphosphatidylinositol-glycan (PI-G)-anchored membrane proteins in cell-free systems: Cleavage of the nascent protein and addition of the PI-G moiety depend on the size of the COOH-terminal signal peptide. *Proc. Natl. Acad. Sci. USA* **89,** 1350–1353.

Krakow, J. L., Hereld, D., Bangs, J. D., Hart, G. W., and Englund, P. T. (1986). Identification of a glycolipid precursor of the *Trypanosoma brucei* variant surface glycoprotein. *J. Biol. Chem.* **261,** 12147–12153.

Lazarovitz, J., and Roth, M. A. (1988). A single amino acid change in the cytoplasmic domain allows the influenza virus hemaggluntinin to be endocytosed through coated pits. *Cell* **53,** 743–752.

Lebel, D., and Beattie, M. (1988). The major protein of pancreatic zymogen granule membranes (GP-2) is anchored via covalent bonds to phosphatidylinositol. *Biochem. Biophys. Res. Commun.* **154,** 818–823.

Lemansky, P., Fatemi, S. H., Gorican, B., Meyale, S., Rossero, R., and Tartakoff, A. M. (1990). Dynamics and longevity of the glycolipid-anchored protein, Thy-1. *J. Cell Biol.* **110,** 1525–1531.

Lisanti, M. P., Sargiacomo, M., Graeve, L., Saltiel, A. R., and Rodriguez-Boulan, E. (1988). Polarized apical distribution of glycosyl-phosphatidylinositol-anchored proteins in a renal epithelial cell line. *Proc. Natl. Acad. Sci. USA* **85,** 9557–9561.

Lisanti, M. P., Caras, I. W., Davitz, M. A., and Rodriguez-Boulan, E. (1989). A glycophospholipid membrane anchor acts as an apical targeting signal in polarized epithelial cells. *J. Cell Biol.* **109,** 2145–2156.

Lisanti, M. P., Caras, I. W., Gilbert, T., Hanzel, D., and Rodriguez-Boulan, E. (1990). Vectorial apical delivery and slow endocytosis of a glycolipid-anchored fusion protein in transfected MDCK cells. *Proc. Natl. Acad. Sci. USA* **87,** 7419–7423.

Lisanti, M. P., and Rodriguez-Boulan, E. (1990). Glycophospholipid membrane anchoring provides clues to the mechanism of protein sorting in polarized epithelial cells. *TIBS* **15,** 113–118.

Lisanti, M. P., Rodriguez-Boulan, E., and Saltiel, A. R. (1990). Emerging functional roles for the glycosyl-phosphatidylinositol membrane protein anchor. *J. Membr. Biol.* **117**, 1–10.

Lisanti, M. P., Caras, I. W., and Rodriguez-Boulan, E. (1991). Fusion proteins containing a minimal GPI-attachment signal are apically expressed in transfected MDCK cells. *J. Cell Sci.* **99**, 637–640.

Lisanti, M. P., and Rodriguez-Boulan, E. (1991). Polarized sorting of GPI-linked proteins in epithelia and membrane microdomains. *Cell Biol. Int. Rep.* **15**, 1023–1049.

Low, M. G., and Saltiel, A. R. (1988). Structural and functional roles of glycosyl-phosphatidylinositol in membranes. *Science* **239**, 268–275.

Luhrs, C. A., and Slomiany, B. L. (1989). A human membrane-associated folate binding protein is anchored by a glycosyl-phosphatidylinositol tail. *J. Biol. Chem.* **264**, 21446–21449.

Masterson, W. J., Doering, T. L., Hart, G. W., and Englund, P. T. (1989). A novel pathway for glycan assembly: Biosynthesis of the glycosyl-phosphatidylinositol anchor of the trypanosome variant surface glycoprotein. *Cell* **56**, 793–800.

Masterson, W. J., Raper, J., Doering, T. L., Hart, G. W., and Englund, P. T. (1990). Fatty acid remodeling: A novel reaction sequence in the biosynthesis of trypanosome glycosyl phophatidylinositol membrane anchors. *Cell* **62**, 73–80.

Mayor, S., and Menon, A. K. (1990). Structural analysis of the glycoinositol phospholipid anchors of membrane proteins. *Methods* **3**, 297–305.

Mayor, S., Menon, A. K., Cross, G. A. M., Ferguson, M. A. J., Dwek, R. A., and Rademacher, T. W. (1990). Glycolipid precursors for the membrane anchor of *Trypanosoma brucei* variant surface glycoproteins. I. Glycan structure of the phosphatidylinositol-specific phospholipase C sensitive and resistant glycolipids. *J. Biol. Chem.* **265**, 6164–6173.

Mayor, S., Menon, A. K., and Cross, G. A. M. (1991). Transfer of glycosyl-phosphatidylinositol membrane anchors to polypeptide acceptors in a cell-free system. *J. Cell. Biol.* **114**, 61–71.

Mayor, S., Menon, A. K., and Cross, G. A. M. (1992). Galactose-containing glycosylphosphatidylinositols in *Trypanosoma brucei*. *J. Biol. Chem.* **267**, 754–761.

McConville, M. J. (1991). Glycosylated-phosphatidylinositols as virulence factors in *Leishmania*. *Cell Biol. Int. Rep.* **15**, 779–799.

Menon, A. K., Mayor, S., Ferguson, M. A., Duszenko, M., and Cross, G. A. M. (1988). Candidate glycophospholipid precursor for the glycosylphosphatidylinositol membrane anchor of *Trypanosoma brucei* variant surface glycoproteins. *J. Biol. Chem.* **263**, 1970–1977.

Menon, A. K., Mayor, S., and Schwartz, R. T. (1990a). Biosynthesis of glycosyl-phosphatidylinositol lipids in *Trypanosoma brucei*: Involvement of mannosyl-phosphoryldolichol as the mannose donor. *EMBO J.* **9**, 4249–4258.

Menon, A. K., Schwarz, R. T., Mayor, S., and Cross, G. A. M. (1990b). Cell-free synthesis of glycosyl-phosphatidylinositol precursors for the glycolipid membrane anchor of *Trypanosoma brucei* variant surface glycoproteins: Structural characterization of putative biosynthetic intermediates. *J. Biol. Chem.* **265**, 9033–9042.

Menon, A. K. (1991). Biosynthesis of glycosyl-phosphatidylinositol. *Cell Biol. Int. Rep.* **15**, 1007–1021.

Mensa-Wilmot, K., Doering, T. L., Raper, J., Buxbaum, L., Hart, G. W., and Englund, P. T. (1991). Glycosyl phosphatidylinositol membrane anchors in African trypanosomes. *In* "Molecular and Immunological Aspects of Parasitism" (C. C. Wang, ed.) pp. 43–54. Am. Assoc. Advan. Sci., Washington, D.C.

Metz, C. N. (1991). Characterization of the plasma glycosylphosphatidylinositol-specific phospholipase D (GPI-PLD). *Cell Biol. Int. Rep.* **15**, 875–882.

Micanovic, R., Gerber, L. D., Berger, J., Kodukula, K., and Udenfriend, S. (1990a). Selectivity of the cleavage/attachment site of phosphatidylinositol-glycan-anchored membrane proteins determined by site-specific mutagenesis and Asp-484 of placental alkaline phosphatase. *Proc. Natl. Acad. Sci. USA* **87**, 157–161.

Micanovic, R., Kodukula, K., Gerber, L. D., and Udenfriend, S. (1990b). Selectivity at the cleavage/attachment site of phosphatidylinositol glycan anchored membrane proteins is enzymatically determined. *Proc. Natl. Acad. Sci. USA* **87**, 7939–7943.

Mitchell, R. N., Shaw, A. C., Weaver, Y. K., Leder, P., and Abbas, A. K. (1991). Cytoplasmic tail deletion converts membrane immunoglobulin to a phosphatidylinositol-linked form lacking signaling and efficient antigen internalization functions. *J. Biol. Chem.* **266**, 8856–8860.

Moran, P., and Caras, I. W. (1991). A nonfunctional sequence converted to a signal for glycosylphosphatidylinositol membrane anchor attachment. *J. Cell Biol.* **115**, 329–336.

Moran, P., Raab, H., Kohr, W. J., and Caras, I. W. (1991). Glycophospholipid membrane anchor attachment: Molecular analysis of the cleavage/attachment site. *J. Biol. Chem.* **266**, 1250–1257.

Mostov, K., Apodaca, G., Aroeti, B., and Okamoto, C. (1992). Plasma membrane protein sorting in polarized epithelial cells. *J. Cell. Biol.* **116,** 577–583.

Nagata, A., Hirota, N., Sakai, T., Fujimoto, M., and Komada, T. (1991). Molecular nature and possible presence of a membranous glycan-phosphatidylinositol anchor of CA125 antigen. *Tumor Biol.* **12**, 279–286.

Nitsch, L., Tramontano, D., and Ambesi-Impiombato, F. S. (1985). Morphological and functional polarity of an epithelial thyroid cell line. *Eur. J. Cell Biol.* **38**, 57–66.

Noda, M., Yoon, K., Rodan, G. A., and Koppel, D. E. (1987). High lateral mobility of endogenous and transfected alkaline phosphatase: A phosphatidylinositol-anchored membrane protein. *J. Cell Biol.* **105**, 1671–1677.

Nuoffer, C., Jeno, P., Conzelmann, A., and Riezman, H. (1991). Determinants for glycophospholipid anchoring of the *Saccharomyces cerevisiae* GAS1 protein to the plasma membrane. *Mol. Cell Biol.* **11**, 27–37.

Orlean, P., Albright, C., and Robbins, P. W. (1988). Cloning and sequencing of the yeast gene for dolichol phosphate mannose synthase, an essential protein. *J. Biol. Chem.* **263**, 17499–17507.

Orlean, P. (1990). Dolichol phosphate mannose synthase is required *in vivo* for glycosylphosphatidylinositol membrane anchoring, O-mannosylation and N-glycosylation of protein in *Saccharomyces cerevisiae. Mol. Cell Biol.* **10**, 5796–5805.

Paul, E., Leblond, F. A., and Lebel, D. (1991). In resting conditions, the pancreatic granule membrane protein GP-2 is secreted by cleavage of its glycosylphosphatidylinositol anchor. *Biochem. J.* **277**, 879–881.

Phelps, B. M., Primakoff, P., Koppel, D. E., Low, M. G., and Myles, D. G. (1988). Restricted lateral diffusion of PH-20, a PI-anchored sperm membrane protein. *Science* **240**, 1780–1782.

Powell, S. K., Cunningham, B. A., Edelman, G. M., and Rodriguez-Boulan, E. (1991a). Targeting of transmembrane and GPI-anchored forms of N-CAM to opposite domains of a polarized epithelial cell. *Nature* **353**, 76–77.

Powell, S. K., Lisanti, M. P., and Rodriguez-Boulan, E. (1991b). Thy-1 expresses two signals for apical localization in epithelial cells. *Am. J. Physiol.* **260**, C715–C720.

Puoti, A., Desponds, C., Fankhauser, C., and Conzelmann, A. (1991). Characterization of a glycophospholipid intermediate in the biosynthesis of glycophosphatidylinositol anchors accumulating in the Thy-1 negative lymphoma line SIA-b. *J. Biol. Chem.* **266**, 21051–21059.

Ramamoorthy, R., Donelson, J. E., Paetz, K. E., Maybodi, M., Roberts, S. C., and Wilson, M. E. (1992). Three distinct RNAs for the surface protease gp63 are differentially expressed during development of *Leishmania donovani chagasi* promastigotes to an infectious form. *J. Biol. Chem.* **267**, 1888–1895.

Rifkin, M. R., and Landsberger, F. R. (1990). Trypanosome variant surface glycoprotein transfer to target membranes: A model for the pathogenesis of trypanosomiasis. *Proc. Natl. Acad. Sci. USA* **87**, 801–805.

Roberts, W. L., Myher, J. J., Kuksis, A., Low, M. G., and Rosenberry, T. L. (1988a). Lipid analysis of the glycoinositol phospholipid membrane anchor of human erythrocyte acetylcholinesterase. *J. Biol. Chem.* **263**, 18766–18775.

Roberts, W. L., Santikarn, S., Reinhold, V. N., and Rosenberry, T. L. (1988b). Structural characterization of the glycoinositol phospholipid membrane anchor of human erythrocyte acetylcholinesterase by fast atom bombardment mass spectrometry. *J. Biol. Chem.* **263**, 18776–18784.

Robinson, P. J., Millrain, M., Antoniou, J., Simpson, E., and Mellor, A. L. (1989). A glycophospholipid anchor is required for Qa-2 mediated T cell activation. *Nature* **342**, 85–87.

Robinson, P. J. (1991a). Signal transduction by GPI-anchored membrane proteins. *Cell Biol. Int. Rep.* **15**, 761–767.

Robinson, P. J. (1991b). Phosphatidylinositol membrane anchors and T-cell activation. *Immunol. Today* **12**, 35–41.

Roditi, I., Schwarz, H., Pearson, T. W., Beecroft, R. P., Liu, M. K., Richardson, J. P., Buhring, H. J., Pleiss, J., Bulow, R., Williams, R. O., and Overath, P. (1989). Procyclin gene expression and loss of the variant surface glycoprotein during differentiation of *Trypanosoma brucei*. *J. Cell. Biol.* **108**, 737–746.

Romero, G. (1991). Inositolglycans and cellular signalling. *Cell Biol. Int. Rep.* **15**, 827–852.

Rosenberry, T. L. (1991). A chemical modification that makes glycoinositol phospholipids resistant to phospholipase C cleavage: Fatty acid acylation of inositol. *Cell Biol. Int. Rep.* **15**, 1133–1149.

Rothberg, K. G., Ying, Y. S., Kamen, B. A., and Anderson, R. G. W. (1990a). Cholesterol controls the clustering of the glycophospholipid-anchored membrane receptor for 5-methyltetrahydrofolate. *J. Cell. Biol.* **111**, 2931–2938.

Rothberg, K. G., Ying, Y. S., Kolhouse, J. F., Kamen, B. A., and Anderson, R. G. W. (1990b). The glycophospholipid-linked folate receptor internalizes folate without entering the clathrin-coated pit endocytic pathway. *J. Cell Biol.* **110**, 637–649.

Rothberg, K. G., Heuser, J. E., Donzell, W. C., Ying, Y. S., Glenney, J. R., and Anderson, R. G. W. (1992). Caveolin, a protein component of caveolae membrane coats. *Cell* **68**, 673–682.

Salmon, J. E., Brogle, N. L., Edberg, J. C., and Kimberly, R. P. (1991). Fcg receptor III induces actin polymerization in human neutrophils and primes phagocytosis mediated by Fcγ receptor II. *J. Immunol.* **146**, 997–1004.

Saltiel, A. R. (1991). The role of glycosyl-phosphoinositides in hormone action. *Bioenergetics Biomembr.* **23**, 29–41.

Scallon, B. J., Scigliano, E., Freedman, V. H., Miedel, M. C., Pan, Y.-C. E., Unkeless, J. C., and Kochan, J. P. (1989). A human immunoglobulin G receptor exists in both polypeptide-anchored and phosphatidylinositol-glycan-anchored forms. *Proc. Natl. Acad. Sci. USA* **86**, 5079–5083.

Schneider, P., Ferguson, M. A. J., McConville, M. J., Mehlert, A., Homans, S. W., and Bordier, C. (1990). Structure of the glycosyl-phosphatidylinositol membrane anchor of the *Leishmania major* promastigote surface protease. *J. Biol. Chem.* **265**, 16955–16964.

Shak, S., Davitz, M. A., Wolinsky, M. L., Nussenzweig, V., Turner, M. J., and Gurnett, A. (1988). Partial characterization of the cross-reacting determinant, a carbohydrate epitope shared by decay accelerating factor and the variant surface glycoprotein of the African *Trypanosoma brucei*. *J. Immunol.* **140**, 2046–2050.

Simons, K., and van Meer, G. (1988). Lipid sorting in epithelial cells. *Biochemistry* **27**, 6197–6202.

Simons, K., and Wandinger-Ness, A. (1990). Polarized sorting in epithelia. *Cell* **62**, 207–210.

Singh, N., Singleton, D., and Tartakoff, A. M. (1991). Anchoring and degradation of glycolipid-anchored membrane proteins by L929 versus by LM-TK- mouse fibroblasts: Implications for anchor biosynthesis. *Mol. Cell. Biol.* **11**, 2362–2374.

Stadler, J., Keenan, T. W., Bauer, G., and Gerisch, G. (1989). The contact site A glycoprotein of *Dictyostelium discoidem* carries a phospholipid anchor of a novel type. *EMBO J.* **8**, 371–377.

Stahl, N., Baldwin, M. A., Burlingame, A. L., and Prusiner, S. B. (1990a). Identification of glycoinositol phospholipid linked and truncated forms of the scrapie prion protein. *Biochemistry* **29**, 8879–8884.

Stahl, N., Borchelt, D. R., and Prusiner, S. B. (1990b). Differential release of cellular and scrapie prion proteins from cellular membranes by phosphatidylinositol-specific phospholipase C. *Biochemistry* **29**, 5405–5412.

Stahl, N., and Prusiner, S. B. (1991). Prions and prion proteins. *FASEB J.* **5**, 2799–2807.

Stefanova, I., Horejsi, V., Ansotegui, I. J., Knapp, W., and Stockinger, H. (1991). GPI-anchored cell-surface molecules complexed to protein tyrosine kinases. *Science* **254**, 1016–1018.

Stevens, V. L., and Raetz, C. R. H. (1991). Defective glycosyl phosphatidylinositol biosynthesis in extracts of three Thy-1 negative lymphoma cell mutants. *J. Biol. Chem.* **266**, 10039–10042.

Su, B., Waneck, G. L., Flavell, R. A., and Bothwell, A. L. M. (1991). The glycosyl phospatidylinositol anchor is critical for Ly-6A/E-mediated T cell activation. *J. Cell Biol.* **112**, 377–381.

Sugiyama, E., DeGasperi, R., Urakaze, M., Chang, H.-M., Thomas, L. J., Hyman, R., Warren, C. D., and Yeh, E. T. H. (1991). Identification of defects in glycosylphosphatidylinositol anchor biosynthesis in the Thy-1 expression mutants. *J. Biol. Chem.* **266**, 12119–12122.

Takami, N., Misumi, Y., Kuroki, M., Matsuoka, Y., and Ikehara, Y. (1988). Evidence for carboxyl-terminal processing and glycolipid-anchoring of human carcinoembryonic antigen. *J. Biol. Chem.* **263**, 12716–12720.

Takami, N., Oda, K., and Ikehara, Y. (1992). Aberrant processing of alkaline phosphatase precursor caused by blocking the synthesis of glycosylphosphatidylinositol. *J. Biol. Chem.* **267**, 1042–1047.

Teng, M.-H., Hedayati, S., Alexander, A. A., Barkin, R., Basch, R. S., and Buxbaum, J. N. (1989). A Thy-1 negative lymphoma cell variant defective in the formation of glycosyl-phosphatidylinositol membrane protein anchors. *Mol. Immunol.* **26**, 391–402.

Thomas, J. R., Dwek, R. A., and Rademacher, T. W. (1990). Structure, biosynthesis, and function of glycosylphosphatidylinositols. *Biochemistry* **29**, 5413–5422.

Thompson, T. E., and Tilack, T. W. (1985). Organization of glycosphingolipids in bilayers and plasma membranes of mammalian cells. *Annu. Rev. Biophys. Chem.* **14**, 361–386.

Tisdale, E. J., Schimenti, J. C., and Tartakoff, A. M. (1991). Sodium butyrate causes reexpression of three membrane proteins on glycolipid-anchoring mutants. *Somatic Cell Mol. Genet.* **17**, 349–357.

Turco, S. J. (1988). The lipophosphoglycan of leishmania. *Parasitol. Today* **4**, 255–257.

Turner, M. J., Cardoso de Almeida, M. L., Gurnett, A. M., Raper, J., and Ward, J. (1985). Biosynthesis, attachment and release of variant surface glycoproteins of the African trypanosome. *Curr. Top. Microbiol. Immunol.* **117**, 23–55.

Udenfriend, S., Micanovic, R., and Kodukula, K. (1991). Structural requirements of a nascent protein for processing to a PI-G anchored form: Studies in intact cells and cell-free systems. *Cell Biol. Int. Rep.* **15**, 739–759.

Udenfriend, S., Kodukula, K., and Amthauer, R. (1992). Cell-free processing of nascent proteins destined to be linked to the plasma membrane by a phosphatidylinositol-glycan anchor. *Cell. Mol. Biol.* **38**, 11–16.

Walsh, F. S., and Doherty, P. (1991). Glycosylphosphatidylinositol anchored recognition molecules that function in axonal fasciculation, growth, and guidance in the nervous system. *Cell Biol. Int. Rep.* **15**, 1151–1166.

Walter, E. I., Roberts, W. L., Rosenberry, T. L., Ratnoff, W. D., and Medof, M. E. (1990). Structural basis for variations in the sensitivity of human decay accelerating factor to phosphatidylinositol-specific phospholipase C cleavage. *J. Immunol.* **144**, 1030–1036.

Wilson, J. M., Fasel, N., and Kraehenbuhl, J-P. (1990). Polarity of endogenous and exogenous glycosyl-phosphatidylinositol-anchored membrane proteins in Madin–Darby canine kidney cells. *J. Cell. Sci.* **96**, 143–149.

Zamze, S. E, Ferguson, M. A. J., Collins, R., Dwek, R. A., and Rademacher, T. W. (1988). Characterization of the cross-reacting determinant (CRD) of the glcycosyl-phosphatidylinositol membrane anchor of *Trypanosoma brucei* variant surface glycoprotein. *Eur. J. Biochem.* **176**, 527–534.

Zhang, T., Crise, B., Su, B., Hou, Y., Rose, J. K., Bothwell, A., and Jacobson, K. (1991). Lateral diffusion of membrane-spanning and glycosylphosphatidylinositol-linked proteins: Toward establishing rules governing the lateral mobility of membrane proteins. *J. Cell Biol.* **115**, 75–84.

The Use of Glycosylation Inhibitors to Study Glycoconjugate Function

Alan D. Elbein

Department of Biochemistry and Molecular Biology, University of Arkansas for Medical Sciences, Little Rock, Arkansas 72205

I. Introduction

Complex carbohydrates have been shown to be intricately involved in numerous biological systems including, but not limited to, fertilization (1), lysosomal enzyme targeting (2), uptake of or removal of glycoproteins from the blood (3), cell–cell comunication as well as cell adhesion (4,5), host–parasite interactions (6), and numerous other physiological processes (7,8). The fact that these many different systems have complex carbohydrates as a common theme has given rise to the term "glycobiology," which has been defined as "a study of the structure and function of the carbohydrate modifications on different types of macromolecules" (9).

Complex carbohydrates are found attached both to proteins and to lipids. The glycoproteins produced by eucaryotic cells can have either N-linked oligosaccharides or O-linked oligosaccharides, and frequently both types of structures are found on the same protein (10). In addition, the number of oligosaccharide chains attached to the protein can vary from only one to many chains. In the N-linked types (described in more detail below), the carbohydrate is attached to the amide nitrogen of asparagine residues that occur in the tripeptide sequences, Asn–X–Ser (Thr)(11). However, this sequon is not sufficient for glycosylation to occur since many –Asn–X–Ser(Thr)– sequences that are found in proteins are not glycosylated. It appears that the conformation of the protein around this sequence and perhaps its exposure are critical factors in addition of carbohydrate. On the other hand, in the more common types of O-linked structures such as the mucins, the oligosaccharides are attached by O-glycosidic bonds that usually involve GalNAc linked to the OH groups of serine or threonine residues, but there is no known amino acid sequence necessary for glycosylation (12).

In terms of lipids, the best known and most common types of glycolipids are the sphingolipids, which contain oligosaccharide chains composed of some of the following sugars: glucose, galactose, GlcNAc, GalNAc, sialic acid, and fucose. These sugars may be present in various configurations and combinations and are linked to a ceramide that is made up of fatty acid and sphingosine. These compounds are of considerable importance to medicine since certain gangliosides have been shown to be receptors for diphtheria and cholera toxins, while some cerebrosides are involved in bacterial attachment to animal cells. Since a number of current reviews have covered these compounds in detail, they will not be considered here (13,14). A more recently described type of glycolipid is the glycosyl phosphatidylinositol (GPI) type, in which a glycan of mannose and GlcNAc (and sometimes other sugars such as galactose) is attached to the membrane phospholipid phosphatidylinositol. These GPI molecules appear to be important in anchoring many proteins to the membrane (15,16), and some of the inhibitors of N-linked oligosaccharide assembly also affect the synthesis of the glycan portion of these anchors. They will be considered in more detail in a later section.

Although a considerable amount of information is available concerning the structures of N-linked oligosaccharides and GPI glycans, there is relatively little data concerning the function of the carbohydrate portion of these molecules. It appears that protein (or lipid) glycosylation has two major roles: to modulate or change the biochemical properties of these molecules, including such parameters as bioactivity, folding, stability or turnover, immunoreactivity, and so on; and to serve as a determinant in molecular recognition events, such as lysosomal enzyme targeting, parasite or symbiote binding or adhesion to host cells, attachment of toxins or other biologically important ligands to cell surfaces, and so on (17).

In most cases, it has been somewhat difficult to demonstrate a functional role for the carbohydrate portion of the glycoprotein. One obvious way to do this would be to obtain the particular glycoprotein without its carbohydrate and then compare the biological and biochemical properties of the unglycosylated form with those of the normal glycoprotein. There are a number of ways to produce glycoproteins without their associated carbohydrate, especially in the case of the N-linked glycoproteins. For example, one can remove the N-linked oligosaccharides with endoglycosidases, such as endoglucosaminidase H (Endo H), which cleaves "high-mannose" oligosaccharides between the two internal GlcNAc residues (18), or peptide glycosidase F, which hydrolyzes most N-linked oligosaccharides at the GlcNAc to asparagine bond (19). The problem with this approach is that many N-linked oligosaccharides are resistant to cleavage by these enzymes when the protein is in its native state. Therefore, it is necessary to denature the protein in order to obtain complete removal of the N-linked oligosaccharides, making this method untenable for functional studies. Another problem with this approach is that removing the carbohydrate after the protein has been synthesized and has achieved its proper conformation may be quite different physiologically than having the protein synthesized in the cell in the absence of carbohydrate.

A second possibility is to produce the glycoprotein of interest in a cell system in the presence of the antibiotic tunicamycin (20). This compound inhibits the GlcNAc-1-P transferase that catalyzes the first step in the "dolichol cycle," thereby preventing the formation of lipid-linked oligosaccharides (21). This inhibition usually results in a complete blockage in N-linked glycosylation (22). The difficulty with this approach is that many glycoproteins may be improperly folded or become insoluble when they are synthesized without any N-linked carbohydrate. As a result, physiological studies again become of questionable value.

A third alternative for preventing glycosylation is to use site-directed mutagenesis to modify the Asn–X–Ser sequon(s) that is glycosylated to one that is a "non-acceptor" sequon (23,24). The beauty of this method is that one can modify one glycosylation site at a time (assuming that there is more than one) and determine how an alteration at that site affects function. Or one can modify any number of these sites at the same time and compare the various proteins. Two potential obstacles are (1) the gene coding for the protein of interest must be available and (2) one must make the assumption that the change in the amino acid sequence does not itself affect protein conformation. This methodology has provided some valuable insights into the role of asparagine-linked glycosylation. However, if all N-linked oligosaccharide sequons are modified, then the same problem of improper folding or solubility that one finds with tunicamycin would also apply here. Another possibility is to modify glycosylation rather than prevent it, so that the resulting N-linked glycoprotein contains an altered oligosaccharide structure. One can then determine if and how this alteration affects the biological activity of a given glycoprotein. There are several ways to modify N-linked oligosaccharide structures. For example, the protein of interest can be produced in a mutant cell line that has a defect in one of the biosynthetic enzymes. Cell lines that are missing various glycosyltransferases or processing glycosidases have been isolated, and several excellent reviews have described the various mutants and the techniques for their utilization (25,26). In this case also, one must have the gene for the glycoprotein of interest in order to introduce it into the mutant cell line for synthesis of the appropriate protein.

The carbohydrate structure can also be modified with selected inhibitors that block the activity of key enzymes in the processing pathway, so that synthesis is stopped at a premature step and the various types of complex chains cannot be formed. Such inhibitors and their sites of action are the main topics of this chapter. One disadvantage on the use of such inhibitors in *in vivo* studies is that all N-linked glycoproteins are affected, and therefore it may be difficult to relate changes in physiology to an effect on the specific glycoprotein of interest. In fact, one criticism that can be levied at the use of inhibitors in most biological systems (and also as chemotherapeutic agents) is that absolute specificity is seldom if ever achieved. Of course in the above case, if the altered glycoprotein can be isolated from the inhibited (or mutated) cell and its activity compared with the normal protein, either *in vitro* or in another system, then this criticism can be overcome.

One final method for altering carbohydrate structure is to disrupt or alter the normal transport of glycoproteins from the ER through the various Golgi stacks to their ultimate location. Since it is now clear that certain of the oligosaccharide trimming reactions occur in the ER, whereas others occur in the cis-Golgi and still others in the medial Golgi, etc., the basis of this approach is to prevent the protein from being transported to that region where the enzymatic activity is located. If one can somehow prevent the glycoprotein from reaching the specific compartment that houses that activity, then that reaction should not occur. There are several valuable compounds that have such properties and these drugs will be discussed in a later section. These types of inhibitors have been extremely useful tools for helping us to understand the mechanism(s) of protein targeting.

In terms of the glycan portion of the GPI anchor, a number of recent studies have demonstrated that various inhibitors that act on the N-linked oligosaccharides also inhibit some of the reactions in anchor formation. These inhibitors are providing useful information on the biosynthetic pathway and should also be helpful for functional studies. However, at this stage, we still do not know enough about the synthesis of the glycan portion of the GPI anchors, nor do we really understand their function. This chapter will deal with various types of inhibitors that affect GPI anchor formation, as well as those that act on the N-linked glycoproteins.

II. Structure and Biosynthesis of N-Linked Oligosaccharides

The most common types of glycoproteins that are present in eucaryotic cells, as cell surface proteins, membrane proteins, and secretory proteins, are those having oligosaccharides attached to the amide nitrogen of asparagine (N-linked or asparagine-linked) (10). The oligosaccharide portion of these molecules may have a variety of different structures that fall into one of the general groups shown in Fig. 1. These are referred to as high-mannose, complex, or hybrid types of oligosaccharides. It can be seen from the figure that each of these structures contains the same basic core region (shown by the boxed-in area) composed of the oligosaccharide sequence Manα1,6(Manα1,3) Manβ1,4GlcNAcβ1,4GlcNAc. In the high-mannose structures, this core region is elongated by as many as six additional mannose residues (i.e., a total of nine mannose units) attached in α1,2, α1,3, and α1,6 linkages, as shown by the typical structure presented in Fig. 1. However, some high-mannose oligosaccharides may have fewer than six additional mannoses, since some trimming can occur during biosynthesis (see Section V).

On the other hand, in the complex types of oligosaccharides, the pentasaccharide core region can be extended with a number of

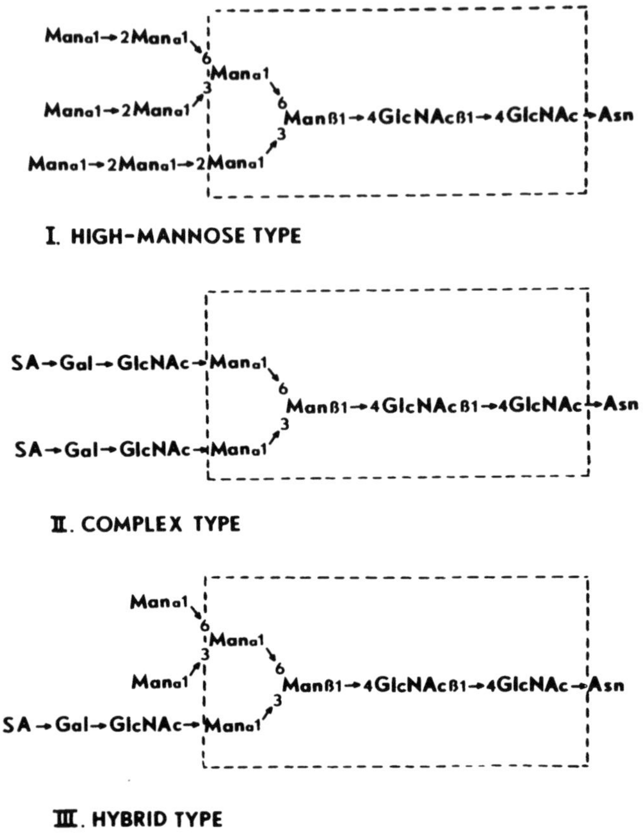

FIG. 1. Generalized structures of the various types of N- linked oligosaccharides. As indicated here, these structures are classified as high-mannose (top), complex (middle), and hybrid (bottom) structures.

different substitutions. Shown in the figure is a typical biantennary type of oligosaccharide, containing two of the trisaccharide sequences, NeuAcα2,6Galβ1,4GlcNAcβ1,2. On the other hand, some complex chains may have three (triantennary) or four (tetraantennary) of these trisaccharide sequences, as well as various other substitutions such as α1,6-linked L-fucose on the innermost GlcNAc or sulfate residues attached to this GlcNAc (10). Various other alterations may also occur, such as α1,3-linked fucose units on the outer GlcNAc residues, polylactosamine chains, a β1,4-bisecting GlcNAc residue attached to the β-linked mannose and so on, as well as various glycosidic linkages such as α2,3-linked sialic acid rather than α2,6-linked sialic acid, etc. Although the exact number of different types of complex chains is not clear, it has been estimated that there are probably several hundred known structures (27).

The final type of N-linked oligosaccharide is the hybrid structure, which still contains a Manα1,6 (Manα1,3) branch on the α1,6-mannose arm, but also has a type of complex chain on the α1,3-branch. Most hybrid chains also apparantly have a bisecting GlcNAc attached in β1,4-linkage to the β-linked mannose, and this substitution may play a role in the control of further processing reactions.

The biosynthesis of all of these N-linked oligosaccharides involves two independent and sequential pathways that occur in the endoplasmic reticulum (ER) and Golgi apparatus of the cell. In the first series of reactions, membrane and secretory proteins that are undergoing synthesis on membrane-bound polysomes become glycosylated by transfer of a common oligosaccharide having the structure $Glc_3Man_9(GlcNAc)_2$ from its lipid-linked saccharide form, i.e., $Glc_3Man_9(GlcNAc)_2$-PP-dolichol, to specific –Asn–X–Ser (Thr)– sequences on the protein (28). This pathway is referred to as the "dolichol cycle" and is depicted in Fig. 2. A series of glycosyltransferases that are associated with the ER membrane sequentially add the sugars GlcNAc, mannose, and glucose to the lipid carrier dolichyl-P to give the above-mentioned lipid-linked oligosaccharide donor.

To study this pathway in animal cell culture systems, one can add one of the radioactive sugars, i.e., [^{3}H]glucosamine, [^{3}H]galactose (as a tag for the glucose residues), or [^{3}H]mannose, to the medium and examine the incorporation of the label into lipid-linked saccharides. These intermediates are easy to extract from the cells by a sequential extraction procedure, first with $CHCl_3$:CH_3OH:H_2O (1:1:1), which extracts the lipid-linked monosaccharides and smaller sized lipid-linked oligosaccharides, and then extraction of the remaining residue with $CHCl_3$:CH_3OH:H_2O (10:10:3) to obtain the lipid-linked oligosaccharides that contain the larger sized oligosaccharides. The intact lipid-linked monosaccharides can usually be identified by TLC, whereas the larger oligosaccharides from the lipid-linked oligosaccharides are usually released by mild acid hydrolysis and characterized by gel filtration on Bio-Gel P-4 or by HPLC (29,30).

Thus, the first enzyme in the biosynthesis of N-linked oligosaccharides transfers GlcNAc-1-P from UDP–GlcNAc to dolichyl-P to form dolichyl-PP-GlcNAc (31,32). This enzyme is the site of action of the very widely used glycosylation inhibitor tunicamycin (21) (discussed later). The GlcNAc-1-P transferase has been purified to homogeneity from bovine mammary glands (33), and the gene for this enzyme was recently cloned and expressed is COS cells (34). The next GlcNAc

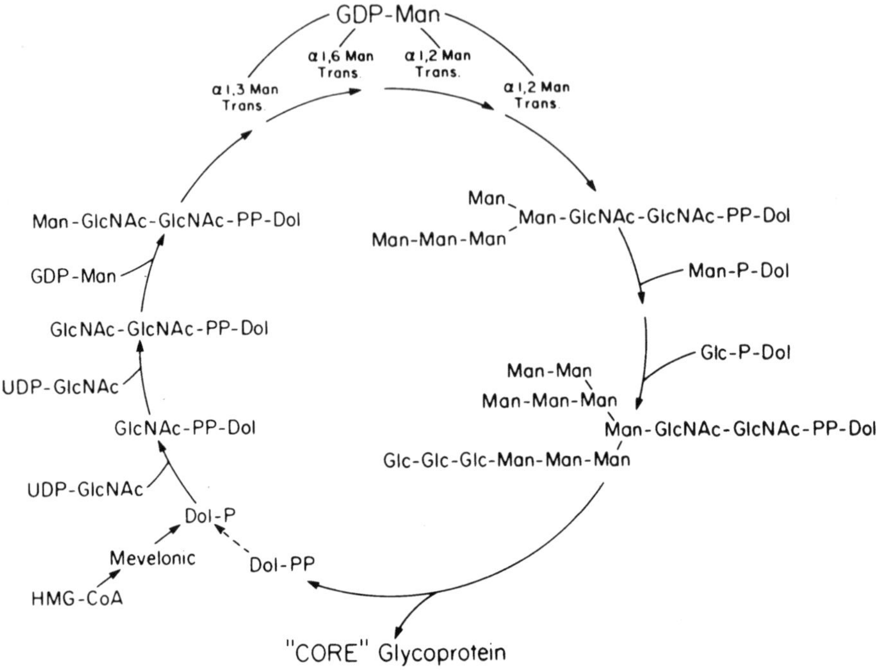

FIG. 2. Reactions involved in the assembly of the lipid-linked oligosaccharide that donates oligosaccharide to protein. This reaction sequence occurs in the ER of cells and involves the sequential transfer of sugars to dolichyl-P by a series of membrane-bound glycosyltransferases.

transferase adds the second GlcNAc, also from UDP–GlcNAc, to dolichyl-PP-GlcNAc to form GlcNAcβ1,4GlcNAc-PP-dolichol. This enzyme has recently been purified to homogeneity from pig aorta (35).

The next five sugars to be added are mannose residues and are donated via GDP–mannose. As indicated below, two lines of evidence have demonstrated that GDP–mannose, rather than dolichyl-P-mannose, is the mannosyl donor for these additions: (a) A mutant cell line that is missing the dolichyl-P-mannose synthase can still form the Manα1,2Manα1,2Manα1,3(Manα1,6)Manβ1,4GlcNAcβ1,4GlcNAc-PP-dolichol but is not able to further elongate it (36); and (b) amphomycin, an inhibitor of the *in vitro* formation of dolichyl-P-mannose (37), does not prevent the formation of the above $Man_5(GlcNAc)_2$-PP-dolichol, but it does block further elongation of this intermediate (38). The β-mannosyl transferase that adds the first mannose to the dolichol intermediates has recently been purified to homogeneity from pig aorta, and an azido-labeled [^{32}P]GDP–mannose was used to identify the subunits of this protein (39). Several of the α-mannosyltransferases have also been partially purified from liver (40–43), and these enzymes require GDP–mannose rather than dolichyl-P-mannose as the mannosyl donor. However, these proteins have not yet been purified to homogeneity so that their detailed properties or their orientation in the ER membrane is not known.

Once the $Man_5(GlcNAc)_2$-PP-dolichol is formed, it can be further glycosylated by the addition of four more mannose residues and then three glucose units (44). These mannose residues all are believed to be donated via dolichyl-P-mannose (because of the topology of the oligosaccharide in the ER lumen and the results with mutants missing the dolichyl-P-mannose synthase) (36), whereas the glucosyl donor is dolichyl-P-glucose (45). Most of these enzymes have not been isolated or studied in any detail, and therefore little is known about their properties. However, the enzyme that adds the first mannose from dolichyl-P-mannose (i.e., the sixth mannose to the lipid) has been partially purified from aorta extracts and its substrate specificity was determined. For activity, this enzyme required dolichyl-P-mannose as the mannosyl donor and $Man_5(GlcNAc)_2$-PP-dolichol as the mannosyl acceptor (46). Likewise, the enzyme that adds the first glucose from dolichyl-P-glucose to $Man_9(GlcNAc)_2$-PP-dolichol was purified to near homogeneity from pig aorta and its substrate specificity and oligosaccharide acceptor specificity were determined. While a number of oligosaccharide structures linked to dolichol could serve as glucose acceptors, they all required the Manα1,2Manα1,2Manα1,3– branch for this activity (47). The enzyme(s) that adds the last two glucoses has not yet been purified.

The role of the glucose units in the biosynthesis of N-linked oligosaccharides is presumably to facilitate the transfer of oligosaccharide from lipid to protein. That is, the oligosaccharyltransferase that catalyzes this final step in the dolichol cycle recognizes or prefers glucose-containing lipid-linked oligosaccharides as the oligosaccharide donor (48,49). In fact, studies with microsomes showed that glucose-containing lipid-linked oligosaccharides were much more efficient oligosaccharide donors than were glucose-free lipid-linked oligosaccharides. Thus, this last step of the dolichol cycle occurs in the ER and the oligosaccharide is transferred to specific asparagine residues on the protein, as the polypeptide is being elongated on ER-bound polysomes (50). The oligosaccharyltransferase has recently been purified to apparant homogeneity from liver (51).

Once the oligosaccharide chain(s) is transferred to the protein, and while the protein is still undergoing synthesis in the ER, two membrane-bound and highly specific glucosidases remove all three glucoses from the $Glc_3Man_9(GlcNAc)_2$. The processing pathway for N-linked oligosaccharides is outlined in Fig. 3. Thus, glucosidase I removes the outermost α1,2-linked glucose to give a $Glc_2Man_9(GlcNAc)_2$–protein (52), whereas glucosidase II removes the next two α1,3-linked glucoses to give the $Man_9(GlcNAc)_2$–protein (53). Glucosidase I has been purified from calf (54) and porcine (55) liver, from bovine mammary glands (56), from mung bean seedlings (57), and from *Saccharomyces cerevesiae* (58). This enzyme has a pH optimum in the neutral range (6.4–6.8), does not work with *p*-nitrophenyl-α-D-glucopyranoside, and is quite specific for α1,2-glucosidic linkages. The mammalian enzyme is a tetramer with subunits of about 85 kDa, whereas the yeast enzyme has subunits of about 95 kDa. In *S. cerevesiae*, the *gls1* mutation results in cells that lack glucosidase I and produce glycoproteins that maintain the $Glc_3Man_9(GlcNAc)_2$ structure (59). However, this alteration in structure has no effect on secretion of the glycoproteins, in contrast to mammalian cells in which such an alteration had profound effects (see section on glucosidase inhibitors). A CHO mutant cell line has been isolated by virtue of its resistance to L-PHA,

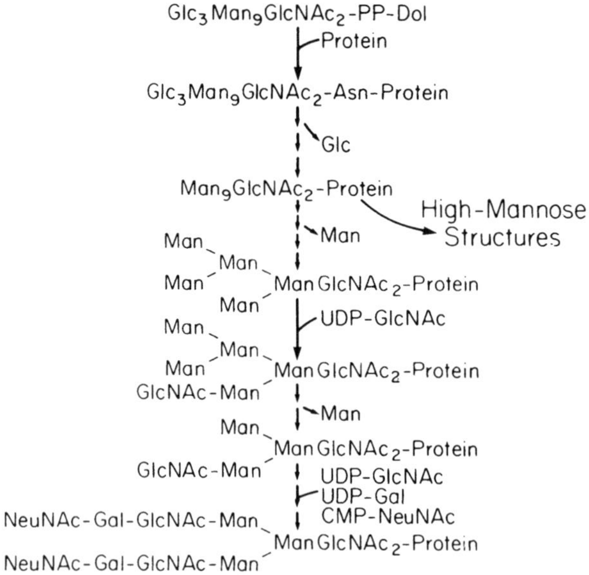

FIG. 3. Reactions involved in the processing of the protein- bound oligosaccharide. A series of glycosidases and glycosyltransferases, localized in the ER and Golgi, processes the oligosaccharide chains to give rise to the various types of N-linked structures seen in Fig. 1.

and this mutant cell, called Lec23, has been shown to lack glucosidase I (60). These cells were found to synthesize the G protein of VSV having hybrid and high-mannose oligosaccharides, as well as complex chains.

Glucosidase II was initially identified in microsomal fractions of rat liver (61), calf liver (62), and yeast (63) and was shown to remove both of the α1,3-linked glucoses from $Glc_2Man_9(GlcNAc)_2$ to form $Man_9(GlcNAc)_2$. However, removal of the outermost α1,3-linked glucose is quite rapid ($t_{1/2}$ = 5 min) as compared with cleavage of the innermost glucose ($t_{1/2}$ = 20–30 min). Although glucosidase II also can catalyze the cleavage of aryl-α-glucosides such as p-nitrophenyl-α-glucopyranoside, it appears to be fairly specific for α1,3-linked glucoses since its activity is inhibited by the α1,3-linked glucose disaccharide nigerose, but not by other α-linked glucose disaccharides (i.e., kojibiose, maltose, isomaltose, or trehalose) (64). Glucosidase II from pig kidney was reported to have a subunit molecular mass of 100 kDa and to be a high-mannose glycoprotein (65), whereas the enzyme from mung beans was composed of two identical glycoprotein subunits of 110 kDa, also with high mannose oligosaccharides (some of which probably also contain some glucose residues) (64). In some animal systems, however, the glucosidase II had quite different subunit molecular weights of around 65 kDa (56,66). The localization of the enzyme has also been studied and may vary, depending on the tissue, from being concentrated in rough ER (RER) and smooth ER (SER) in pig hepatocytes (67) to post-Golgi apparatus structures in pig kidney tubular cells (65).

The best substrates for assaying these processing glucosidases are radiolabeled $Glc_3Man_9GlcNAc$ and $Glc_2Man_9GlcNAc$. These substrates can be prepared by incubating animal cells (or virus-infected cells) in the presence of the processing inhibitors castanospermine (discussed later), to prevent the removal of glucose residues, and deoxymannojirimycin, to prevent the removal of α1,2-linked mannoses. [^{3}H]Galactose is then added to the cell cultures to label the glucose moieties in the newly synthesized glycoproteins. The glycopeptides resulting from pronase digestion of the glycoproteins are isolated by gel filtration on columns of Bio-Gel P-4, and these glycopeptides are then treated with Endo H to obtain the [^{3}H]glucose labeled $Glc_3Man_9GlcNAc$. This oligosaccharide can then be treated with a glucosidase I preparation that is free of glucosidase II and α-mannosidases to obtain [^{3}H]Glc_2Man_9Glc NAc. For assay of the processing glucosidase activities, the previously mentioned radiolabeled substrates are incubated with the appropriate enzyme preparations, and the release of labeled glucose is measured by a Concanavalin A–Sepharose binding assay (68).

The removal of three glucose residues from the $Glc_3Man_9(Glc$ $NAc)_2$–protein results in the formation of a high-mannose Man_9

$(GlcNAc)_2$–protein, which can then be further processed to other high-mannose or complex types of oligosaccharides (as seen in Figs. 1 and 3). Early studies on the processing pathway indicated that there were two or three "neutral" α-mannosidase activities involved in trimming the N-linked oligosaccharide, and these activities were distinct from those of the lysosomal α-mannosidase (69). As a result of increasing interest in this area of research, a number of new α-mannosidase activities have been described and current work is aimed at trying to understand how these various enzymes function in the removal of α-linked mannoses from eucaryotic cell glycoproteins. In fact, one of the important reasons for having a number of specific mannosidase inhibitors (see Section V) is to be able to use them to distinguish among various mannosidase activities, as well as to determine what happens to processing and function of a given glycoprotein when a specific α-mannosidase is inhibited. A well-written review deals with the various α-mannosidases (70), so these enzymes will be only briefly covered here.

An ER α-mannosidase that removes a single α1,2-linked mannose from $Man_9(GlcNAc)_2$ to give a specific $Man_8(GlcNAc)_2$ isomer has been purified from yeast (71). This enzyme is inhibited by the processing inhibitor deoxymannojirimycin, requires Ca^{2+} for activity, and does not work with aryl-α-mannosides. The gene for the enzyme has been isolated and codes for a 63-kDa protein with three *N*-glycosylation sites, a Ca^{2+} binding consensus sequence, and a noncleavable signal sequence near the N terminus that acts as a transmembrane type II domain (72). A Ca^{2+} requiring α1,2-mannosidase was also purified from rabbit liver and this enzyme shows considerable homology (38% identity and 58% similarity) to the yeast enzyme (73). However, the rabbit liver enzyme can remove most if not all of the α1,2-linked mannoses from $Man_9(GlcNAc)_2$, and it requires either nonionic detergents or phospholipids for activity (74). Two other Ca^{2+} requiring α-mannosidases have been purified from calf (75) and pig (76) livers, and these enzymes differ from the rabbit enzyme in some of their properties. These enzymes also do not work with the aryl-α-mannopyranosides as substrates. The pig liver enzyme was localized by immunological techniques and found to be distributed in RER, SER, transitional elements of the RER, and vesicles that engage in transportation between the ER and the Golgi, but it was not in the Golgi (77).

Actually, the first neutral α-mannosidase to be reported was named mannosidase I and this enzyme was isolated from the Golgi apparatus of eucaryotic cells (78). This α-mannosidase has catalytic properties very similar to those of the Ca^{2+}-dependent mannosidases, but it apparantly does not need Ca^{2+}, although this is not certain because the requirement was not tested in the presence of EDTA (79). The mannosidase I was resolved into two distinct proteins by chromatographic methods and these two forms were termed mannosidase IA and IB. IA was purified to homogeneity and shown to be a tetramer composed of 56 kDa glycoprotein subunits (80). Presumably IA and IB have similar substrate specificities, but it is not clear if they are derived from the same gene.

Once the four α1,2-linked mannose units are removed, a medial Golgi GlcNAc transferase (called GlcNAc transferase I) adds a GlcNAc from UDP–GlcNAc to the mannose that is attached to the β-linked mannose in α1,3-linkage (81). This GlcNAc transferase was purified over 100,000-fold to homogeneity from liver (82) and the enzyme has been sequenced and cloned (83). The GlcNAc is apparantly a recognition signal for another highly specific α-mannosidase, called mannosidase II, that cleaves the α1,3-linked and α1,6-linked mannoses from the GlcNAc–$Man_5(GlcNAc)_2$–protein to give a GlcNAcβ1,2Manα1,3[Manα1,6]Manβ1,4GlcNAcβ1,4GlcNAc-protein (84). In spite of this substrate specificity, mannosidase II does act on the aryl-α-mannopyranosides such as *p*-nitrophenyl-α-D-mannopyranoside. Mannosidase II was purified to homogeneity from rat liver (85,86) and from mung bean seedlings (87). Both enzymes migrate on SDS gels as proteins of about 125 kDa, although the sequence for the murine enzyme predicts a MW of 132 kDa for the deglycosylated protein. Both enzymes are also glycoproteins and the murine enzyme has been shown to be a type II transmembrane glycoprotein with a single transmembrane region and a luminally oriented catalytic site (88).

There are several other α-mannosidases that are distinct from the above enzymes based on their sensitivity to processing inhibitors and to catalysis of high-mannose oligosaccharides. However, the relative role of these enzymes in the glycoprotein processing pathway remains to be established. One of these enzymes is a cytosolic and neutral α-mannosidase from rat liver (89). Although it was originally thought that this enzyme was a proteolytic product of an ER α-mannosidase, it now appears that both the cytosolic and the ER α-mannosidase are related to an α-mannosidase that was isolated from rat brain and that can cleave α1,2-, α1,3-, and α1,6-linked mannose residues (90). This last enzyme can cleave a $Man_9(GlcNAc)_2$ structure to a $Man_3(GlcNAc)_2$. The difference between the ER liver enzyme and the brain enzyme is that the former enzyme is also active with aryl-α-mannosides, whereas the latter is not. Two other α-mannosidases that can cleave α1,2-, α1,3-, and α1,6-linked mannoses have been purified from rat sperm (91) and rat liver (92).

Another type of hydrolytic enzyme that is also proposed to be involved in processing is an endo-α-mannosidase that can cleave a Glcα1,3Man disaccharide from $Glc_1Man_9(GlcNAc)_2$ (or a Glc_3Man from $Glc_3Man_9(GlcNAc)_2$, etc.) (93). It is interesting to note that the $Man_8(GlcNAc)_2$ that results from the action of this enzyme is a different isomer from the $Man_8(GlcNAc)_2$ that is formed by the action of the ER α-mannosidase or the yeast α-mannosidase (94). This endomannosidase appears to be a resident of the Golgi apparatus and may function to trim oligosaccharides on glycoproteins that have escaped the ER still retaining one or more glucose residues.

As shown in Fig. 3, following the trimming of all of the glucose residues and a number of mannose units from the N-linked oligosaccharides, various glycosyltransferases act to add different sugars, such as GlcNAc, galactose, sialic acid, fucose, etc., to the core N-linked structure to form many different types of complex chains (95). Since there are no inhibitors available for these glycosyltransferases, they will not be considered further in this review.

III. Structure and Biosynthesis of Glycosyl Phosphatidylinositol Anchors

Glycosyl phosphatidylinositols are a recently discovered class of glycolipids that have been shown to anchor many proteins (and other molecules such as heparan sulfates) to biological membranes via covalent linkages. Recently, the chemical nature of these molecules has been deduced, but the complete structure is still elucidated only for a few of these anchors. The characterization of these glycolipids has involved compositional analyses using GC–MS in combination with specific chemical treatments and sequential digestion with exoglycosidases, analysis of glycosidic linkages by methylation analyses, and gel filtration and TLC of intact glycans and their degradation products (96). The best known and most thoroughly detailed GPI anchors are those of *Trypanosoma brucei* and rat Thy-1. Although these structures show some differences, they do have a conserved linear region shown in

FIG. 4. General structure of the glycosylphosphatidylinositol that serves as a membrane anchor for many cell surface proteins.

Fig. 4 to be composed of 6-*O*-ethanolamine-P-6-Manα1,2Manα1, 6Manα1,4GlcNα1,6-inositol (97). In the case of the variable surface antigen (i.e., VSG) of trypanosomes, there are a number of galactose residues attached to the first mannose, whereas in Thy-1 there may be a GalNAc and/or an ethanolamine-P attached to this mannose (98). Also these structures may differ in the number of fatty acid residues attached to the inositol and also in the structure of the fatty acids.

Although some of the details of the biosynthesis of the glycan portion of these anchors are known, there are still considerable gaps in our understanding of the pathway. Nevertheless, the steps in biosynthesis are proposed to occur as outlined in Fig. 5. The sequence begins with the transfer of a GlcNAc from UDP–GlcNAc to phosphatidylinositol to form GlcNAc-PI, which then undergoes deacetylation to form GlcN-PI (99). These reactions have been demonstrated *in vivo* by labeling studies with ^{3}H-glucosamine and in cell-free extracts incubated with UDP–GlcNAc. Three mannose residues are then added to the GlcN-PI and these mannoses are all proposed to be derived from dolichyl-P-mannose, rather than from GDP–mannose (see Section IX for experimental evidence). None of these enzymes has been purified and virtually nothing is known about how other sugars such as Gal or GlcNAc are added. In addition, it is not clear when in the biosynthetic pathway the fatty acids are added to the PI, although there is now evidence that the GlcN-PI must be acylated to serve as an acceptor for the first mannose (100). Once the mannose units have been added, phosphatidylethanolamine serves as a donor of ethanolamine-P, which becomes attached via a phosphodiester bond to the 6 position of the terminal mannose. The signal for GPI addition to the nascent protein resides in a short hydrophobic carboxy-terminal peptide that is removed and replaced by the GPI anchor (101). This addition of protein to the GPI apparantly occurs on internal membranes, but the details of these biosynthetic steps are not known at this time.

IV. Inhibitors of Lipid-Linked Saccharide Synthesis

A. *Tunicamycin*

Probably the most widely used and most valuable inhibitor of N-linked glycosylation is tunicamycin. This compound is a nucleoside antibiotic that is produced by the bacterium *Streptomyces lysosuperificus* and was initially found to inhibit the replication of a number of enveloped viruses (102). The structure of tunicamycin was elegantly

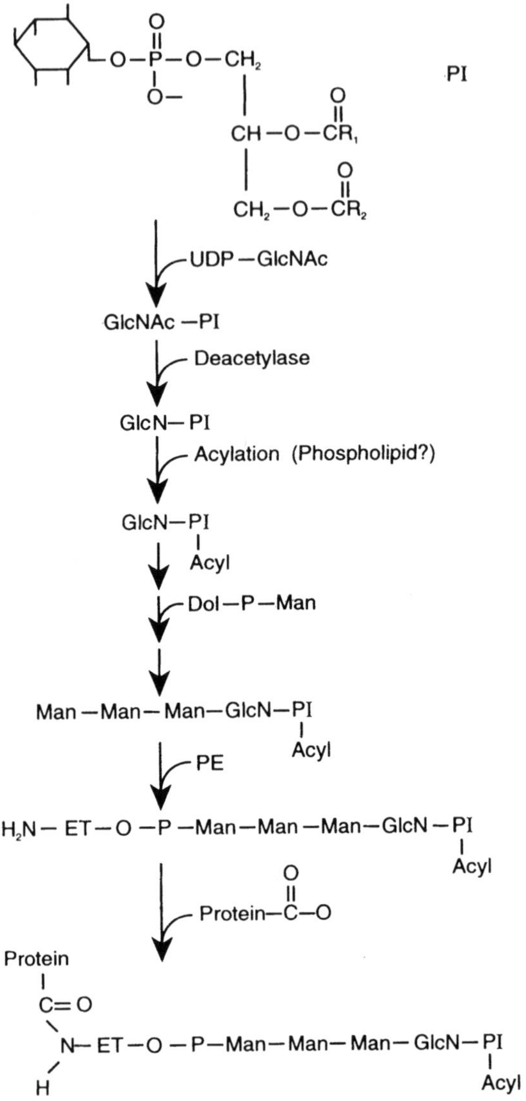

FIG. 5. Outline of the general pathway for biosynthesis of the glycan portion of the GPI anchors.

elucidated by Tamura and co-workers (103) and is shown in Fig. 6. The antibiotic contains a uracil moiety to which is attached, by an N-glycosidic bond, an 11 carbon aminodeoxydialdose called tunicamine. The tunicamine is itself substituted by having a GlcNAc attached to

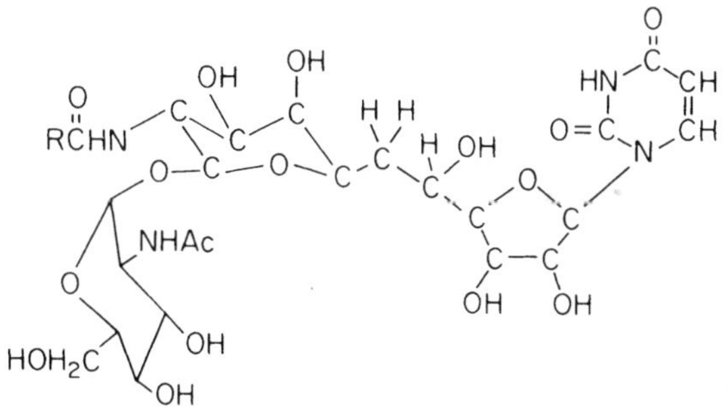

I : R =	$CH_3)_2CH(CH_2)_7CH{=}CH$-	VI : R =	$(CH_3)_2CH(CH_2)_{11}$-
II :	$(CH_3)_2CH(CH_2)_8CH{=}CH$-	VII :	$(CH_3)_2CH(CH_2)_{10}CH{=}CH$-
III :	$CH_3(CH_2)_{10}CH{=}CH$-	VIII :	$CH_3(CH_2)_{12}CH{=}CH$-
IV :	$CH_3(CH_2)_{11}CH{=}CH$-	IX :	$CH_3(CH_2)_{13}CH{=}CH$-
V :	$(CH_3)_2CH(CH_2)_9CH{=}CH$-	X :	$(CH_3)_2CH(CH_2)_{11}CH{=}CH$-

FIG. 6. The structure(s) of the various tunicamycin components. *R* refers to the different fatty acids, designated I to X, that may be attached to the tunicamine portion of the molecule.

its anomeric carbon in an O-glycosidic bond and a fatty acid linked in amide bond to the amino group. The fatty acid may vary in structure from C_{12} to C_{16}, may be saturated or unsaturated, and may be branched or unbranched. In fact, tunicamycin is actually produced as a complex that can be separated into as many as 10 to 15 peaks by HPLC (22,104), and the basis of this separation is the fatty acid moiety. Interestingly enough, all of these peaks have biological activity, but those components with the longer chain fatty acids appear to be more effective inhibitors in cell culture (22), possibly because they are taken up by the cells at a faster rate.

The site of action of tunicamycin was initially demonstrated with a microsomal enzyme preparation from calf liver, and the antibiotic was shown to be a potent inhibitor of the enzyme that transfers GlcNAc-1-P from UDP–GlcNAc to dolichyl-P (32,105). This reaction produces dolichyl-PP-GlcNAc, which is the first intermediate in the dolichol pathway. Tunicamycin had the same site of action in plant microsomes (106) and in microsomes from chick oviduct (107). In addition, tunicamycin is a very potent inhibitor with partially purified preparations of the UDP–GlcNAc:dolichyl-P GlcNAc-1-P transferase from several animal tissues (22,32,108). This inhibitor also affects several microbial enzymes that are involved in the synthesis of peptidoglycan, i.e., the enzyme that forms undecaprenyl-PP-*N*-acetylmuramoyl-pentapeptide

(109), and the enzyme involved in production of undecaprenyl-PP-GlcNAc (110).

However, tunicamycin had no effect on other GlcNAc transferases such as the enzyme that adds the next GlcNAc to dolichyl-PP-GlcNAc (111), the enzymes that add terminal GlcNAc residues to the N-linked oligosaccharide core (107), the phosphotransferase that adds GlcNAc-1-P to high-mannose chains on lysosomal enzymes (112), and the transferase that adds GlcNAc to serine residues on nuclear proteins (113). It also had no effect on the formation of undecaprenyl-PP-galactose or undecaprenyl-PP-GalNAc (114), but at high concentrations it did inhibit formation of dolichyl-P-glucose (115). Summing up these data, it appears that tunicamycin inhibits phosphotransferases that catalyze the translocation of a GlcNAc-1-P from its UDP derivative to a polyprenyl-P, but the sugar must be chirally related to GlcNAc. Tunicamycin is probably a competitive inhibitor of the GlcNAc-1-P transferase, but it binds so tightly to this enzyme that it has not been possible to reverse the inhibition by adding high concentrations of UDP–GlcNAc (105). It was suggested that this compound was a tight-binding reversible inhibitor and might be a substrate-product transition-state analog (32).

Tunicamycin has been used in many cell culture systems to try and assess the role of N-linked glycosylation in the function of various glycoproteins. Probably the best test of whether tunicamycin is really effective in preventing glycosylation in cell culture is to determine whether it blocks the *in vivo* incorporation of labeled glucosamine into dolichyl-PP-GlcNAc or the incorporation of [2-^{3}H]mannose into lipid-linked oligosaccharides. To do this, one can incubate cell cultures in the presence of various concentrations of tunicamycin (usually 0.1 to 10 μg/ml) for 30 min to 1 hr and then add the labeled glucosamine or mannose to these cells. After an appropriate period of labeling (determined experimentally for each cell type), the cell pellet is extracted with $CHCl_3$:CH_3OH:H_2O (1:1:1) and then $CHCl_3$:CH_3OH:H_2O (10:10:3) to obtain the lipid-linked monosaccharides and the lipid-linked oligosaccharides (116). One important point here is that the effective concentration of inhibitor may vary widely depending on the cell system being used, and the optimum concentration needs to be determined in each case. In fact, it has been demonstrated that tunicamycin-resistant mutants can be isolated and these mutant cells have amplified levels of the GlcNAc-1-P transferase (117). Thus, it seems reasonable to assume that the level of this enzyme, and therefore the susceptibility of the cell to tunicamycin, may vary from cell to cell or from tissue to tissue.

Something else to consider when using tunicamycin for functional studies is that this compound may inhibit protein synthesis (118). For this reason, any effects on glycoprotein function should be thoroughly examined to be certain that they result from the absence of carbohydrate, rather than from the loss of certain proteins. For example, in some of the various studies, at tunicamycin concentrations of 0.1 to 10 μg/ml, [^{3}H]leucine incorporation into protein was inhibited from 10 to 60%, while glycosylation was inhibited from 70 to 90%. But again, the important point is that some cell types are much more sensitive to inhibitors than are other cell types. Perhaps one control for some of these studies would be to use other inhibitors (i.e., cycloheximide or puromycin) that are known to block protein synthesis but do not affect carbohydrate synthesis or addition, or some other inhibitors that modify the structure of the carbohydrate chain, and compare the effects of these compounds with those of tunicamycin on the function of the particular glycoprotein.

The results of producing glycoproteins in cells in the presence of tunicamycin vary considerably depending on the glycoprotein in question. In terms of structural studies, tunicamycin at the appropriate concentration seems to prevent glycosylation of all N-linked glycoproteins, regardless of the biological system. However, in terms of the protein portion of the molecule, some glycoproteins are still synthesized at a normal rate and are still secreted even when they are not glycosylated (119). Thus, no effect of this inhibitor on the secretion of procollagen from chick embryo cells, secretion of a number of serum proteins from rat hepatocytes, interferon secretion by leukocytes or L cells, secretion of the enzyme β-*N*-acetylhexosaminidase by fibroblasts, and so on was observed (120). The conclusion here is that a number of different types of cells and glycoproteins do not require their carbohydrate for either folding or conformation and appear to function normally in the absence of added carbohydrate.

Some proteins, however, are greatly affected by the absence of carbohydrate. Thus, a considerable decrease in the rate of secretion of IgE and IgA by mouse and rat plasma cells was observed in the presence of tunicamycin, as was the secretion of IgM in hybridoma cells (121). One rather strange and interesting result that emerged from these studies was the observation that tunicamycin inhibited IgM secretion in hybridoma cells, but did not affect the secretion of IgG. These results led the authors to hypothesize that the differential effects of this inhibitor were due to factors intrinsic to the respective heavy-chain polypeptides themselves, rather than to some property of the cells (122). It would seem likely that such factors are related to the conformation of the respective proteins and to the influence of the carbohydrate on protein structure and/or folding.

Not surprisingly, tunicamycin also has dramatic effects on some membrane receptors, such as the acetylcholine receptor of muscle cells (123), the insulin receptor (124), the receptor for low-density lipoproteins (LDL) (125), and the receptor for nerve growth factor (126), to cite only a few examples. In general, with all of these receptors, tunicamycin significantly decreased the number of receptor molecules at the cell surface, but did not change the total number of receptors in the cells or the binding affinity of those receptors that were present in the plasma membrane. In some of these studies, the effects of tunicamycin were reversible, and in one study with the insulin receptor, it appeared that the nonglycosylated receptor could become glycosylated after removal of the drug (127). That would suggest that in some cases, proteins might be glycosylated post-translationally. Since there is some precedence for surface glycoproteins being endocytosed and recycled back through the Golgi (and perhaps the ER) (128), it is conceivable that glycosylation could occur on some exposed asparagine sequons.

Probably one of the reasons for the great variation in tunicamycin effects on protein function is that some glycoproteins require their N-linked oligosaccharides for stability (or longevity), and in the absence of glycosylation these proteins are rapidly degraded. That is, the carbohydrate may protect or somehow prevent the degradative enzymes from attacking glycosylated proteins. Such has been suggested to be the case for fibronectin, acetylcholine receptor, influenza viral hemagglutinin, and various immunoglobulins, just to cite a few examples (129). In fact, in the influenza system, it was observed that when viral replication was examined in the presence of tunicamycin, the viral envelope glycoproteins could not be detected. However, when a protease inhibitor, such as TCPK or leupeptin, was added to the cells along with the tunicamycin, the viral proteins in their nonglycosylated form could be detected (130). These studies and others cited above all indicate that the influence of the N-linked oligosaccharides on the protein in question depends on its amino acid sequence, which in turn determines folding and conformation. Thus, the carbohydrate may play a much greater role in affecting or maintaining the conformation of some proteins than that of others.

One study that provides strong support for the hypothesis that the amino acid sequence of the protein determines what type of and how important a role the carbohydrate plays in the function of the glycoprotein was done using various temperature-sensitive mutants of vesicular stomatitis virus (VSV) that were grown in the presence of tunicamycin. With one strain of VSV (called the San Juan strain), tunicamycin prevented viral replication by more than 90% when the virus was grown at either 38 or 30°C. However, with another strain (called the Orsay strain), the drug inhibited viral replication by 90% when the virus was grown at 37°C, but there was only 30 to 50% inhibition at 30°C. Furthermore, when this virus was grown in tunicamycin at 30°C, the nonglycosylated viral envelope protein (G protein) was detected at the host cell surface, indicating that it was synthesized and targeted in the normal fashion. However, at 37°C, this protein could not be detected at the cell surface. Finally, studies on the physical properties of these various nonglycosylated proteins also indicated that these proteins differed markedely in their solubility in detergents such as Triton X-114 at the different temperatures (131). The probable explanation for such results is that the carbohydrate has a significant role in providing a conformation that has increased solubility as well as stability.

B. *Tunicamycin-Related Antibiotics*

Five other antibiotics have been described that are now known to be structurally and functionally related to tunicamycin. These antibiotics are streptoviridin (132), mycospocidin (133), antibiotic 24101 (134), antibiotic MM 19290 (135), and corynetoxin (136). All of these compounds have the same general structure as tunicamycin, but they differ in the nature of the fatty acids attached to the tunicamine. Thus, in some cases, they may be identical to the various tunicamycin homologs discussed earlier. For example, the streptoviridin components have shorter chain fatty acids than the tunicamycins, as evidenced by the fact that their molecular weights range from 790 to 814, whereas those of tunicamycins are from 802 to 858 (22,132). On the other hand, HPLC comparison of the mycospocidin components with those from the tunicamycin complex showed a number of identical components (133). The corynetoxins are an interesting group of tunicamycin-related compounds that appear to be involved in or responsible for ryegrass toxicity. In this case, the developing seed heads of annual ryegrass may be invaded by a nematode which then causes a gall to be produced. Some of the nematodes carry the bacterium *Corynebacterium rathayi,* which produces the tunicamycin analog corynetoxin. The corynetoxins are apparantly toxic to sheep that eat the infected ryegrass (136).

All of these compounds appear to have the same action as tunicamycin, i.e., to inhibit the GlcNAc-1-P transferase. Detailed studies on the mechanism of action were done with streptoviridin and this inhibitor was compared with tunicamycin on the crude and partially purified preparations of the enzyme, as well as in cell culture systems (115). As indicated above for the various tunicamycin homologs, this group of analogs also showed differences in inhibitory activity, and those compounds with the longer chain fatty acids were more effective inhibitors in cell culture systems than those with shorter chain fatty acids. In addition, studies with cell-free extracts or partially purified

GlcNAc-1-P transferase preparations showed that the streptoviridins had a lower affinity for the enzyme than did the tunicamycin complex (115). In fact, with this inhibitor it was possible to demonstrate competitive inhibition with regard to UDP–GlcNAc, because the affinity of the GlcNAc-1-P transferase for streptoviridin was considerably lower than that for tunicamycin.

C. Lipopeptide Antibiotics

A number of other antibiotics and other compounds have been found to be useful inhibitors of some of the steps in the lipid-linked saccharide pathway. Many of these inhibitors, especially the antibiotics, were initially discovered because they block the formation of bacterial cell wall polymers such as peptidoglycan or lipopolysaccharide. Because the formation of these polymers also involves polyprenyl-linked saccharide intermediates, these compounds appeared to be likely candidates to test on the N-linked glycosylation pathway. For example, the antibiotics amphomycin and tsushimycin inhibit the transferase that adds phospho-*N*-acetylmuramoyl-pentapeptide to the growing peptidoglycan chains (137), whereas bacitracin inhibits the enzyme that catalyzes the dephosphorylation of undecaprenyl-pyrophosphate (138), the carrier lipid in cell wall biosynthesis. These antibiotics also inhibit the formation of N-linked oligosaccharide assembly as discussed later.

Amphomycin is a hydrophobic undecapeptide containing either 3-isododecanoic or 3-anteisododecanoic acids attached to the N-terminal aspartic acid in an amide linkage (139). The structure of amphomycin is presented in Fig. 7A. In cell-free extracts of aorta tissue, amphomycin blocked the formation of dolichyl-P-mannose from GDP–mannose, but it did not affect the transfer of mannose to lipid-linked oligosaccharides from either dolichyl-P-mannose or GDP–mannose (37,38). Thus, when microsomes were incubated with GDP–[^{14}C]mannose in the presence of enough amphomycin to completely inhibit the formation of dolichyl-P-[^{14}C]mannose, these extracts accumulated ^{14}C-Man$_5$(GlcNAc)$_2$-PP-dolichol, but they were not able to elongate this lipid to larger sized oligosaccharide-lipids. On the other hand, when dolichyl-P-[^{14}C]mannose was used as the substrate with these extracts, mannose was transferred to oligosaccharide-lipids that contained six or more mannose residues. These studies suggested that the first five mannose residues in the lipid-linked oligosaccharides were donated from GDP–mannose, whereas the next four mannoses came from dolichyl-P-mannose (37). Similar results were obtained with oviduct tissue (140) and embryonic liver (141). The isolation of a CHO mutant cell line missing the dolichyl-P-mannose synthase clearly established that doli-

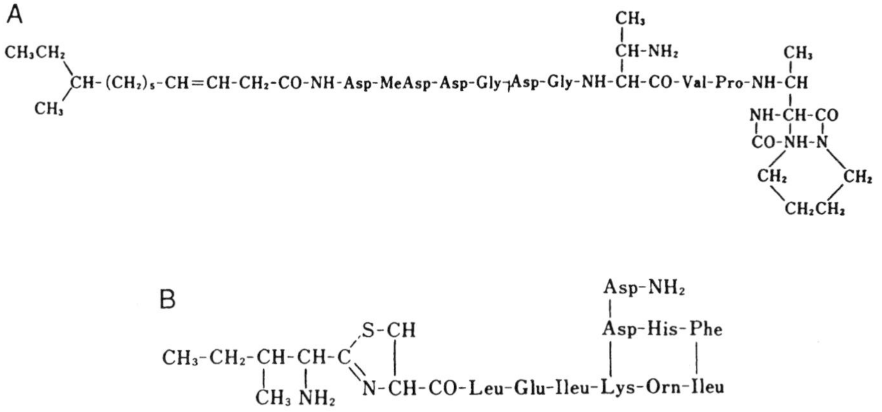

FIG. 7. Lipopeptide antibiotics that inhibit lipid-linked saccharide formation. (A) Structure of amphomycin. (B) Structure of bacitracin.

chyl-P-mannose was the mannosyl donor for the last four mannose residues (36).

In cell-free extracts of brain tissue, amphomycin inhibited the production of dolichyl-P-glucose, dolichyl-P-mannose, and dolichyl-PP-GlcNAc, leading these investigators to suggest that amphomycin might form a complex with dolichyl-P and dolichyl-P derivatives (142). Some evidence for such a complex between dolichyl-P and amphomycin has been obtained, although it is not clear how this relates to the inhibition of lipid-linked saccharide formation. Amphomycin has also been used in various other experiments, such as one study that demonstrated that dolichyl-P was involved in sugar translocation across the ER membrane (143), and in one case to show that dolichyl-P-mannose, rather than GDP–mannose, was involved in the elongation of $Man_7(GlcNAc)_2$-PP-dolichol (144). Another lipopeptide antibiotic, called tsushimycin, has a mechanism of action similar to that of amphomycin (145). Unfortunately, these peptide antibiotics do not appear to be effective in cell cultures of eucaryotic cells, probably because they do not cross the plasma membrane.

Bacitracin is another lipopeptide antibiotic that has been useful in studying both bacterial cell wall biosynthesis and the synthesis of lipid-linked saccharides. The structure of this compound is presented in Fig. 7B. Bacitracin was shown to inhibit the synthesis of bacterial cell wall peptidoglycan by inhibiting the dephosphorylation of undecaprenyl-pyrophosphate, which is necessary for the reutilization of this carrier molecule (138). In animal cells, the specific site of inhibition by this antibiotic is not clear because it has been reported to act at several different points, including some of the mannosyltransferases that add mannose to the trisaccharide-lipid Manβ(GlcNAc)$_2$-PP-dolichol, the GlcNAc-1-P transferase that forms GlcNAc-PP-dolichol, and the GlcNAc transferase that forms GlcNAc-GlcNAc-PP-dolichol (146). It seems likely that the action of this inhibitor is similar to that of amphomycin; i.e., it may bind various dolichyl derivatives under the right conditions and therefore may interfere with a variety of reactions in the "dolichol cycle."

D. *Other Inhibitors of Lipid-Linked Saccharide Formation*

Several other antibiotics that affect the synthesis of bacterial cell walls are also effective on the reactions described here. Generally, these compounds have not been well defined structurally or functionally, but some of their inhibitory actions have been reported. The structure of diumycin is not known, but it was reported to inhibit the formation of dolichyl-P-mannose in a yeast membrane preparation, as well as the transfer of mannose from dolichyl-P-mannose to serine or threonine residues (147). In a membrane preparation from *Acanthamoeba*, diumycin blocked the transfer of mannose and GlcNAc to lipid-linked monosaccharides, but it did not affect the synthesis of dolichyl-P-glucose (148). In fact, diumycin (and another antibiotic, showdomycin) has been used to add support to the hypothesis that dolichyl-P-mannose is a positive regulator of the GlcNAc-1-P transferase that forms GlcNAc-PP-dolichol (143). Thus, when either of these drugs was added to a microsomal preparation, it blocked the formation of dolichyl-P-mannose from GDP–mannose and therefore prevented the stimulation of GlcNAc-PP-dolichol formation by GDP–mannose.

Flavomycin is an antibiotic of the moenomycin group of compounds. This drug interferes with the formation of lipid-linked saccharides in microsomes from pig brain (149). Unfortunately, no other studies have been reported with this compound, so additional studies will be needed before we can understand its mechanism of action. The same applies to diumycin.

Showdomycin is a nucleoside antibiotic whose structure is shown in Fig. 8. This compound has a maleimide group that reacts with sulfhydryls on proteins, and this reaction is probably the basis of the showdomycin inhibition of lipid-linked saccharide formation. In cell-free extracts of aorta, the synthesis of dolichyl-P-mannose and dolichyl-P-glucose were both inhibited by this drug, but the formation of dolichyl-PP-GlcNAc was not affected (150). However, in *Volvox* extracts,

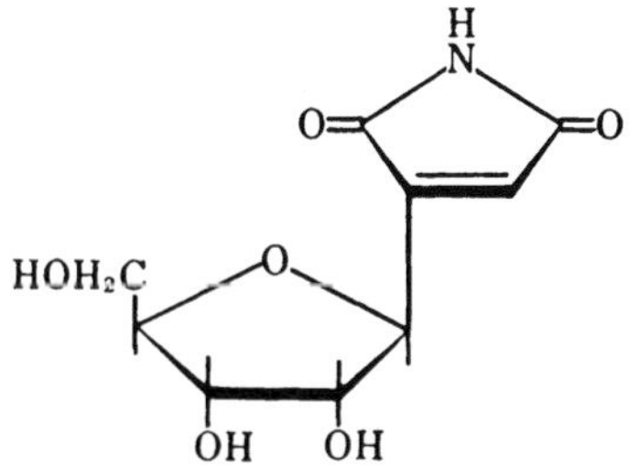

FIG. 8. Structure of showdomycin.

the syntheses of dolichyl-P-glucose and dolichyl-PP-GlcNAc were equally sensitive to showdomycin in the presence of Triton X-100, and this inhibition was reversed by the addition of dithiothreitol (151). As indicated earlier, showdomycin has also been used to study the regulation of lipid-linked saccharide formation by dolichyl-P-mannose.

E. Inhibition by Sugar Analogues

Probably two of the earliest inhibitors of viral envelope formation to be described were glucosamine (152) and 2-deoxyglucose (153). It is now known that glucosamine interferes with the formation of lipid-linked oligosaccharides and blocks protein glycosylation. This inhibition, however, requires intact cells and involves the metabolism of glucosamine probably to an intermediate such as UDP–glucosamine (154), which may compete with UDP–GlcNAc for various GlcNAc transferases. When MDCK cells were incubated in the presence of 1 m*M* glucosamine, there was a complete shift in the composition of the lipid-linked oligosaccharides from mostly $Glc_3Man_9(GlcNAc)_2$-PP-dolichol in normal cells to $Man_{6\text{-}8}(GlcNAc)_2$-PP-dolichol in inhibited cells. This shift to smaller sized oligosaccharide-lipids became even more pronounced when the concentration of glucosamine was raised to 10 m*M* (155). The glucosamine effect could be reversed by removal of this amino-sugar from the culture medium, but the exact site or mechanism of action of this compound is still unknown.

Interestingly enough, another amino-sugar, mannosamine, also inhibits the formation of lipid-linked oligosaccharides, but the mechanism of action of this compound is quite different from that of glucosamine. The inhibition by mannosamine was dose-dependent and resulted in a shift in the size of the oligosaccharide moieties of the lipid-linked oligosaccharides from $Glc_3Man_9(GlcNAc)_2$-PP-dolichol to lipids with $Man_3(GlcNAc)_2$ to $Man_9(GlcNAc)_2$. The major species were oligosaccharides having five or six mannose residues. These studies demonstrated that mannosamine itself was incorporated into the lipid-linked oligosaccharides, mostly in place of the mannose that is in α1,3-linkage to the β-linked mannose. This substitution appears to prevent the addition of the α1,2-linked mannose to this α1,3-linked mannosamine (156). In the presence of mannosamine, a series of basic oligosaccharides linked to dolichol were formed, and at least some of these oligosaccharides were still transferred to protein. However, it is not clear whether these altered glycopeptides can be processed by the normal pathway or they accumulate in the cells as mannosamine and mannose containing structures.

In terms of glucose analogs, such as 2-deoxyglucose, it is now clear that this sugar is metabolized in cells and is converted to UDP–2-deoxyglucose and GDP–2-deoxyglucose, as well as to dolichyl-P-2-deoxyglucose. The inhibition of glycoprotein synthesis by this deoxy-sugar appears to be due to the GDP–2-deoxymannose (glucose), because mannose itself can reverse the inhibition (157). Probably when deoxyglucose is incorporated into the lipid-linked oligosaccharides instead of the mannose at the 3-branch, it acts like mannosamine and prevents further extension of this chain because additional mannoses are linked in α1,2-bonds. These 2-deoxyglucose-containing oligosaccharides appeared not to be transferred to protein in contrast to the situation with the mannosamine-containing oligosaccharides (from the lipid-linked oligosaccharides). A number of fluoro-sugars such as 2-deoxy-2-fluoro-D-mannose, and 4-deoxy-4-fluoro-D-mannose have also been shown to be inhibitors of glycosylation (158). It seems likely that the mechanism of action of these analogs is similar to that of 2-deoxyglucose.

F. Effects of Glucose Starvation and Alterations in Energy Charge

Some animal cells, especially Chinese hamster ovary (CHO) cells, show a phenomenon that has been called “glucose starvation.” Thus, when glucose is removed from the medium, these cells can no longer produce the typical $Glc_3Man_9(GlcNAc)_2$-PP-dolichol that is the natural oligosaccharide donor for N-linked glycosylation. Instead, such cells produce a truncated lipid-linked oligosaccharide that is a $Man_5(GlcNAc)_2$-PP-dolichol (i.e., Manα1,2Manα1,2Manα1,3[Manα1,6]Manβ1,4GlcNAcβ1,4GlcNAc-) and then glucosylate this intermediate to produce a $Glc_3Man_5(GlcNAc)_2$-PP-dolichol (159–161). It appears that under glucose starvation conditions, the cells are not able to synthesize dolichyl-P-mannose, but the reason for this lesion is not clear because these cells still do make dolichyl-P-glucose and appear to glucosylate the oligosaccharides. In several cell lines, the effect of glucose starvation was seen within 20 min after the removal of glucose, as long as the cells were at low to moderate density, but at high density no effect was seen (162). The glucose starvation phenomenon could be overcome by the addition of glucose, but not the addition of galactose, glutamine, pyruvate, inositol, or glycerol. In rat hepatoma cells that were starved for glucose, a lower molecular weight form of α_1-acid glycoprotein was produced and this decrease in molecular weight was due not only to the truncated oligosaccharides, but also to fewer oligosaccharide chains on the protein (163).

Carbonyl-cyanide *m*-chlorophenylhydrazone (CCCP) is an uncoupler of oxidative phosphorylation that causes cells to be depleted of energy. When this compound is given to various animal cells, it causes a similar situation to that seen with glucose starvartion; i.e., cells are unable to produce dolichyl-P-mannose, although they still synthesize dolichyl-PP-GlcNAc and dolichyl-P-glucose. These cells form a $Glc_3Man_5(GlcNAc)_2$-PP-dolichol and transfer this oligosaccharide to protein (164). On the other hand, in thyroid slices incubated with CCCP, $Man_{9\text{-}8}(GlcNAc)_2$-PP-dolichol intermediates were produced, and a decrease in N-linked glycosylation was observed (165). In this study, other inhibitors of respiration such as N_2 or antimycin A caused the same effects. Since some uncouplers of oxidative phosphorylation have been reported to disrupt the recycling of the glucose transporter, these compounds may limit the entry of glucose into the cells, thus mimicking the glucose starvation effect. The puzzle here is the same as that mentioned earlier; i.e., Why is the formation of dolichyl-P-mannose inhibited, but not that of dolichyl-P-glucose? Perhaps it could have something to do with the postulation that the dolichyl-P-mannose synthase is a regulatory enzyme and the protein itself undergoes phosphorylation under some conditions.

V. Inhibitors of N-Linked Glycoprotein Processing

A. *Glucosidase Inhibitors*

A number of naturally occurring, sugarlike compounds in which the ring oxygen is replaced by a nitrogen have been isolated, and these alkaloids have been found to be potent inhibitors of various glycosidases. Apparantly this nitrogen in the ring makes these compounds inactive from a metabolic standpoint, but they are still recognized by specific glycosidases and therefore function as useful enzyme inhibitors. Because the enzymes involved in the early glycoprotein processing reactions are glycosidases, these inhibitors have played a significant role in studying the sequence of reactions of N-linked oligosaccharides, as well as the requirement for glycoprotein processing in many animal cell lines. Many of these inhibitors have been isolated from plant sources, but the characterization of their structures and the demonstration of their biological activity have stimulated the chemical synthesis of a number of other analogues. The plant inhibitors are all alkaloids and fall into various chemical classes designated piperidines, pyrrolidines, indolizidines, and pyrrolizidines (166).

Castanospermine is an indolizidine alkaloid that is found in the seeds of the Australian chestnut tree, *Castanospermum australe* (167). Its structure is shown in Fig. 9. This alkaloid is a potent inhibitor of β-glucosidase (168) and also inhibits the glycoprotein processing glucosidases (I and II), as well as various other α-glucosidases such as sucrase, maltase, and lysosomal α- glucosidase (169). Because this compound

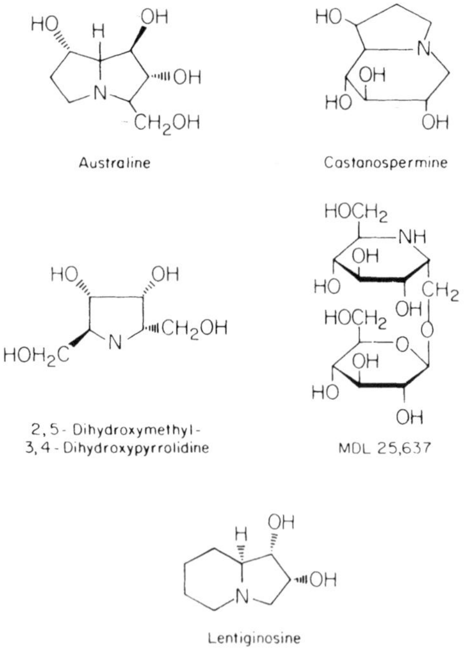

FIG. 9. Structures of various glucosidase inhibitors. Castanospermine and 2,5-dihydroxymethyl-3,4-dihydroxypyrrolidine are inhibitors of both glucosidase I and glucosidase II, whereas australine is a much better inhibitor of glucosidase I than glucosidase II and MDL is a more potent inhibitor of glucosidase II. Lentiginosine, shown for comparison, is an inhibitor of amyloglucosidase but is ineffective against the processing glucosidases.

is such a strong inhibitor of α-glucosidases (such as sucrase and maltase) and therefore prevents the normal digestion of starch and sucrose, the seeds of this plant are toxic to animals and cause severe diarrhea (170). When various cultured animal cells are grown or incubated in the presence of this alkaloid, the processing of N-linked glycoproteins is blocked at the first step in the pathway (see Fig. 3), and the asparagine-linked glycoproteins have oligosaccharides mostly of the $Glc_3Man_{7\text{-}9}(GlcNAc)_2$ structure. Another glucosidase inhibitor is the polyhydroxylated piperidine analog nojirimycin, which corresponds to glucose in the pyranose ring form (see Fig. 9). This compound was originally isolated from *Streptomyces* and was found to be a potent inhibitor of α- and β-glucosidases (171). Chemical reduction of nojirimycin gave the more stable deoxynojirimycin, which is also a glucosidase inhibitor and which has subsequently been isolated from bacteria and plants (172). Deoxynojirimycin causes the same general effects in cell culture as does castanospermine, although it is more active toward glucosidase II than toward glucosidase I, and it does cause other alterations such as changes in the structures of the lipid-linked saccharides (173). Another glucosidase I inhibitor that also is found in plants is the pyrrolidine alkaloid 2,5-dihydroxymethyl-3,4-dihydroxypyrrolidine (174). This compound, although considerably less effective than the six-membered ring structures described above (e.g., compare the structure in Fig. 9 with those of castanospermine and deoxynojirimycin), causes similar changes in the structures of the N-linked oligosaccharides.

The effect on glycoprotein targeting and function of preventing the removal of glucoses from the N-linked oligosaccharides is quite dramatic. For example, when hepatocytes (Hep-G2 cells) are incubated in the presence of deoxynojirimycin, the rate of secretion of α_1-antitrypsin was substantially diminished, whereas the rate of secretion of other N-linked glycoproteins such as ceruloplasmin and the C3 component of complement was only marginally altered. Based on some preliminary cell fractionation studies, it was suggested that the presence of glucose on the oligosaccharides of certain glycoproteins might retard the movement of these glycoproteins from the ER to the Golgi apparatus (175). Similar results were obtained with fibroblasts and smooth muscle cells when the effect of castanospermine on the LDL receptor was examined. In this case, there was a significant decrease in the binding and endocytosis of LDL when arterial smooth muscle cells were upregulated to produce LDL receptors in the presence of castanospermine. The reason for this decreased binding was shown to be due to a reduced number of receptors at the cell surface rather than to a change in the affinity of receptor for its ligand. Furthermore, the total number of LDL receptors in the cell did not change in the presence of castanospermine, indicating that this drug did not inhibit the synthesis of the receptor, nor did it affect its degradation. Cell fractionation studies designed to isolate various membrane fractions demonstrated a shift in the distribution of LDL receptors within the cell compartments and indicated a decrease in receptors at the cell surface and a corresponding increase in receptors in the ER–Golgi membrane fractions (176).

In the case of IM-9 lymphocytes, castanospermine was used to study the biosynthesis and processing of the insulin receptor. When these cells were treated with this alkaloid, they showed a 50% decrease in cell-surface insulin receptors as demonstrated by the binding of ^{125}I-insulin, as well as by Na^{125}I-lactoperoxidase labeling and by cross-linking studies with ^{125}I-insulin. These studies demonstrated that processing, i.e., removal of glucose from the N-linked oligosaccharides, was not necessary for the cleavage of the insulin proreceptor, i.e., for protein maturation. However, as indicated above, the presence of glucose appears to slow down the transport of glycoproteins from the ER to the Golgi apparatus (177). These data suggest the presence of a glucose receptor in the ER that "holds up" or binds glucose-containing glycoproteins, or the data may indicate that some proteins undergo a conformational change when glucose is removed, and the newly altered protein is transferred from the ER to the Golgi at a more rapid rate.

In the case of synthesis of the E_2 glycoprotein of coronavirus, castanospermine or deoxynojirimycin caused a drop in virus formation by 2 logs and also caused a dramatic decrease in the appearance of E_2 at the cell surface. Interestingly enough, the E_2 that was produced in the presence of these drugs was still acylated with a fatty acid, as was the control E_2, but the glycoprotein accumulated intracellularly in a compartment that was distinct from the Golgi apparatus (178). Could this compartment be the ER?

In another case, post-translational addition of palmitic acid was also not affected by glucosidase I inhibitors. The Na channel of rat brain neurons is composed of α- and β-subunits that form a complex during maturation. The α-subunit is post-translationally modified by the addition of a palmitic acid, and the incorporation of labeled palmitate into this glycoprotein is blocked by the glycosylation inhibitor tunicamycin. However, the α-subunit formed in the presence of castanospermine is of lower molecular weight, does not contain sialic acid, and has a lower sulfate content, but it is similar to control α-subunit in its content of palmitic acid (179). The drug also did not prevent the covalent assembly of α- and β-subunits, nor did it affect the biological activity of the channel with regard to its binding of saxitoxin.

The external glycoprotein of the AIDS-associated virus HIV is called GP120 and is a heavily glycosylated protein that is involved in the mechanism of attachment of HIV to the CD4 receptor on T lymphocytes and other susceptible cells. The GP120 interacts with target molecules on susceptible cells to cause the fusion of the cells, which is referred to as the formation of syncytia. Both glucosidase I inhibitors (i.e., castanospermine and deoxynojirimycin) caused a drastic inhibition of virus replication as well as of syncytium formation (180–182). Because of these interesting results, castanospermine and deoxynojirimycin, as well as other glucosidase I and II inhibitors and analogs of these various inhibitors, are being tested as potential anti-AIDS compounds.

As shown in Fig. 9, there are other compounds besides castanospermine and deoxynojirimycin, that also have activity against the α-glucosidases and that function as glycoprotein processing inhibitors. One of these compounds is the pyrrolidine alkaloid mentioned earlier and referred to as 2,5-dihydroxymethyl-3,4-dihydroxypyrrolidine (DMDP), which has been found in several plants of the *Leguminosae* family. In cell culture, DMDP gives the same type of oligosaccharide structure, i.e., $Glc_3Man_{7\text{-}9}(GlcNAc)_2$, as does castanospermine, although it is much less active as an inhibitor (i.e., higher concentrations are required) (174). Nevertheless, it is an interesting inhibitor because it demonstrates that a six-membered ring structure is not absolutely required for inhibitory activity, and knowledge of such structures helps us to define the particular structural requirements that are necessary for such biological activity. Along these lines, two other structures that are also interesting glucosidase inhibitors, partly because they are unusual and partly because they show more selectivity than those described thus far, are shown in Fig. 9.

Australine is a tetrahydroxypyrrolizidine alkaloid that was isolated from the seeds of the tree *C. australe* (as is castanospermine). This compound has a unique substitution pattern (i.e., (1*R*,2*R*,3*R*,7*S*,7α*S*)-3-hydroxymethyl-1,2,7-trihydroxypyrrolizidine). This bicyclic alkaloid is a reasonably good, competitive inhibitor of the fungal amyloglucosidase (an aryl-glucosidase) and it also inhibits the glycoprotein processing enzyme glucosidase I. However, in contrast to castanospermine and deoxynojirimycin, australine is a very poor inhibitor of glucosidase II (183). This compound is the first glucosidase inhibitor to distinguish between these two processing glucosidases. Thus, additional compounds such as australine that show increased specificity will be valuable to help us understand how these two glucosidases differ in their mechanism of catalysis and also to determine the chirality and spacial configurations that are necessary for activity of various glycosidases.

Another inhibitor whose structure is shown in Fig. 9 is 2,6-diamino-2,6-imino-7-O-(β-D-glucopyranosyl)-D-glycero-L-guloheptitol (MDL). This compound was chemically synthesized to mimic a disaccharide that would act as a transition state analog of the intestinal enzyme sucrase (184). As anticipated, MDL inhibited rat intestinal maltase, sucrase, isomaltase, glucoamylase, and trehalase at micromolar concentrations. Interestingly enough, when tested on the glycoprotein processing glucosidases, it was the opposite of australine in its specificity. That is, MDL was a much more effective inhibitor of glucosidase II than it was of glucosidase I (185). In cell culture experiments, MDL caused the accumulation of glycoproteins having mostly Glc_2Man_{7-9} $(GlcNAc)_2$ structures. The finding that australine and MDL are more specific inhibitors than castanospermine and deoxynojirimycin will hopefully stimulate a more intense search for other naturally occurring inhibitors with improved selectivity for glucosidases and mannosidases, and will also entice chemists to use their imagination to synthesize more novel compounds.

B. *Mannosidase Inhibitors*

The first glycoprotein processing inhibitor to be reported was swainsonine (186), which is described in more detail below. This indolizidine alkaloid was initially found to be an inhibitor of lysosomal and arylmannosidases (187), and was later shown to specifically inhibit the processing mannosidase II (188). However, for ease of understanding, I will consider the various mannosidase inhibitors in the order in which they occur in the processing pathway rather than in the order in which they were first reported.

The discovery of deoxynojirimycin and castanospermine as glycoprotein processing inhibitors indicated that a good glycosidase inhibitor should possess the following properties: (a) a pyranose ring structure with a nitrogen in the ring replacing the oxygen and (b) the stereochemistry of the hydroxyl groups in keeping with those of the particular sugar that is recognized by the glycosidase of interest. Thus, it was reasonable to assume that a structure similar to deoxynojirimycin but with the mannose configuration (i.e., the 2-epimer) should be a good inhibitor of mannosidases. The 2-epimer of deoxynojirimycin was synthesized chemically and was indeed found to be a potent inhibitor of the glycoprotein processing enzyme mannosidase I (189). Surprisingly enough, deoxymannojirimycin did not inhibit jack bean or lysosomal α-mannosidase, nor did it inhibit the processing mannosidase II. In fact, this inhibitor demonstrates that it is dangerous to screen for new inhibitors using aryl-glycosidases as the source of enzymes or even using only a few of the processing enzymes, because the most useful and desirable processing inhibitors will be specific for a particular processing glycosidase and may not inhibit these other activities.

The structures of deoxymannojirimycin, swainsonine, and other known mannosidase inhibitors are shown in Fig. 10. Since the time when many of the studies on deoxymannojirimycin were done, several other neutral α-mannosidases that have different specificities and are enzymes distinct from mannosidase I or mannosidase II have been identified. Some of these enzymes have been found to be resistant to inhibition by deoxymannojirimycin, but there are not enough mannosidase inhibitors yet known to be able to specifically inhibit some of these activities. In addition, most of these other enzymes have not yet been purified, so they have not been tested against the various processing inhibitors. In animal cells, deoxymannojirimycin inhibited the Golgi mannosidase IA/B and caused the accumulation of glycoproteins having high-mannose oligosaccharides, mostly of the $Man_{8\text{-}9}(GlcNAc)_2$ structure (191). This inhibitor, however, did not prevent the secretion of IgD or IgM in cell culture, although the glucose analog deoxynojirimycin did (192). Thus the presence of glucose on the oligosaccharides does have a significant effect on the intracellular transport of at least some glycoproteins, but the presence of high-mannose oligosaccharides rather than the normal complex chains does not appear to be a problem

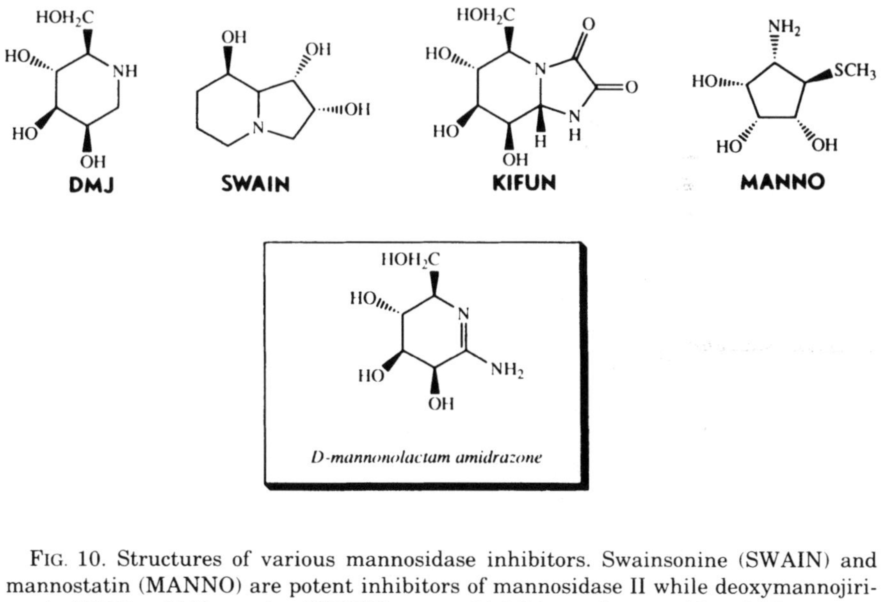

FIG. 10. Structures of various mannosidase inhibitors. Swainsonine (SWAIN) and mannostatin (MANNO) are potent inhibitors of mannosidase II while deoxymannojirimycin (DMJ) and kifunensine (KIFUN) are good inhibitors of mannosidase I. The mannonolactam amidrazone is a very potent general mannosidase inhibitor that acts on arylmannosidases, as well as mannosidase I and mannosidase II.

in terms of protein secretion. Deoxymannojirimycin was also without effect on the formation and appearance of the G protein of vesicular stomatitis virus, the hemagglutinin of influenza virus, and the HLA A,B, and C antigens, as well as on α_1-acid glycoprotein and α_1-protease inhibitor (192).

Deoxymannojirimycin was used in one study to determine if membrane glycoproteins were recycled through the Golgi apparatus during the endocytic process. In this experiment, membrane glycoproteins were synthesized and labeled with [^{3}H]mannose in the presence of deoxymannojirimycin so that the glycoproteins would all have high-mannose structures. Then the inhibitor and labeled sugar were removed, and the oligosaccharide structure of the transferrin receptor was determined at various times and in conditions under which this receptor would be endocytosed. The initial oligosaccharides of the transferrin receptor were high-mannose structures because of the presence of deoxymannojirimycin during their synthesis, but during the chase, a small percentage of the recycled receptor underwent processing and attained complex types of oligosaccharide structures, indicating that some of the endocytosed glycoprotein was recycled through the cis-Golgi compartments (193). However, the amount of glycoprotein that actually underwent modification during the recycling was quite small, indicating that, while some glycoprotein does recycle through the Golgi, this is probably not a major route.

The role of endoplasmic reticulum α-mannosidase was also examined in UT-1 cells, a cell line that overexpresses HMG CoA reductase. The HMG CoA reductase resides in the ER of cells and is a glycoprotein that has high mannose oligosaccharides mostly of the $Man_8(GlcNAc)_2$ and $Man_6(GlcNAc)_2$ structures. Previous studies indicated that "ER" α-mannosidase was not inhibited by deoxymannojirimycin. Thus, in the presence of deoxymannojirimycin, the initial trimming of the oligosaccharide chains of the reductase still occurred, but in this case, the major oligosaccharide structure on the protein was a $Man_8(GlcNAc)_2$ and the smaller sized oligosaccharides were not detected. These studies indicated that the ER α-mannosidase was involved in the removal of the first mannose, but that other α-mannosidases were necessary for the removal of other mannose residues (194).

Another mannosidase inhibitor is 1,4-dideoxy-1,4-imino-D-mannitol (DIM). This inhibitor has a furanose ring structure rather than the pyranose rings of most of the other inhibitors. DIM was synthesized chemically from benzyl-α-D-mannopyranose and was found to be a good inhibitor of jack bean α-mannosidase (195). It also inhibited the processing of N-linked glycoproteins in cell culture when it was added to the culture medium of MDCK cells. In those experiments, the major oligosaccharides formed on the cell glycoproteins were $Man_9(GlcNAc)_2$ structures in keeping with the hypothesis that DIM inhibited the Golgi mannosidase I (196). However, it is not known whether DIM also inhibits the ER mannosidase, because this activity does not appear to be present in MDCK cells. Furthermore, there is no indication that this inhibitor has been tested against this ER mannosidase or other mannosidases, although it is active against mannosidase I. Thus, *in vitro* studies with DIM and a partially purified preparation of mannosidase I from rat liver showed that this compound inhibits the release of $[^3H]$mannose from $[^3H]$mannose-labeled $Man_9GlcNAc$. DIM is not nearly as effective as a mannosidase inhibitor as is swainsonine or kifunensine. However, it is much more readily available in large amounts because the chemical synthesis is quite simple. Therefore, DIM and its isomers should be valuable for animal studies, as well as for other experiments requiring large amounts of mannosidase inhibitor.

Kifunensine is an alkaloid, produced by the actinomycete *Kitasatosporia kifunense* and corresponding in structure to the cyclic oxamide derivative of 1-aminodeoxymannojirimycin (197). Kifunensine is a very weak inhibitor of jack bean α-mannosidase ($IC_{50} = 1.2 \times 10^{-4}$ M), but it is a potent inhibitor of the glycoprotein processing mannosidase I ($IC_{50} = 2–5 \times 10^{-8}$ M). Thus, this inhibitor is about 100 times more effective against this enzyme than is deoxymannojirimycin. On the other hand, kifunensine had no effect on mannosidase I (198). Using rat liver microsomes as a source of mannosidase activities, kifunensine strongly inhibited the Golgi mannosidase I but it was probably without effect on the proposed ER α-mannosidase. In cell culture with MDCK cells, kifunensine gave rise to glycoproteins having $Man_9(GlcNAc)_2$ structures just as did deoxymannojirimycin, but the amount of kifunensine needed for this effect was only about one-fiftieth as much as deoxymannojirimycin.

Recently, a broad-spectrum α-mannosidase inhibitor was synthesized chemically as a mimic of the mannopyranosyl cation, which is the putative intermediate in the hydrolysis of mannopyranosides (199). This compound, whose structure is shown in Fig. 10, is called mannonolactam amidrazone. This compound was found to be a more general mannosidase inhibitor than any of the other currently known compounds and those shown in Fig. 10. Thus, the amidrazone not only inhibits the Golgi mannosidase I ($IC_{50} = 4$ μM) and mannosidase II ($IC_{50} = 90–100$ nM), but also is the first compound that has been found to be a potent inhibitor of the soluble or ER α-mannosidase ($IC_{50} = 1$ μM). In addition, the amidrazone also inhibited both the aryl-α-mannosidase and the aryl-β-mannosidase, although it was much more effective on the enzyme recognizing the α-linkage ($IC_{50} = 400$ nM for jack bean α-mannosidase compared with 150 μM for β-mannosidase). In cell culture studies, the amidrazone completely prevented the formation of complex types of oligosaccharides and instead caused the formation of oligosaccharides having $Man_9(GlcNAc)_2$ and $Man_8(GlcNAc)_2$ structures. It is hypothesized that the reason that this compound is so effective against all of these mannosidases is that it is the first monosaccharide analog that mimics the true half-chair conformation of the cationic intermediate that is believed to be involved in the catalysis of the α-mannose glycosidic bond. This compound should serve as a useful model to synthesize other more specific mannosidase inhibitors.

As indicated above, the first glycoprotein processing inhibitor to be identified was swainsonine. The initial interest in this compound was due to the fact that it is found in plants that are toxic to grazing animals. In Australia, the major plant having this alkaloid is called *Swainsona canescens* (200), while in the United States, various plants of the genus *Astragalus* contain swainsonine (201, 202). These plants present a major hazard to the livestock industry and cause severe neurological and skeletal problems in animals that feed on them. Early studies on the inhibition of aryl-α-mannosidases by swainsonine suggested that its potent inhibitory activity ($IC_{50} = 1 \times 10^{-7}\ M$) resulted from the structural similarity of the protonated form of swainsonine to that of the mannosyl cation, since the glycosyl ion intermediate is probably formed during natural hydrolysis by glycosidases (203).

Lymphnodes of sheep that had injested these toxic plants contained high levels of mannose-rich oligosaccharides (204), which led to the postulation that swainsonine causes a "lysosomal-like" storage disease that is similar to that of human α-mannosidosis, both morphologically and biochemically. That swainsonine is actually the causative agent of these symptoms was indicated by feeding pigs either the whole toxic plant (i.e., *Astragalus*) or swainsonine and showing that both sets of animals exhibited the same symptoms (205). In these animals, liver Golgi mannosidase II activity was significantly decreased, but strangely enough, a number of lysosomal hydrolases such as α-mannosidase, α-fucosidase, and β-hexosaminidase increased in activity. In addition, the activity of various hydrolases in the plasma was also elevated. In sheep fed locoweed, various high-mannose oligosaccharides were found in the urine and the two major oligosaccharides were identified as $Man_4(GlcNAc)_2$ and $Man_6(GlcNAc)_6$ (206). The relative abundance of the various oligosaccharides changed over the course of the feeding, but when the feeding was discontinued, the amount of urinary oligosaccharides declined rapidly and reached a baseline within about 12 days. Similar types of high-mannose oligosaccharides also accumulated in rats and guinea pigs fed swainsonine, but there were some differences in the structures, especially the presence of one or two GlcNAc residues at and near the reducing terminus (207). These differences probably reflect the presence or absence of an endoglucosaminidase in some of these animals.

Swainsonine was the first compound that was shown to inhibit glycoprotein processing (186). In the initial studies, swainsonine was added to the culture medium of MDCK cells and was found to cause a great decrease in the amount of [2-^{3}H]mannose incorported into complex types of glycoproteins and a corresponding increase in incorporation of labeled sugar into oligosaccharides that were sensitive to digestion by endoglucosaminidase H. In influenza-virus-infected MDCK cells, swainsonine caused the viral hemagglutinin, an N-linked glycoprotein with several complex chains, to migrate more rapidly on SDS gels, and these changes were shown to be due to changes in the structure of the N-linked oligosaccharides (208). Later studies with purified processing enzymes showed that the site of action of swainsonine was the Golgi mannosidase II and that this alkaloid was inactive toward the processing mannosidase I (188). In keeping with this site of action, this indolizidine alkaloid also caused the formation of hybrid types of structures when it was placed in the culture medium of cells producing various glycoproteins such as the VSV G protein (209), fibronectin (210), and BHK cell surface glycoproteins (211). Swainsonine has been used with a number of cell culture systems to try and determine the role of specific oligosaccharide structures in glycoprotein function. In most of these studies, swainsonine was found to have little effect on the glycoprotein in question, indicating that a partial complex oligosaccharide (i.e., hybrid chain) is sufficient for functional activity, at least in tissue culture. For example, swainsonine did not impair the synthesis or export of thyroglobulin in porcine thyroid cells (212), nor did it affect surfactant glycoprotein A in Type II epithelial cells (213), H2-DK histocompatability antigen in macrophages (214), or von Willebrand protein in epithelial cells (215). It also did not affect the targeting or function of various membrane receptors, including insulin receptor (216), epidermal growth factor receptor (217), or the receptor for asialoglycoproteins (218). The inhibitor did, however, block the receptor-mediated uptake of mannose-terminated glycoproteins by macrophages. This blockage appeared to be due to the formation of hybrid types of glycoproteins on the macrophage surface, which then reacted with and bound the mannose receptors (219). Swainsonine proved to be useful in demonstrating the sequence of addition of the various sugars in the N-linked oligosaccharide assembly pathway. Thus, when fucosylation (220) and sulfation (221) of the influenza viral hemagglutinin was examined in cells grown in the presence of swainsonine or castanospermine, it was found that the hemagglutinin did contain [^{3}H]-fucose and [^{35}S]sulfate when the virus was raised in the presence of swainsonine, but not when virus was produced in the presence of castanospermine (or deoxymannojirimycin). These studies suggest that the fucosyltransferase and the sulfotransferase recognize the GlcNAc–$Man_5(GlcNAc)_2$ structure as the appropriate acceptor substrate.

Some proteins do show a loss of function when they are produced in the presence of swainsonine. Thus, the glucocorticoid stimulation of resorptive cells, which appears to involve the attachment of osteoblasts and related cells to bone, is inhibited or blocked by swainsonine (222). This drug also reduced the interaction of *Tyrpanosoma cruzi* with peritoneal macrophages when either the host cells or the parasites were treated with swainsonine (223). Also a dramatic decline in the ability of B16 melanoma cells to colonize the lungs of experimental animals was observed when the cells were incubated in swainsonine (224). Treatment of lymphocytes with the mitogen concanavalin A stimulates them to proliferate, but this proliferation can be blocked by an immunosuppresive factor found in the serum of mice that contain the tumor sarcoma 180. The suppression caused by this factor is somehow overcome by swainsonine, indicating that this alkaloid might have some chemotherapeutic value in immunosuppressive diseases (225). The turnover or degradation of endocytosed glycoproteins is also inhibited by swainsonine and this leads to the accumulation of these glycoproteins in the lysosomes of tissue culture cells incubated with this drug. The block here appears to be due to the inhibition of lysosomal mannosidase (and perhaps other enzymes), indicating that the carbohydrate portion of these glycoproteins must be degraded before the protein (226). All of these studies suggest that swainsonine may have important effects on some cell systems and therefore it or some related compounds may have some chemotherapeutic value. However, additional studies will be required with these inhibitors.

Another inhibitor of mannosidase II that was recently isolated from a fungus is called mannostatin (227). This compound is interesting because it is the only one of these inhibitors presently known that has an exocyclic nitrogen rather than a nitrogen in the ring. In addition, as seen in Fig. 10, mannostatin has a thiomethyl group and a five-membered ring. Mannostatin was found to be a potent inhibitor of mannosidase II (IC_{50} = 100 nM) and jack bean mannosidase. In cell culture studies, mannostatin gave rise to the same types of hybrid structures that are observed in cells incubated in the presence of swasinsonine (228). However, the N-acetylated derivative of mannostatin was inactive against any of the mannosidases. Although mannostatin is not any different in its site of action, or any more active than swainsonine, it is an interesting addition to the repertoire of inhibitors because it shows a heretofore unanticipated structure as a glycosidase inhibitor. This opens up a number of new possibilities for chemical synthesis of related structures or perhaps structures that combine some aspects of mannostatin and some aspects of other alkaloids.

A number of laboratories have been developing methods for the synthesis of the above alkaloids or their derivatives, as well as of other new inhibitors (229). So far, with the exception of mannostatin, all of the inhibitors have had a nitrogen in the ring in place of the oxygen, and at least three assymetric hydroxyl groups. The substitution of the ring oxygen by a nitrogen apparantly renders the compound metabolically inert but does not prevent the recognition of the compound by glycosidases or perhaps by other carbohydrate recognizing enzymes. Presumably the hydroxyl groups in these inhibitors have the same chirality as the sugar that is a substrate for the glycosidase to be inhibited. Interestingly enough, these inhibitors do not necessarily need to have a six-membered ring structure to mimic the catalytic intermediate that is involved in the substrate recognition or hydrolytic step. In fact, swainsonine best fits as a mannose analog when it is in a staggered or half-chair conformation (203), and other pyrrolidine or pyrrolizidine ring structures may also assume similar conformations. Molecular modeling studies with these compounds will help to understand the mechanism of inhibition by some of these compounds. Such modeling studies have been done with some of the mannosidase inhibitors compared with the mannopyranosyl cation as the active intermediate of jack bean mannosidase (230). However, the synthesis or isolation of other and different inhibitors of these enzymes will add to this information.

VI. Effects of Inhibitors of Protein Synthesis, or Compounds That Modify Protein Structure, on N-Linked Glycosylation

The synthesis of lipid-linked oligosaccharides in animal tissues or by cultured animal cells has been shown to be inhibited by several inhibitors of protein synthesis (231). For example, when MDCK cells were grown in the presence of cycloheximide or puromycin, both inhibitors of protein synthesis but by different mechanisms, there was a dose-dependent inhibition in the incorporation of [2-^{3}H]mannose into the lipid-linked oligosaccharides. However, these inhibitors did not affect mannose incorporation into dolichyl-P-mannose, nor did they block mannose incorporation from GDP–[^{14}C]mannose into lipid-linked saccharides in cell-free extracts prepared from MDCK cells or from porcine aorta. These results could be explained by one of the following postulations: (a) limitation in the amount of dolichyl-P available to serve as a carrier of the sugars or (b) feedback inhibition in the synthesis of lipid-linked saccharides caused by the accumulation of lipid-linked oligosaccharide resulting from lack of acceptor protein. However, since the synthesis of dolichyl-P-mannose was not inhibited, the feedback mechanism seemed more likely (232).

Additional studies showed that the inhibition by cycloheximide could not be overcome by adding exogenous dolichyl-P, even though addition of dolichyl-P to normal cells did stimulate the incorporation of mannose into lipid-linked saccharides. However, the specific site of inhibition (or feedback inhibition) by these inhibitors is not known. Actinomycin, an inhibitor of RNA synthesis, was also found to inhibit the incorporation of mannose into lipid-linked oligosaccharides, but again the exact site of inhibition was not determined (233). Based on the above results, as well as additional studies on the levels of various intermediates in normal and inhibited cells, it was postulated that the mechanism of inhibition involved a feedback type of control that was caused by elevated levels of GTP in the cells (234).

In N-linked glycosylation, the consensus sequence that accepts the oligosaccharide chain is H_3N . . . Asn–X–Ser (Thr) . . . , where *X* can be any amino acid except proline or aspartic acid. This sequence is, however, not sufficient for glycosylation to occur, and the conformation of the protein around this glycosylation site is apparantly also critical. Statistical studies on a number of asparagine-linked glycoproteins have indicated that most of the asparagine residues that do become glycosylated are located in regions of the polypeptide chain that favor the formation of β-turns.

The importance of the threonine residue in the tripeptide sequence was examined by using a threonine analog, β-hydroxynorvaline, and determining its effects on cotranslational glycosylation in extracts of ascites cells. This analog inhibited the glycosylation of the α-subunit of human chorionic gonadotropin and the β subunit of bovine leutinizing hormone. This inhibition could be overcome by adding threonine, strongly suggesting that the analog was inhibiting by competing with threonine for incorporation into protein (235). In cultured fibroblasts, β-hydroxynorvaline caused the formation of a variety of cathepsin D molecules that differed from each other by having one, two, or no oligosaccharide chains on the protein. The nonglycosylated form of cathepsin was a minor component and it was rapidly degraded within 45 min of its synthesis. However, those forms having one or two oligosaccharide chains behaved normally, were segregated to the lysosomes, and underwent proteolytic maturation in the usual fashion (236). Similar types of results were observed with respect to the effects of this analog on the formation of α_1-acid glycoprotein (237).

Another amino acid analog that has interesting effects on glycosylation is threo-β-fluoroasparagine. This analog is toxic to asparagine-dependent cells when aspartic acid is included in the culture medium. The actual mechanism of toxicity has not been determined but it is believed to be due to the incorporation of the analog into protein. At 1 m*M* concentrations, this analog strongly inhibits glycosylation but this inhibition is blocked by adding asparagine, further strengthening the idea that the analog is incorporated. However, the erythro-analog was not an inhibitor of glycosylation (238). Interestingly enough, when the above two amino acid analogs (i.e., β-hydroxynorvaline and threo-β-fluoroasparagine) were used to inhibit protein synthesis, they also inhibited the incorporation of radioactive mannose into lipid-linked oligosaccharides as demonstrated with cycloheximide and puromycin (232). It seems likely that the toxicity of fluoroasparagine is due to its preventing the cells from glycosylating their N-linked glycoproteins in much the same manner as tunicamycin.

VII. Effects of Inhibitors of Polyprenol Synthesis on N-Linked Glycosylation

As indicated earlier, dolichyl-P serves as an obligatory carrier in the glycosylation of N-linked glycoproteins. Dolichyl-P is produced by the same initial pathway as that responsible for the synthesis of polyprenol units and, therefore, one would anticipate that inhibitors of polyprenol formation should also inhibit dolichyl-P formation. Furthermore, such inhibitors should also have a potent effect on the glycosylation of asparagine-linked glycoproteins. Since the formation of polyprenols utilizes the rate-limiting step catalyzed by the enzyme hydroxymethylglutaryl-Co A reductase (HMG CoA reductase), inhibitors of this enzyme such as 25-hydroxycholesterol and compactin (or mevanolin) should also inhibit glycosylation.

When smooth muscle cells, derived from aorta, are grown in the presence of 25-hydroxycholesterol, the incorporation of [^{14}C]acetate into cholesterol and into dolichyl-P is inhibited by up to 90%, as is N-linked glycosylation. However, mevalonate incorporation into these lipids is unaffected, as expected. Mevalonate, in fact, when added to inhibited cells, can overcome the inhibition with respect to N-linked glycosylation, presumably because it allows dolichyl-P to be formed (239). On the other hand, in mouse L cells, hydroxycholesterol also affected the synthesis of dolichol and cholesterol, but in this system, there were conditions that gave great fluctuations in cholesterol synthesis but only slight effects on dolichyl-P formation. The authors postulated that there were other points in the pathway that must control the levels of these two lipids (i.e., cholesterol and dolichol) (240). Cholestyramineis a compound that when fed to animals causes an increase in the activity of HMG CoA reductase and a great stimulation in the incorporation of acetate into cholesterol. However, it did not increase the incorporation of acetate into dolichyl-P (241). On the other hand, feeding animals a diet high in cholesterol would be expected to suppress cholesterol synthesis at the level of the HMG CoA reductase. However, this treatment caused a significant increase in the incorporation of mevalonate into dolichol and dolichol-linked oligosaccharides as well as increased activity of some of the glycosyltransferases that are required for glycoprotein synthesis (242). The explanation for these data is not clear, but perhaps inhibiting the cholesterol branch of the polyprenol pathway results in the accumulation of certain intermediates that are key players in other branches of the biosynthetic pathway.

Compactin is a fungal product that is a competitive inhibitor of HMG CoA reductase (243). Compactin, and a number of chemically synthesized derivatives of this structure (i.e., mevinolin), have been studied extensively because of their enormous chemotherapeutic potential in lowering serum cholesterol levels. In one study, when rat hepatocytes were cultured in compactin for 24 hr, the synthesis of dolichyl-P was decreased by 75–90%, suggesting that dolichyl-P was mainly synthesized by the *de novo* pathway and that the CTP-mediated phosphorylation of dolichol was of limited use. However, another study (244) indicated that the dolichol kinase and the dolichyl-P phosphatase played key roles in regulating the cellular levels of dolichyl-P. When compactin, at concentrations of 1–5 μM, was fed to sea urchin embryos, it induced abnormal gastrulation. This effect appeared to be due to an inhibition in dolichyl-P formation because these embryos also showed a decreased capacity to synthesize lipid-linked oligosaccharides and glycoproteins. Furthermore the inhibitory effect of compactin on development could be overcome by supplementing the embryos with dolichyl-P, but the addition of cholesterol or coenzyme Q could not overcome the inhibition (245). Similar results were obtained with mouse embryos, in which compactin and an oxygenated sterol arrested development at the 32-cell stage and left the blastomeres decompacted. Here also, N-glycosylation was effected and this affect could be overcome by adding mevalonate (246).

VIII. Inhibitors of Golgi Trafficking Affect N-Linked Glycosylation

During their synthesis and processing, the N-linked glycoproteins are transported from the initial site of synthesis in the endoplasmic reticulum of the cell through the various Golgi stacks (i.e., from cis-Golgi, through medial Golgi, to trans-Golgi) to their final destination in the plasma membranes, in lysosomes or in other locations in the cell. While the signals that direct these proteins to specific sites are known in some cases (e.g., the mannose-6-phosphate receptor for lysosomal targeting has been elegantly described (247)), in other cases the signals are just beginning to become known. Nevertheless, a number of compounds have been identified that perturb the transport or migration of proteins through the various compartments of the cell. Most of these pertubants are ionophores that affect the concentrations of various ions in some cellular compartments and may also affect the internal pH of these compartments. For example, the secretion of immunoglobulins by plasma cells was markedly inhibited by monensin, and striking alterations in the ultrastructural appearance of the Golgi apparatus were seen (248). The suggestion was made that depletion of Ca^{2+} levels by this ionophore rendered the Golgi vesicles incapable of fusing, which resulted in a disruption in the flow of glycoproteins from the ER to the Golgi (249).

The transport of other proteins, such as fibronectin and procollagen, is also inhibited by ionophores (250), as is the transport of various viral envelope glycoproteins (251). In these experiments, membrane vacuoles could be visualized by electron microscopy, and large accumulations of procollagen and fibronectin could be seen in these vacuoles by immunofluoresence. In one interesting study, monensin was used as a tool to isolate those cisternae that contained accumulations of viral capsids because they sedimented at a higher density that normal cisternae (252). In the presence of monensin, terminal glycosylation is inhibited, i.e., addition of complex sugars such as galactose, sialic acid, fucose, etc., but the incorporation of amino acids into protein and the hydroxylation of proline were not affected. Thus, in monensin-treated cells, newly synthesized proteins accumulate in intracellular vesicles that appear to be derived from the Golgi apparatus, and some of the post-translational modifications are blocked. Apparantly there exists an acidic compartment in the Golgi that is necessary for the secretory process (253).

A recently discovered and very useful antibiotic, called brefeldin A, also inhibits the secretion of glycoproteins in hepatocytes but does not affect protein synthesis. Brefeldin inhibited the terminal glycosylation of α_1-protease inhibitor and haptoglobulin, as well as the conversion of proalbumin to albumin. Both of these modifications appear to occur in the trans-Golgi (254). On the other hand, the proteolytic processing of haptoglobulin occurs in the endoplasmic reticulum and this modification was not affected by brefeldin (255).

The effect of brefeldin on the Golgi complex and endoplasmic reticululm of retrovirus-transformed murine erythroleukemia (MEL) cells was studied. Within 5 min of addition of this compound, most stacked cisternae were converted to vesicles that were scattered throughout the centrosphere region, and by 30 min, the Golgi complexes were completely disassembled. Over the next few hours, the number of vesicles in the Golgi area decreased concomitantly with an expansion of a predominantly smooth endoplasmic reticulum that consisted of a network of dilated tubules in continuity with regular RER cisternae, annulate lamellae, and the nuclear envelope. By electron microscopy, viral glycoproteins appeared to accumulate on these membranes and immature virions budded preferentially into the cisternal space. These findings suggest that brefeldin blocks the intracellular transport of newly synthesized cellular and viral proteins immediately distal to the distinct compartment of the ER. The effects of brefeldin were completely reversible and, within a short time after removal of the drug, Golgi cisternae bergan to reappear (256).

In a cell free system, brefeldin A prevented the assembly of non-clathrin-coated vesicles from Golgi cisternae. Furthermore, when coated vesicle assembly is blocked, extensive tubule networks form. These connect previously separate cisternae and stacks into a single topological unit that allows the mixing of the contents of Golgi cisternae (257). These studies are consistent with *in vivo* studies that showed that the earliest detectable effect of brefeldin was the displacement from the Golgi stacks of a 110-kDa protein (258, 259), and this protein was found to be identical to one of the major coat protein subunits (COPS) of Golgi non-clathrin-coated vesicles, β-COP (260). Thus, a simple explanation for the effect of this interesting and very useful compound is that it acts by blocking the budding of coated transport vesicles. Brefeldin A is now being widely used in cell biology and biochemistry and its use will undoubtedly provide important insight into mechanisms of protein transport and Golgi assembly and disassembly.

IX. Inhibitors of the Glycan Portion of Membrane Anchors

As indicated earlier, many membrane proteins are anchored to the cell surface by a glycosylphosphatidylinositol (GPI) anchor in which the protein is linked from its carboxyl-terminal region to the glycan via a phosphoethanolamine bridge. Since the glycan portion of the molecule is essential to this anchoring mechanism, inhibitors of glycan formation should be of considerable interest for determining the role of this anchoring mechanism, as well as for understanding the function of the anchored proteins. As discussed in a previous section, the mannosyl residues of the anchor are donated via dolichyl-P-mannose, so inhibitors that block the formation of this glycolipid should affect anchor formation. Thus, amphomycin was effective in cell-free experiments to demonstrate that the mannosyl donor for the three mannose units of the glycan portion of the anchor was indeed dolichyl-P-mannose (261). The sugar analog, 2-fluoro-2-deoxyglucose, which was previously shown to inhibit dolichyl-P-mannose formation and lipid-linked oligosaccharides (154), also altered the synthesis of the GPI-anchored protein alkaline phosphatase in JEG-3 cells and caused the accumulation of a proform of the enzyme. This proform was gradually processed into a secretory form by proteolytic removal of the carboxy-terminal hydrophobic peptide (262). Thus, both of these studies indicated that inhibition of the synthesis of dolichyl-P-mannose can block protein–anchor assembly.

Another recently reported inhibitor of protein–anchor formation is mannosamine. Mannosamine was previously shown to be an inhibitor of lipid-linked saccharide formation and to cause the production of truncated lipid-linked oligosaccharides in MDCK cells incubated with micromolar concentrations of this compound (156). When mannosamine was included in the growth medium of MDCK cells, it blocked the incorporation of glycosylphosphatidylinositol into GPI-anchored proteins. Thus, in polarized MDCK cell cultures, this amino-sugar markedely reduced the appearance at the cell surface of a recombinant GPI-anchored protein and converted this apically oriented membrane protein to an unpolarized secretory protein. It also blocked the surface appearance of other endogenous GPI-anchored proteins. In this system, mannosamine also inhibited the incorporation of [^{3}H]ethanolamine into GPI-anchored proteins and GPI intermediates, suggesting that this amino-sugar inhibits the assembly of the glycan portion of the anchor (263).

The effect of mannosamine on the biosynthesis of intermediates involved in anchor formation was also examined in MDCK cell cultures by incubating these cells with [2-^{3}H]mannose in the presence of various amounts of mannosamine. This amino-sugar inhibited the incorporation of mannose into GPI intermediates in a dose-dependent and reversible manner. Mannosamine was shown to be incorporated into the glycan as mannosamine and probably mostly in the second mannose position. Thus, it appears to block further addition of mannose because the next mannose is attached to the 2-hydroxyl of this second mannose. Various smaller sized intermediates were detected in cells incubated in the presence of this drug. As a result, ethanolamine-P could not be added, nor could the protein be transferred to this anchor. This inhibition was specific for mannosamine, and other amino-sugars such as galactosamine and trehalosamine were not inhibitory (264). Further studies with mannosamine or with other inhibitors as they become known should help us to understand the physiological significance of this anchoring mechanism.

X. Summary and Future Directions

Inhibitors are certainly not a panacea for all biochemical studies, and as indicated earlier there are potential problems in their utilization, such as lack of specificity or interference with other key reactions. Nevertheless, they can provide valuable insights into understanding specific steps in a biochemical pathway as well as in determining the physiological function of various biochemical molecules. For example, in terms of the biosynthesis of the N-linked oligosaccharides, inhibitors of the lipid-linked oligosaccharide pathway, such as amphomycin, have helped to establish that the mannosyl donor for the last four mannose residues of the $Glc_3Man_9(GlcNAc)_2$-PP-dolichol was dolichyl-P-mannose, rather than GDP–mannose. Other inhibitors of this pathway, such as tunicamycin, have provided evidence to suggest that the carbohydrate portion of the asparagine-linked glycoproteins may play a number of different roles, including enhancing the stability of the glycoprotein, affecting the conformation of the protein, providing a recognition signal for targeting the protein to specific sites in the cell or to uptake from the circulation, providing a receptor molecule for cell–cell adhesion or interactions and so on. The actual role of the carbohydrate depends on the amino acid sequence of the protein in question.

The second group of important inhibitors are those that affect glycosidases. These types of enzymes are widespread in nature and are involved in many natural processes such as cell division and cell separation, turnover of macromolecules, N-linked glycoprotein processing, and so on. Thus, inhibitors of these enzymes have great potential for use as biochemical tools or even as therapeutic agents, or they may also be lethal as shown for the glucosidase and mannosidase inhibitors castanospermine and swainsonine. Again these latter compounds have gained widespread use for studies on the role of carbohydrate in N-linked glycoprotein function.

However, all of these studies point out the need for additional inhibitors as well as for structurally different inhibitors. That is, the studies with glycosidase inhibitors have shown that even the best of these compounds is still somewhat nonspecific and inhibits more than one enzyme. For example, castanospermine inhibits the processing enzymes glucosidase I and glucosidase II, as well as lysosomal α-glucosidase, and other glucosidases. Swainsonine, which is somewhat more specific, still inhibits mannosidase II and lysosomal α-mannosidase. Can we find, or synthesize, more specific inhibitors that act only on one enzyme? Of course, the answer to that question is not known at this time since the mechanism of catalysis has not been demonstrated for most of these enzymes. However, even if some of these enzymes share a common catalytic mechanism, it should still be possible to narrow the site of action of these inhibitors to groups of similarly acting enzymes. Since there are now several inhibitors that appear to distinguish between glucosidase I and glucosidase II, these two enzymes must differ in their catalytic mechanism or in their active site. Thus, such studies provide hope that more specificity can be obtained with the right inhibitors.

How can we find better inhibitors? It seems that there are a number of approaches that could be used. One of the most logical, of course, would be to work out and understand the catalytic mechanism of each of the enzymes to be inhibited and then design inhibitors based on this knowledge. While this approach is certainly possible, it is not likely to be feasible in the immediate future. Most of the enzymes involved in the biosynthesis of N-linked oligosaccharides or membrane anchors appear to be present in cells in relatively low abundance, and therefore even the purification of these enzymes to homogeneity is difficult. Actually, several of these proteins have now been purified to homogeneity, and perhaps with modern approaches of molecular biology, they can be cloned and produced in sufficient amounts for crystallization.

A second approach is to design new inhibitors based on the structures of already known prototypes. In this regard, various laboratories are currently conducting modeling studies with the known inhibitors to provide the basis for developing analogs of already-known compounds. While this is a viable and logical way to develop new inhibitors, it does have one major shortcoming, which is that this approach cannot provide completely new and novel structures that could lead to even better inhibitors than those that we already have available. In that respect, perhaps we need to take a lesson from nature since evolution has had millions of years to develop all sorts of unusual chemical structures. It is unlikely that penicillin, streptomycin, or bleomycin, to name a few, could have been chemically synthesized *de novo* without first having been isolated and characterized from natural sources. Thus, the third approach to better inhibitors that should not be ignored is to look for new inhibitory compounds from natural sources, because this seems like the most likely way to find new and previously unknown structures.

REFERENCES

1. Wasserman, P. M. (1987). Early events in mammalian fertilization. *Annu. Rev. Cell. Biol.* **3,** 109–140.
2. Kornfeld, S., and Mellman, I. (1989). The biogenesis of lysosomes. *Annu. Rev. Cell Biol.* **5,** 483–525.
3. Ashwell, G., and Harford, J. (1982). Carbohydrate specific receptors of the liver. *Annu. Rev. Biochem.* **51,** 531–554.
4. Roth, S. (1973). A molecular model for cell interactions. *Q. Rev. Biol.* **48,** 541–563.
5. Rosen, S. D. (1989). Lymphocyte homing: Progress and prospects. *Curr. Opin. Cell Biol.* **1,** 913–919.
6. Firon, N., Ofek, I., and Sharon, N. (1983). Carbohydrate specificity of the surface lectins of *E. coli, K. pneumoniae,* and *S. typhimurium. Carbohydr. Res.* **120,** 235–242.
7. Feizi, T., and Child, R. A. (1987). Carbohydrates as antigenic determinants of glycoproteins. *Biochem. J.* **245,** 1–11.
8. Stanley, P. (1992). Glycosylation engineering. *Glycobiology* **2,** 99–107.
9 Rademacher, T. W., Parekh, R. B., and Dwek, R. A. (1988). Glycobiology. *Annu. Rev. Biochem.* **57,** 785–838.
10. Kornfeld, R., and Kornfeld, S. (1976). Comparative aspects of glycoproteins structure. *Annu. Rev. Biochem.* **45,** 217–237.
11. Kobata, A. (1984). The carbohydrates of glycoproteins. *In* "Biology of Carbohydrates" (V. Ginsburg and P. W. Robbins, eds.), pp. 87–161. Wiley, New York.
12. Carraway, K. L., and Hill, S. R. (1991). Cell surface mucin-type glycoproteins and mucin-like domains. *Glycobiology* **1,** 131–138.
13. Hakomori, S. (1989). Aberrant glycosylation in tumors and tumor-associated carbohydrate antigens. *Adv. Cancer Res.* **52,** 257–331.
14. Karlsson, K. (1989). Animal glycosphingolipids as membrane attachment sites for bacteria. *Annu. Rev. Biochem.* **58,** 309–350.
15. Cross, G. A. M. (1990). Glycolipid anchoring of plasma membrane proteins. *Annu. Rev. Cell Biol.* **6,** 1–39.
16. Ferguson, M. A. J., and Williams, A. F. (1988). Cell surface anchoring of proteins via glycosylphosphatidylinositol structures. *Annu. Rev. Biochem.* **57,** 285–320.
17. Holt, G. D. (1991). Identifying glycoconugate binding domains: Building on the past. *Glycobiology* **1,** 329–336.
18. Maley, F., Trimble, R. B., Tarentino, A. L., and Plummer, T. H. (1989). Characterization of glycoproteins and their associated oligosaccharides through the use of endoglycosidases. *Anal. Biochem.* **180,** 195–204.
19. Alexander, S., and Elder, J. W. (1989). Endoglycosidases from *Flavobacterium meningosepticum* and application to biological problems. *In* "Methods in Enzymology" (V. Ginsburg, ed.), Vol. 179, pp. 505–520. Academic Press, San Diego.
20. Tamura, G. (ed.) (1982). "Tunicamycins." Japan Scientific Society Press, Tokyo.
21. Tkacz, J. S., and Lampen, J. P. (1975). Tunicamycin inhibition of polyisoprenyl *N*-acetylglucosaminyl pyrophosphate formation in calf liver microsomes. *Biochem. Biophys. Res. Commun.* **65,** 248–253.

22. Keenan, R. W., Hamill, R. L., Occolowitz, J. L., and Elbein, A. D. (1981). Biological activities of isolated tunicamycin and streptovirudin fractions. *Biochemistry* **70,** 2968–2973.
23. Santos-Agudo, J. Biro, P. A., Fukrmann, U., Strominger, J. A., and Barbosa, J. A. (1987). Amino acid sequences in the α_1 domain and not glycosylation are important in IILA-A2/βZ-microglobulin association and cell surface association. *Mol. Cell Biol.* **7,** 982–990.
24. Grinnell, B. W., Walls, J. D., and Gerlitz, B. (1991). Glycosylation of human protein C affects its secretion, processing, functional activities and activation by thrombin. *J. Biol. Chem.* **226,** 9778–9785.
25. Stanley, P. (1984). Glycosylation mutants of animal cells. *Annu. Rev. Genet.* **18,** 525–552.
26. Stoll, J., Robbins, A. R., and Krag, S. (1982). A mutant of Chinese hamster ovary cells with altered mannose-6-P receptor activity is unable to synthesize mannosylphosphoryl-dolichol. *Proc. Natl. Acad. Sci. USA* **79,** 2296–2300.
27. Wagh, P., and Bahl, O. (1981). Sugar residues on proteins. *Crit. Rev. Biochem.* **10,** 307–377.
28. Hubbard, S. C., and Ivatt, R. (1981). Synthesis and processing of asparagine-linked oligosaccharides. *Annu. Rev. Biochem.* **50,** 555–583.
29. Chambers, J., and Elbein, A. D. (1975). Glycoprotein synthesis in aorta: Biosynthesis and characterization of lipid-linked sugars and oligosaccharides. *J. Biol. Chem.* **250,** 6904–6913.
30. Elbein, A. D. (1979). The role of lipid-linked saccharides in the biosynthesis of glycoproteins. *Annu. Rev. Plant Physiol.* **30,** 239–260.
31. Kaushal, G. P., and Elbein, A. D. (1985). Purification and properties of UDP–GlcNAc:dolichol-P GlcNAc-1-P-transferase. *J. Biol. Chem.* **260,** 16303–16309.
32. Keller, R. K., Boon, D. Y., and Crum, F. C. (1979). *N*-Acetylglucosamine-1-P transferase from hen oviduct: Solubilization, characterization and inhibition by tunicamycin. *Biochemistry* **18,** 3946–3952.
33. Shailubhai, K., Dong-Yu, B., Saxena, E. S., and Vijay, I. K. (1988). Purification and characterization of UDP–GlcNAc:dolichyl-P GlcNAc-1-P transferase involved in the biosynthesis of asparagine-linked glycoproteins in mammary gland. *J. Biol. Chem.* **263,** 15964–15972.
34. Lehrman, M. A. (1991). Biosynthesis of GlcNAc-PP-dolichol, the committed step of asparagine-linked oligosaccharide assembly. *Glycobiology* **1,** 553–562.
35. Kaushal, G. P., and Elbein, A. D. (1992). Purification and properties of the GlcNAc transferase that adds the second GlcNAc to GlcNAc–GlcNAc-PP-dolichol. Manuscript in preparation.
36. Chapman, A., Fujimoto, K., and Kornfeld, S. (1980). The primary glycosylation defect in class E thy-1-negative mutant mouse lymphoma cells is an inability to synthesise dolichol-P-mannose. *J. Biol. Chem.* **255,** 4441–4446.
37. Kang, M. S., Spencer, J. P., and Elbein, A. D. (1978). Amphomycin inhibits mannose incorporation into lipid-linked saccharides. *Biochem. Biophys. Res. Commun.* **82,** 568–575.
38. Kang, M. S., Spencer, J. P., and Elbein, A. D. (1978). Amphomycin inhibition of mannose and GlcNAc incorporation into lipid-linked saccharides. *J. Biol. Chem.* **253,** 8860–8868.
39. Kaushal, G. P., and Elbein, A. D. (1992). Purification and properties of β-mannosyl transferase that produces Manβ1,4GlcNAcβ1, 4GlcNAc-PP-dolichol. Manuscript in preparation.

40. Jensen, J. W., and Schutzbach, J. S. (1981). The biosynthesis of oligosaccharide-lipids: Partial purification and characterization of mannosyltransferase II. *J. Biol. Chem.* **256,** 12899–12904.
41. Schutzbach, J. S., Springfield, J. D., and Jensen, J. W. (1980). The biosynthesis of oligosaccharide lipids: Formation of an α1,2-mannosyl–mannose linkage. *J. Biol. Chem.* **255,** 4170–4175.
42. Prakosh, C., Katial, A., Kang, M. S., and Vijay, I. K. (1984). Solubilization of mannosyltransferase activities for the biosynthesis of mammary glycoproteins. *Eur. J. Biochem.* **139,** 87–93.
43. Spencer, J. P., and Elbein, A. D. (1981). Transfer of mannose from GDP–mannose to lipid-linked oligosaccharides by soluble mannosyltransferase. *Proc. Natl. Acad. Sci. USA* **77,** 2524–2529.
44. Kornfeld, R., and Kornfeld, S. (1985). Assembly of asparagine-linked oligosaccharides. *Annu. Rev. Biochem.* **54,** 631–664.
45. Behrens, N. H., Parodi, A. J., and Leloir, L. F. (1971). Glucose transfer from dolichol monophosphate glucose: The product formed with endogenous microsomal acceptor. *Proc. Natl. Acad. Sci. USA* **68,** 2857–2860.
46. Sharma, C. B., Kaushal, G. P., Pan, Y. T., and Elbein, A. D. (1990). Purification and characterization of dolichyl-P-mannose:$Man_5(GlcNAc)_2$-PP-dolichol mannosyltransferase. *Biochemistry* **29,** 8901–8907.
47. D'Souza, C. M., Sharma, C. B., and Elbein, A. D. (1992). Purification and properties of the first glucosyltransferase that forms $Glc_1Man_9(GlcNAc)_2$-PP-dolichol from dolichyl-P-glucose and $Man_9(GlcNAc)_2$-PP-dolichol. Manuscript in preparation.
48. Turco, S. J., Stetson, B., and Robbins, P. W. (1977). Comparative rates of transfer of lipid-linked oligosaccharides to endogenous glycoprotein acceptors *in vitro*. *Proc. Natl. Acad. Sci. USA* **74,** 4411–4416.
49. Spiro, M. J., Spiro, R. G., and Bhoyroo, V. D. (1979). Glycosylation of proteins by oligosaccharide-lipids: Studies on a thyroid enzyme involved in oligosaccharide transfer and the role of glucose in this reaction. *J. Biol. Chem.* **254,** 7668–7679.
50. Rothman, J. E., and Lodish, H. F. (1977). Synchronized transmembrane insertion and glycosylation of a nascent membrane protein. *Nature* **269,** 775–780.
51. Kelleher, D. J., Krebach, G., and Gilmore, R. (1992). Oligosaccharyl-transferase activity is associated with a protein complex composed of ribophorins I and II and a 48 kd protein. *Cell* **69,** 55–65.
52. Elting, J. J., Chen, W. C., and Lennarz, W. J. (1980). Characterization of a glucosidase involved in an initial step in the processing of oligosaccharide chains. *J. Biol. Chem.* **255,** 2325–2331.
53. Grinna, L. S., and Robbins, P. W. (1980). Substrate specificities of rat liver microsomal glucosidases which process glycoproteins. *J. Biol. Chem.* **255,** 2255–2258.
54. Hettkamp, H., Legler, G., and Bause, E. (1984). Purification by affinity chromatography of glucosidase I, an ER hydrolase involved in processing of asparagine-linked oligosaccharides. *Eur. J. Biochem.* **142,** 85–90.
55. Bause, E., Schweden, J., Gross, A., and Orthen, B. (1989). Purification and characterization of trimming glucosidase I from pig liver. *Eur. J. Biochem.* **183,** 661–669.
56. Shailubhai, K., Pratta, M. A., and Vijay, I. K. (1987). Purification and characterization of glucosidase I involved in N-linked glycoprotein processing in bovine mammary gland. *Biochem. J.* **247,** 555–562.
57. Szumilo, T., Kaushal, G. P., and Elbein, A. D. (1986). Purification and properties of glucosidase I from mung bean seedlings. *Arch. Biochem. Biophys.* **247,** 261–271.
58. Bause, E., Schweden, J., and Jaenicke, L. (1986). Purification and characterization of trimming glucosidase I from *Saccharomyces cerevesiae*. *FEBS Lett.* **206,** 308–312.

59. Esmon, B., Esmon, P., and Scheckman, R. (1984). Early steps in processing yeast glycoproteins. *J. Biol. Chem.* **259,** 10322–10327.
60. Stanley, P., Sallustro, S., Krag, S., and Dunn, B. (1990). Lectin resistant CHO cell: Selection of 7 new mutants resistant to ricin. *Somat. Cell Mol. Genet.* **16,** 211–233.
61. Grinna, L., and Robbins, P. W. (1979). Glycoprotein biosynthesis: Rat liver microsomal glucosidases which process oligosaccharides. *J. Biol. Chem.* **254,** 8814–8818.
62. Michael, J. M., and Kornfeld, S. (1980). Partial purification and characterization of the glucosidase involved in the processing of asparagine-linked oligosaccharides. *Arch. Biochem. Biophys.* **199,** 249–258.
63. Saunier, B., Kilker, D., Tkacz, J. S., Quaroni, A., and Herscovics, A. (1982). Inhibition of N-linked complex oligosaccharide formation by deoxynojirimycin, an inhibitor of processing glucosidases. *J. Biol. Chem.* **257,** 14155–14161.
64. Kaushal, G. P., Pastuszak, I., Hatanaka, K., and Elbein, A. D. (1990). Purification to homogeneity and properties of glucosidase II from mung bean seedlings and suspension cultured soybean cells. *J. Biol. Chem.* **265,** 16271–16279.
65. Brada, D., Kerjaschki, P., and Roth, J. (1990). Cell type-specific post-Golgi apparatus localization of a "resident" ER glycoprotein, glucosidase II. *J. Cell Biol.* **110,** 309–318.
66. Burns, D. M., and Touster, O. (1982). Purification and characterization of glucosidase II, an endoplasmic reticulum hydrolase involved in glycoprotein biosynthesis. *J. Biol. Chem.* **257,** 9991–10,000.
67. Lucocq, J., Brada, D., and Roth, J. (1986). Immunolocalization of the oligosaccharide trimming enzyme, glucosidase II. *J. Cell Biol.* **102,** 2137–2146.
68. Szumilo, T., and Elbein, A. D. (1985). A simple and reliable assay for glycoprotein processing glucosidases. *Anal. Biochem.* **151,** 32–40.
69. Chester, M., Lundblad, A., and Masson, P. (1975). Relationship between different forms of human α-mannosidase. *Biochem. Biophys. Acta* **391,** 341–348.
70. Trimble, R. B., Moremen, K., and Herscovics, A. (1992). Glycosidases of the endoplasmic reticulum and Golgi apparatus. *In* "Guidebook to the Secretory Pathway" (T. Stevens, P. Novick, and J. Kornblatt, eds.). Sambrook & Tooze, Dallas.
71. Jelinek-Kelly, S., and Herscovics, A. (1988). Glycoprotein biosynthesis in *Saccharomyces cerevesiae:* Purification of the mannosidase which removes one specific mannose residue from $Man_9(GlcNAc)_2$. *J. Biol. Chem.* **263,** 14757–14763.
72. Camirand, A., Heysen, A., Grondin, B., and Herscovics, A. (1991). Glycoprotein biosynthesis in *Saccharomyces cerevesiae:* Isolation and characterization of the gene encoding a specific processing α-mannosidase. *J. Biol. Chem.* **266,** 15120–15127.
73. Forsee, W. T., and Schutzbach, J. S. (1981). Purification and characterization of a phospholipid dependent α-mannosidase from rabbit liver. *J. Biol. Chem.* **256,** 6577–6582.
74. Forsee, W. T., Palmer, C. F., and Schutzbach, J. S. (1989). Purification and characterization of α1,2-mannosidase involved in processing asparagine-linked oligosaccharides. *J. Biol. Chem.* **264,** 3869–3876.
75. Schweden, J., Legler, G., and Bause, E. (1986). Purification and characterization of a neutral processing mannosidase from calf liver acting on $Man_9(GlcNAc)_2$–oligosaccharides. *Eur. J. Biochem.* **157,** 563–570.
76. Schweden, J., and Bause, E. (1989). Characterization of trimming Man_9–mannosidase from pig liver. *Biochem. J.* **264,** 347–355.
77. Roth, J. Brada, D., Lackee, P., Schweden, J., and Bause, E. (1990). Oligosaccharide trimming Man_9–mannosidase is a resident ER protein and exhibits a more restricted local distribution than glucosidase II. *Eur. J. Cell Biol.* **53,** 131–141.

78. Tabas, I., and Kornfeld, S. (1979). Purification and characterization of rat liver Golgi α-mannosidase capable of processing asparagine linked oligosaccharides. *J. Biol. Chem.* **254,** 11655–11663.
79. Tulsiani, D. R. P., Hubbard, S. C., Robbins, P. N., and Touster, O. (1982). α-D-Mannosidases of rat liver Golgi membranes. *J. Biol. Chem.* **257,** 3660–3668.
80. Tulsiani, D. R. P., and Touster, O. (1988). The purification and characterization of mannosidase IA from rat liver Golgi membranes. *J. Biol. Chem.* **263,** 5408–5417.
81. Oppenheimer, C. L., and Hill, R. L. (1981). Purification and characterization of a rabbit liver α1,3 mannoside β2-*N*-acetyl-glucosaminyltransferase. *J. Biol. Chem.* **256,** 799–804.
82. Bendiak, B., and Schachter, H. (1986). Control of glycoprotein synthesis: Purification of UDP–GlcNAc α-D-mannoside β1,2-GlcNAc transferase II from rat liver. *J. Biol. Chem.* **262,** 5775–5183.
83. Sarkar, M., Hall, E., Nishikawa, Y., Simpson, R. J., Moritz, R. L., Dunn, R., and Schachter, H. (1991). Molecular cloning and expression of cDNA encoding the enzyme that controls conversion of high-mannose to hybrid and complex N-glycans: UDP–*N*-acetylglucosamine: α-3-D-mannoside β-1 2-*N*-acetylglucosaminyltransferase I. *Proc. Natl. Acad. Sci. USA* **88,** 234–238.
84. Dewald, B., and Touster, O. (1973). A new α-D-mannosidase occurring in Golgi membranes. *J. Biol. Chem.* **248,** 7223–7233.
85. Moreman, K. W., and Touster, O. (1985). Biosynthesis and modification of Golgi mannosidase II in Hela and 3T3 Cells. *J. Biol. Chem.* **260,** 6654–6662.
86. Tulsiani, D. R. P., Opheim, D. J., and Touster, O. (1977). Purification and characterization of α-D-mannosidase from rat liver Golgi mambranes. *J. Biol. Chem.* **252,** 3227–3233.
87. Kaushal, G. P., Szumilo, T., Pastuzak, I., and Elbein, A. D. (1990). Purification to homogeneity and properties of mannosidase II from mung bean seedlings. *Biochemistry* **29,** 2168–2176.
88. Moreman, K., and Robbins, P. W. (1991). Isolation, characterization and expression of cDNA encoding murine α-mannosidase II, a Golgi enzyme that controls conversion of high-mannose to complex N-glycans. *J. Cell Biol.* **115,** 1521–1534.
89. Marsh, C. A., and Gourlay, G. C. (1971). Evidence for a non-lysosomal α-mannosidase in rat liver homogenates. *Biochim. Biophys. Acta* **235,** 142–148.
90. Tulsiani, D. R. P., and Touster, O. (1985). Characterization of a novel α-D-mannosidase from rat brain microsomes. *J. Biol. Chem.* **260,** 13081–13087.
91. Tulsiani, D. R. P., Skudlarek, M. D., and Orgebin-Crist, M. C. (1989). Novel α-D-mannosidase of rat sperm plasma membranes: Characterization and potential role in sperm–egg interactions. *J. Cell. Biol.* **109,** 1257–1267.
92. Monis, E., Bonay, P., and Hughes, R. C. (1987). Characterization of a mannosidase acting on α1,3 and α1,6-linked mannose residues of oligomannosidic intermediates of glycoprotein processing. *Eur. J. Biochem.* **168,** 287–294.
93. Lubas, W. A., and Spiro, R. G. (1987). Golgi endo-α-D-mannosidase from rat liver, a novel N-linked carbohydrate unit processing enzyme. *J. Biol. Chem.* **262,** 3775–3781.
94. Bischoff, J., and Kornfeld, R. (1986). The soluble form of rat liver α-mannosidase is immunologically related to the ER membrane α-mannosidaase. *J. Biol. Chem.* **258,** 7907–7910.
95. Beyer, T. A., Sadler, J. E., Rearick, J. I., Paulson, J. C., and Hill, R. W. (1981). Glycosyltransferases and their use in assessing oligosaccharide structure and structure–function relationships. *Adv. Enzymol.* **52,** 23–175.

96. Thomas, J. R., Dwek, R. A., and Rademacher, T. W. (1990). Structure, biosynthesis and function of glycosylphosphatidylinositols. *Biochemistry* **29,** 5413–5422.
97. Menon, A. K., Mayor, S., Ferguson, S. S., Duszenko, M., and Cross, G. (1988). Candidate glycophospholipid precussion for GPI membrane anchor of *Trypanosoma brucei* variant surface glycoprotein. *J. Biol. Chem.* **263,** 1970–1977.
98. Hirose, S., Mohney, R. P., Mulka, S., Lakshmesware, R., Singleton, D., Perry, G., Tartakoff, A. M., and Medof, E. (1992). Derivation and characterization of GPI anchor-defective human K562 cell clones. *J. Biol. Chem.* **267,** 5272–5278.
99. Doering, T. L., Masterson, W. J., Englund, P. J., and Hart, G. W. (1989). Biosynthesis of the GPI membrane anchor of the trypanosome variant surface glycoprotein. *J. Biol. Chem.* **264,** 11168–11173.
100. Urakaze, M., Kamitani, T., DeGespari, R., Sugiyama, E., Chang, W-M., Warren, C. D., and Yeh, E. T. H. (1992). Identification of a missing link in GPI anchor biosynthesis in mammalian cells. *J. Biol. Chem.* **267,** 6459–6462.
101. Caras, I. W. (1991). An internally positioned signal can direct attachment of a GPI membrane anchor. *J. Cell Biol.* **113,** 77–85.
102. Takatsuki, A., Arima, K., and Tamura, G. (1971). Tunicamycin, a new antibiotic, I. Isolation and characterization of tunicamycin. *J. Antibiot.* **24,** 215–223.
103. Ito, T., Kodama, Y., Kawamura, K., Suzuki, K., Takasuki, A., and Tamura, G. (1977). The structure of tunicaminyl uracil, a degradation product of tunicamycin. *Agric. Biol. Chem.* **41,** 2302–2305.
104. Mahoney, W. C., and Duskin, D. (1980). Separation of tunicamycin homologues by reversed high performance liquid chromatography. *J. Chromatogr.* **198,** 506–510.
105. Heifetz, A., Keenan, R. W., and Elbein, A. D. (1979). Mechanism of action of tunicamycin on the UDP–GlcNAc:dolichyl-P GlcNAc-1-P transferase. *Biochemistry* **18,** 2186–2192.
106. Ericson, M. C., Gafford, J., and Elbein, A. D. (1977). Tunicamycin inhibits GlcNAc lipid formation in plants. *J. Biol. Chem.* **252,** 7431–7433.
107. Struck, D. K., and Lennarz, W. J. (1977). Evidence for the participation of saccharide-lipids in the synthesis of the oligosaccharide chain of ovalbumin. *J. Biol. Chem.* **252,** 1007–1013.
108. Takatsuki, A., and Tamura, G. (1982). Inhibition of glycoconjugate biosynthesis by tunicamycin. *In* "Tunicamycin" (G. Tamura, ed.), pp. 35–67. Japan Scientific Society Press, Tokyo.
109. Bettinger, G. E., and Young, F. E. (1975). Tunicamycin, an inhibitor of *Bacillus* peptidoglycan synthesis, a new site of inhibition. *Biochem. Biophys. Res. Commun.* **67,** 16–21.
110. Bracha, R., and Glaser, L. (1976). An intermediate in teichoic acid biosynthesis. *Biochem. Biophys. Res. Commun.* **72,** 1098–1103.
111. Kaushal, G. P., and Elbein, A. D. (1986). Purification and properties of the UDP–GlcNAc:dolichyl-pyrophosphoryl-GlcNAc GlcNAc transferase from mung bean seedlings. *Plant Physiol.* **81,** 1086–1091.
112. Reitman, N. L., and Kornfeld, S. (1981). UDP–*N*-acetylglucosamine:glycoprotein *N*-acetylglucosamine-1-phosphotransferase: Proposed enzyme for the phosphorylation of the high mannose oligosaccharide units of lysosomal enzyme. *J. Biol. Chem.* **256,** 4275–4281.
113. Hart, G. W., Haltiwanger, R. S., Holt, G. D., and Kelly, W. G. (1989). Glycosylation in the nucleus and cytoplasm. *Annu. Rev. Biochem.* **58,** 841–874.
114. Ward, J. B., Wyke, A. W., and Curtis, C. (1980). The effect of tunicamycin on wall-polymer synthesis in *Bacilli*. *Biochem. Soc. Trans.* **8,** 164–166.

115. Elbein, A. D. and Gafford, J., and Kang, M. S. (1979). Inhibition of lipid-linked saccharide synthesis: Comparison of tunicamycin, streptovirudin, and antibiotic 24010. *Arch. Biochem. Biophys.* **196,** 311–318.
116. Elbein, A. D. (1988). The tunicamycins-useful tools for studies on glycoproteins. *Trends Biochem. Sci.* 219–221 (August 1981).
117. Lehrman, M. A., Zhu, X., and Khounlo, S. (1988). Amplification and molecular cloning of the hamster tunicamycin sensitive GlcNAc-1-P transferase gene: The hamster and yeast enzymes share a common peptide sequence. *J. Biol. Chem.* **263,** 19796–19803.
118. Struck, D. K., Siuta, P. R., Lane, M. D., and Lennarz, W. J. (1978). Effect of tunicamycin on the secretion of serum proteins by primary cultures of rat and chick hepatocytes. *J. Biol. Chem.* **253,** 5332–5337.
119. Elbein, A. D. (1991). Inhibitors of the addition, modification or processing of the oligosaccharide chains of the N-linked glycoproteins. *In* "Biology of Carbohydrates" (V. Ginsburg and P. W. Robbins, eds.), Vol. 3, pp. 117–119. JA1 Press, London.
120. Miller, A. L., Kress, B. C., Lewis, L., Stern, R., and Kinnon, C. (1980). Effect of tunicamycin and cycloheximide on the secretion of acid hydrolases from I-cell cultured fibroblasts. *Biochem. J.* **186,** 971–975.
121. Hickman, S., Kulczyki, A., Jr., Lynch, R. G., and Kornfeld, S. (1977). Studies of the mechanism of tunicamycin inhibition of IgA and IgE secretion by plasma cells. *J. Biol. Chem.* **252,** 4402–4408.
122. Sidman, C. (1981). Differing requirement for glycosylation in the secretion of related glycoproteins is determined neither by the producing cell nor by the relative number of oligosaccharide units. *J. Biol. Chem.* **256,** 9374–9376.
123. Prives, J. M., and Olden, K. (1980). Carbohydrate requirement for expression and stability of acetylcholine receptor on the surface of embryonic muscle cells in culture. *Proc. Natl. Acad. Sci. USA* **77,** 5263–5267.
124. Slieker, C. J., and Lane, M. D. (1985). Post-transtional processing of the epidermal growth factor receptor. *J. Biol. Chem.* **260,** 687–690.
125. Chattergee, S., Kwiterovich, P. O., Jr., and Sekerke, C. S. (1979). Effects of tunicamycin on the binding and degradation of low density lipoproteins and glycoprotein synthesis in cultured human fibroblasts. *J. Biol. Chem.* **254,** 3704–3707.
126. Baribault, T., and Neet, K. (1985). Effects of tunicamycin on NGF binding and neurite outgrowth in PC12 cells. *J. Neurosci. Res.* **14,** 49–60.
127. Ronnett, G. V., and Lane, M. D. (1981). Post-translational glycosylation induced activation of aglycoinsulin receptor accumulated during tunicamycin treatment. *J. Biol. Chem.* **256,** 4704–4707.
128. Snider, M. D., and Rogers, O. (1985). Intracellular movement of cell surface receptors after endocytosis: Resialylation of asialo-transferase receptor in human erythroleukemia cells. *J. Cell Biol.* **160,** 826–834.
129. Weintraub, B. D., Stannard, B. S., and Meyers, L. (1983). Glycosylation of thyroid-stimulating hormone in pituitary tumor cells: Influence of high mannose oligosaccharides units on subunit aggregation. *Endocrinology* **112,** 1331–1335.
130. Schwarz, R. T., Rohrschneider, J. M., and Schmidt, M. F. G. (1976). Suppression of glycoprotein formation of semliki forest, influenza, and avian sarcoma virus by tunicamycin. *J. Virol.* **19,** 782–791.
131. Gibson, R., Schlesinger, S., and Kornfeld, S. (1979). The nonglycosylated glycoprotein of vesicular stomatitis virus is temperature-sensitive and undergoes intracellular aggregation at elevated temperatures. *J. Biol. Chem.* **254,** 3600–3607.
132. Elbein, A. D., Occolowitz, J. L., Hamill, R. L., and Eckardt, K. (1981). Streptovirud-

ins of series I and II: Chemical and biological properties. *Biochemistry* **20,** 4210–4216.
133. Tkacz, J., and Wong, A. (1978). Comparison of the biological and chemical properties of tunicamycin and mycospocidin. *Fed. Proc.* **37,** 1766.
134. Murazumi, N., Yamamori, S., Araki, Y., and Ito, E. (1979). Anomeric configuration of *N*-acetylglucosaminyl phosphorylundecaprenols formed in *Bacillus cereus* membranes. *J. Biol. Chem.* **254,** 11791–11793.
135. Kenig M., and Reading C. (1979). Holomycin and an antibiotic (MM 19290) related to tunicamycin. *J. Antibiotics* **32,** 549–554.
136. Vogel, P., Stynes, B. A., Cookley, D., Yoeh, G., and Patterson, D. (1982). Glycolipid toxins from parasitised annual ryegrass: A comparison with tunicamycin. *Biochem. Biophys. Res. Commun.* **105,** 835–840.
137. Tanaka, H., Owia, R., Matsukura, A., Inokoshi, J., and Omura, S. (1987). Studies on bacterial cell wall inhibitors. *J. Antibiot.* **35,** 1216–1221.
138. Wedgewood, J., and Strominger, J. L. (1980). Enzymatic activities in cultured human lymphocytes that dephosphorylate dolichyl pyrophosphate and dolichyl phosphate. *J. Biol. Chem.* **255,** 1120–1123.
139. Bodansky, M., Sigler, G., and Bodansky, A. (1973). Structure of the peptide antibiotic amphomycin. *J. Ann. Chem. Soc.* **95,** 2352–2357.
140. Hayes, G. R., and Lucas, J. (1983). Stimulation of lipid-linked oligosaccharide assembly during oviduct differentiation. *J. Biol. Chem.* **258,** 15095–15100.
141. Kyosseva, S. V., and Zhivkov, V. (1985). Biosynthesis of lipid-linked oligosaccharides in embryonic liver: Formation of mannose containing derivatives. *Int. J. Biochem.* **17,** 813–817.
142. Banerjee, D. K., Scher, M., and Waechter, C. J. (1981). Amphomycin: Effect of the lipopeptide antibiotic and the glycosylation and extraction of dolichyl monophosphate in calf brain membranes. *Biochemistry* **20,** 1561–1568.
143. Kean, E. (1985). Stimulation by dolichol phosphate mannose and phospholipids of the biosynthesis of GlcNAc-PP-dolichol. *J. Biol. Chem.* **260,** 12561–12571.
144. Haselback, A., and Tanner, W. (1982). Dolichyl phosphate-mediated mannosyl transfer through liposomal membranes. *Proc. Natl. Acad. Sci. USA* **79,** 1520–1524.
145. Elbein, A. D. (1981). The effect of tsushimycin on the synthesis of lipid-linked saccharides in aorta. *Biochem. J.* **193,** 477–484.
146. Spencer, J. P., Kang, M. S., and Elbein, A. D. (1978). Inhibition of lipid-linked saccharide synthesis by bacitracin. *Arch. Biochem. Biophys.* **190,** 829–832.
147. Babczinski, P. (1972). Inhibition of lipid-linked mannose and mannoprotein synthesis in yeast by diumycin *in vitro*. *Eur. J. Biochem.* **112,** 53–58.
148. Villemez, C. L., and Carlo, P. L. (1980). Properties of a soluble polyprenyl phosphate: UDP–D-*N*-acetylglucosamine *N*-acetylglucosamine-1-phosphate transferase. *J. Biol. Chem.* **255,** 8174–8178.
149. Bause, E., and Legler, G. (1982). The effect of flavomycin on the synthesis and transfer of lipid-linked saccharides in pig brain. *Biochem. J.* **201,** 481–487.
150. Kang, M. S., Spencer, J. P., and Elbein, A. D. (1979). The effect of showdomycin on glycolipid formation: Inhibition of glucosyl-phosphoryl-dolichol in aorta and simulation of glycosylceramide in yeast. *J. Biol. Chem.* **254,** 10037–10043.
151. Muller, T., Bause, E., and Jaenicke, L. (1981). Glycolipid formation in *Volvox carteri* f *nagariensis:* Efects of tunicamycin and showdomycin. *FASEB Lett.* **128,** 208–212.
152. Kaluza, G., Schmidt, M. F. G., and Scholtissek, C. (1973). Effect of 2-deoxy-D-glucose on the multiplication of semliki forest virus and the reversal of the block by mannose. *Virology* **54,** 179–189.

153. Schwarz, R. T., Schmidt, M. F. G., Answer, V., and Klenk, H. D. (1977). Carbohydrates of influenza virus. I. Glycopeptides derived from viral glycoproteins after labeling of its radioactive sugars. *J. Virol.* **23,** 217–226.
154. Schwarz, R. T., and Datema, R. (1982). The lipid pathway of protein glycosylation and its inhibitors. *Adv. Carbohydr. Chem. Biochem.* **40,** 287–379.
155. Pan, Y. T., and Elbein, A. D. (1982). The formation of lipid-linked oligosaccharides in Madin–Darby canine kidney cells: Changes in oligosaccharide profiles induced by glucosamine. *J. Biol. Chem.* **257,** 2795–2801.
156. Pan, Y. T., DeGespari, R., Warren, C. D., and Elbein, A. D. (1992). Formation of unusual mannosamine-containing lipid-linked oligosaccharides in MDCK cells. *J. Biol. Chem.* **267,** 8991–8999.
157. Datema, R., and Schwarz, R. T. (1978). Formation of 2-deoxyglucose-containing lipid-linked oligosaccharides: Interference with glycosylation of glycoproteins. *Eur. J. Biochem.* **90,** 505–516.
158. McDowell, W., Datema, R., Romero, P. A., and Schwarz, R. T. (1985). Mechanism of inhibition of protein glycosylation by the antiviral sugar analogue 2-deoxy-2-fluoro-D-mannose: Inhibition of synthesis of Man $(GlcNAc)_2$-PP-Dol by the guanosine diphosphate ester. *Biochemistry* **24,** 8145–8152.
159. Rearick, J. F., Chapman, A., and Kornfeld, S. (1981). Glucose starvation alters lipid-linked oligosaccharide biosynthesis in Chinese hamster ovary cells. *J. Biol. Chem.* **256,** 6255–6261.
160. Gershman, H., and Robbins, P. W. (1981). Transitory effects of glucose starvation on the synthesis of dolichol-linked oligosaccharides in mammalian cells. *J. Biol. Chem.* **256,** 7774–7780.
161. Turco, S. J. (1980). Modification of oligosaccharide-lipid synthesis and protein glycosylation in glucose-deprived cells. *Arch. Biochem. Biophys.* **205,** 330–339.
162. Turco, S. J., and Picard, J. L. (1982). Altered G-protein glycosylation in vesicular stomatitis virus-infected glucose-deprived baby hamster kidney cells. *J. Biol. Chem.* **257,** 8674–8679.
163. Docherty, P. A., and Aaronson, N. N., Jr. (1985). Effect of the threonine analog β-hydroxynorvaline on the glycosylation and secretion of α-acid glycoprotein by rat hepatocytes. *J. Biol. Chem.* **260,** 10847–10855.
164. Datema, R., and Schwarz, R. T. (1981). Effect of energy depletion on the glycosylation of viral glycoproteins. *J. Biol. Chem.* **256,** 11191–11198.
165. Spiro, R. G., Spiro, M. J., and Bhoyroo, V. D. (1983). Studies on the regulation of the biosynthesis of glucose-containing oligosaccharide-lipids: Effects of energy deprivation. *J. Biol. Chem.* **258,** 9469–9476.
166. Elbein, A. D., and Molyneux, R. J. (1987). The chemistry and biochemistry of simple indolizidine and related polyhydroxy alkaloids. *In* "Alkaloids: Chemical and Biological Perspectives" (W. Pelletier, ed.), Vol. 5, pp. 1–54. Wiley, New York.
167. Hohenschutz, L. D., Bell, E. A., Jewess, P. J., Leworthy, D. P., and Pryce, R. J. (1981). Castanospermine, a 1,6,7,8-tetrahydroxy-octahydroindolizidine alkaloid from the seeds of *Castonospermum australe. Phytochemistry* **20,** 811–814.
168. Saul, R., Chambers, J. P., Molyneux, R. J., and Elbein, A. D. (1983). Castanospermine, a tetrahydroxylated alkaloid that inhibits β-glucosidase and β-glucocerebrosidase. *Arch. Biochem. Biophys.* **221,** 593–597.
169. Pan, Y. T., Hori, H., Saul, R., Sanford, B. A., Molyneux, R. J., and Elbein, A. D. (1983). Castanospermine inhibits the processing of the oligosaccharide portion of the influenza viral hemagglutinin. *Biochemistry* **22,** 3975–3984.
170. Pan, Y. T., Ghidoni, J., and Elbein, A. D. (1992). The effects of castanospermine and swainsonine on the activity and synthesis of intestinal sucrase. *Arch. Biochem. Biophys.,* in press.

171. Frommer, W., Junge, B., Mueller, L., Schmidt, D., and Truscheit, E. (1979). New enzyme inhibitors from microorganisms. *Plant Med.* **35,** 195–200.
172. Schmidt, D. D., Frommer, W., Muellar, L., and Truscheit, E. (1979). Glucosidase-inhibition aus *Bacillus. Naturwissenschaffen* **66,** 3703–3710.
173. Saunier, B., Kilker, R. P., Tkacz, J. S., Quaroni, A., and Herscovis, A. (1982). Inhibition of N-linked complex oligosaccharide formation by 1-deoxynojirimycin, an inhibitor of processing glucosidases. *J. Biol. Chem.* **257,** 14155–14163.
174. Elbein, A. D., Mitchell, M., Sanford, B. A., Fellows, L. E., and Evans, S. V. (1984). The pyrrolidine alkaloid, 2,5-dihydroxymethyl-3,4-dihydroxypyrrolidine, inhibits glycoprotein processing. *J. Biol. Chem.* **259,** 12409–12413.
175. Lodish, H. F., and Kong, N. (1984). Glucose removal from N-linked oligosaccharides is required for efficient maturation of certain secretory glycoproteins from rough ER to Golgi complex. *J. Cell Biol.* **98,** 1720–1729.
176. Edwards, E. H., Sprague, E. A., Kelley, J. L., Kerbacher, J. J., Schwartz, C. J., and Elbein, A. D. (1989). Castanospermine inhibits the function of the low density lipoprotein receptor. *Biochemistry* **28,** 7679–7687.
177. Arakai, R. F., Hedo, J. A. Collier, E., and Gordon, P. (1987). Effects of castanospermine and 1-deoxynojirimycin on insulin receptor biosynthesis. *J. Biol. Chem.* **262,** 11886–11892.
178. Repp, R., Tamura, T., Boschek, C. B., Wege, H., Schwarz, R. T., and Niemann, H. (1985) The effects of processing inhibitors of N-linked oligosaccharides on the intracellular migration of glycoprotein E2 of mouse hepatitis virus and the mutura-tion of coronavirus particles. *J. Biol. Chem.* **260,** 15873–15879.
179. Schmidt, J. W., and Catterall, W. A. (1987). Palmitylation, sulfation, and glycosylation of the α subunit of the sodium channel: Role of post-translational modification in channel assembly. *J. Biol. Chem.* **262,** 13713–13723.
180. Walker, B. D., Kawalski, M., Goh, W. C., Kozarsky, K., Kreiger, M., Rosen, C., Rohrschneider, S., Haseltine, W. A., and Sodorski, J. (1987). Inhibition of human immunodeficiency virus syncytium formation and virus replication by castanospermine. *Proc. Natl. Acad. Sci. USA* **84,** 8120–8124.
181. Gruters, R. A., Neefjes, J. J., Tersmette, M., de Goede, R. E .Y., Tulp, A., Huisman, H. G., Miedema, F., and Ploegh, H. L. (1987). Interference with HIV induced syncytium formation and viral inefectivity by inhibitors of trimming glycosidases. *Nature (London)* **330,** 74–77.
182. Montefiori, D. C., Robinson, W. E., Jr., and Mitchell, W. M. (1988). Role of protein N-glycosylation in pathogenesis of human immunodeficiency virus type I. *Proc. Natl. Acad. Sci. USA* **85,** 9248–9252.
183. Tropea, J., Molyneux, R. J., Kaushal, G. P., Pan, Y. T., Mitchell, M., and Elbein, A. D. (1989). Australine, a pyrrolizidine alkaloid that inhibits amyloglucosidase and glycoprotein processing. *Biochemistry* **28,** 2027–2034.
184. Liu, P. (1987). Total synthesis of 2,6-dideoxy-2,6-imino-7-O-β-D-glucopyranosy-D-glycero-L-gulo-heptitol hydrochloride: A potent inhibitor of α-glucosidases. *J. Org. Chem.* **52,** 4717–4721.
185. Kaushal, G. P., Pan, Y. T., Tropea, J. E., Mitchell, M., Liu, P., and Elbein, A. D. (1988). Selective inhibition of glycoprotein processing enzymes: Differential inhibition of glucosidases I and II in cell culture. *J. Biol. Chem.* **263,** 17278–17283.
186. Elbein, A. D., Solf, R., Vosbeck, K., and Dorling, P. R. (1981). Swainsonine, an inhibitor of glycoprotein processing. *Proc. Natl. Acad. Sci. USA* **78,** 7393–7397.
187. Dorling, P. R., Huxtable, C. R., and Colegate, S. M. (1980). Inhibition of lysosomal α-mannosidase by swainsonine, an indolizidine alkaloid isolated from *Swainsona canescens. Biochem. J.* **191,** 649–651.

188. Tulsiani, D. P. R., Harris, T., and Touster, O. (1982). Swainsonine inhibits the biosynthesis of complex glycoproteins by inhibition of Golgi mannosidase II. *J. Biol. Chem.* **257,** 7936–7939.
189. Legler, G., and Julich, E. (1984). Synthesis of 5-amino-5-deoxy-D-mannopyranose and 1,5-dideoxy-1,5-imino-D-mannitol, and inhibition of α and β-D-mannosidases. *Carbohydr. Res.* **128,** 61–72.
190. Fuhrmann, V., Bause, E., Legler, G., and Ploegh, H. (1984). Novel mannosidase inhibitor blocking conversion of high-mannose to complex oligosaccharides. *Nature (London)* **307,** 755–758.
191. Elbein, A. D., Legler, G., Tlusty, A., McDowell, W., and Schwarz, R. T. (1984). The effect of deoxymannojirimycin on the processing of influenza viral glycoproteins. *Arch. Biochem. Biophys.* **235,** 579–588.
192. Gross, V. K., Tran-Thi, T. R., McDowell, W., and Schwarz, R. T. (1985). Secretion of high-mannose-type α_1-proteinase inhibitor and α_1-acid glycoprotein by primary cultures of rat hepatocytes in the presence of the mannosidase inhibitor. *Eur. J. Biochem.* **150,** 41–46.
193. Snider, M. D., and Rogers, O. C. (1986). Membrane traffic in animal cells: Cellular glycoproteins return to the site of Golgi mannosidase I. *J. Cell Biol.* **103,** 265–275.
194. Bischoff, J., Liscum, L., and Kornfeld, R. (1986). The use of 1-deoxymannojirimycin to evaluate the role of various α-mannosidases in oligosaccharide processing in intact cells. *J. Biol. Chem.* **261,** 4766–4774.
195. Fleet, G. W. J., Smith, P. W., Evans, S. V., and Fellows, L. E. (1984). Design, synthesis and preliminary evaluation of a potent α-mannosidase inhibitor, 1,4-dideoxy-1,4-imino-D-mannitol. *J. Chem. Soc. Chem. Commun.* 1240–1241.
196. Palamarczyk, G., Mitchell, M., Smith, P. W., Fleet, G. W. J., and Elbein, A. D. (1985). 1,4-Dideoxy-1,4-imino-D-mannitol inhibits glycoprotein processing. *Arch. Biochem. Biophys.* **243,** 35–45.
197. Kayakiri, H., Takaese, S., Shibata, T., Okamoto, M., Terano, H., Hashimoto, M., Tada, T., and Koda, S. (1989). Structure of kifunensine, a new immunomodulator isolated from an actinomycete. *J. Org. Chem.* **54,** 4015–4016.
198. Elbein, A. D., Tropea, J. E., Mitchell, M., and Kaushal, G. P. (1990). Kifunensine, a potent inhibitor of the glycoprotein processing mannosidase I. *J. Biol. Chem.* **265,** 15599–15605.
199. Papandreau, G., and Ganem, B. (1991). Mimicking the glucosidase transition state: Shape/charge considerations. *J. Am. Chem. Soc.* **113,** 8984–8985.
200. Colegate, S. M., Dorling, P. R., and Huxtable, C. R. (1979). A spectrophotometric investigation of swainsonine, an α-mannosidase inhibitor isolated from *Swainsona canescens*. *Aust. J. Chem.* **22,** 2257–2264.
201. Molyneux, R. J., and James, L. F. (1982). Loco intoxication: Indolizidine alkaloids of spotted locoweed (*Astragalus lentiginosus*), *Science* **216,** 190–192.
202. Davis, O., Schwarz, P., Hernandez, T., Mitchell, M., Warnock, B., and Elbein, A. D. (1984). Isolation and characterization of swainsonine from Texas locoweed (*Astragalus mollisimus*). *Plant Physiol.* **76,** 972–975.
203. Broquist, H. (1985). The indolizidine alkaloids, slaframine and swainsonine: Contaminants in animal forages. *Annu. Rev. Nutr.* **5,** 391–409.
204. Dorling, P. R., Huxtable, C. R., and Colegate, S. M. (1978). Lysosomal storage in *Swainsona spp.* toxicosis: An induced mannosidosis. *Neuropathol. Appl. Neuobiol.* **4,** 285–295.
205. Tulsiani, D. P. R., Broquist, H. P., James, L. F., and Touster, O. (1984). The similar effects of swainsonine and locoweed on tissue glycosidases and oligosaccharides of the pig indicate the alkaloid is the principle toxin responsible for the induction of locoism. *Arch. Biochem. Biophys.* **232,** 76–85.

206. Warren, C. D., Sadeh, S., Daniel, P. F., Bugge, B., James, L. F., and Jeanloz, R. W. (1983). Induced mannosidosis-excretion of oligosaccharides by locoweed-intoxicated sheep. *FEBS Lett.* **163,** 99–103.
207. Abraham, D. J., Sidebotham, R., Winchester, B., Dorling, P. R., and Dell, A. (1983). Swainsonine affects the processing of glycoproteins *in vivo. FEBS Lett.* **163,** 110–113.
208. Elbein, A. D., Vosbeck, K., Dorling, P. R., and Horisberger, M. (1982). Swainsonine prevents the processing of the oligosaccharide chains of the influenza viral hemagglutinin. *J. Biol. Chem.* **257,** 1573–1576.
209. Kang, M. S., and Elbein, A. D. (1983). Alteration in oligosaccharide structure of VSV glycoproteins by swainsonine. *J. Virol.* **46,** 60–69.
210. Arumughan, R. G., and Tanzer, M. L. (1983). Abnormal glycosylation of human cellular fibronectin in the presence of swainsonine. *J. Biol. Chem.* **258,** 11883–11889.
211. Foddy, L., Feeny, J., and Hughes, R. C. (1986). Properties of BHK cells treated with swainsonine, an inhibitor of glycoprotein synthesis. *Biochem. J.* **233,** 697–706.
212. Franc, J. L., Housepian, S., Fayet, G., and Bouchilloux, S. (1986). Inhibition of N-linked oligosaccharide processing does not prevent the secretion of thyroglobulin. *Eur. J. Biochem.* **157,** 225–232.
213. Whitsett, J. A., Ross, G., Weaver, T., Rice, W., Scon, C., and Hill, W. (1985). Glycosylation and secretion of surfactant associated glycoprotein A. *J. Biol. Chem.* **260,** 15273–15279.
214. Lee, A. Y., and Doyle, D. (1985). N-linked oligosaccharides of the H-2D^k histocompatibility protein heavy chain influence its transport and cellular distribution. *Biochemistry* **24,** 6238–6245.
215. Wagner, D. D., Mayadas, T., Urban-Pickering, M., Lewis, B. H., and Marder, V. J. (1985). Inhibition of disulfide bonding of Von Willebrand protein by monensin results in small, functionally defective multimers. *J. Cell Biol.* **101,** 112–120.
216. Duronio, V., Jacobs, S., and Cuatracasas, P. (1986). Complete glycosylation of the insulin and insulin-like growth factor I receptors is not necessary for their biosynthesis and function. *J. Biol. Chem.* **261,** 970–975.
217. Soderquist, A. M., and Carpenter, G. (1984). Glycosylation of the epidermal growth factor receptor in A-431 cells. *J. Biol. Chem.* **259,** 12586–12594.
218. Breitfeld, P. P., Plup, D., and Schwartz, A. L. (1984). Influence of the N-linked oligosaccharides on the biosynthesis, intracellular routing and function of the human asialglycoprotein receptor. *J. Biol. Chem.* **259,** 10414–10421.
219. Chung, K-N., Sheperd, V. L., and Stahl, P. (1984). Swainsonine and Castanospermine blockade of mannose glycoprotein uptake by macrophages. *J. Biol. Chem.* **259,** 14637–14641.
220. Schwartz, P., and Elbein, A. D. (1985). The effect of glycoprotein processing inhibitors on fucosylation of glycoproteins. *J. Biol. Chem.* **260,** 14452–14458.
221. Merkle, R., Elbein, A. D., and Heifetz, A. (1985). Effects of swainsonine and castanospermine upon processing of influenza virus sulfated glycoproteins *J. Biol. Chem.* **260,** 1083–1089.
222. Bar-Scavit, Z., Kahn, A. J., Pegg, L. E., Stone, R. R., and Teitelbaum, S. L. (1984). Glucocorticoids modulate macrophage surface oligosaccharides and their binding activity. *J. Clin. Invest.* **73,** 1277–1283.
223. Villata, F., and Kierszenbaum, F. (1985). The effect of swainsonine on the association of *Trypanosoma cruzi* with host cells. *Mol. Biochem. Parasitol.* **16,** 1–10.
224. Humphries, M. T., Matsumoto, K., White, S. L., and Olden, K. (1986). Oligosaccharide modification by swainsonine treatment inhibits pulmonary colonization by B16F10 murine melanoma cells. *Proc. Natl. Acad. Sci. USA* **83,** 1752–1756.

225. Hino, M., Nakayoma, O., Tsurami, Y., Adochi, K., Shibata, T., Terano, H., Kohsaka, M., Aoki, H., and Imanaka, H. (1985). Studies of immunomodulator, swainsonine. *J. Antibiot.(Tokyo)* **38,** 926–935.
226. Winkler, J. R., and Segal, H. L. (1984). Inhibition by swainsonine of the degradation of endocytosed glycoproteins in isolated rat liver parenchymal cells. *J. Biol. Chem.* **259,** 1958–1962.
227. Aoyagi, T., Yamamato, T., Kojiri, K., Moriehima, H., Nagai, M., Hanada, M., Takeuchi, T., and Umezawa, H. (1989). Mannostatins A & B: New inhibitors of α-D-mannosidase, produced by *Streptoverticillium verticillus* var, quintum. *J. Antibiot.* **42,** 883–889.
228. Tropea, J. E., Kaushal, G. P., Pastuszak, I., Mitchell, M., Aoyagi, T., Molyneux, R. J., and Elbein, A. D. (1990). Mannostatin A, a new glycoprotein-processing inhibitor. *Biochemistry* **29,** 10062–10069.
229. Winchester, B., and Fleet, G. W. J. (1992). Amino-sugar glycosidase inhibitors: Versitile tools for glycobiologists. *Glycobiology* **2,** 199–210.
230. Winkler, D. A., and Holam, C. (1989). Design of potential anti-HIV agents: Mannosidase inhibitors. *J. Med. Chem.* **32,** 2084–2089.
231. Spiro, M. J., Spiro, R. G., and Bhoyroo, V. D. (1976). Lipid-saccharide intermediates in glycoprotein biosynthesis. I. Formation of an oligosaccharide-lipid by thyroid slices and evaluation of its role in protein glycosylation. *J. Biol. Chem.* **251,** 6400–6408.
232. Pan, Y. T., and Elbein, A. D. (1990). Control of N-linked oligosaccharide synthesis: Cellular levels of dolichyl-P are not the only regulatory factor. *Biochemistry* **29,** 8077–8684.
233. Hubbard, S. C., and Robbins, P. W. (1980). Synthesis of the N-linked oligosaccharides of glycoproteins: Assembly of the lipid-linked precursor oligosaccharide and its relation to protein synthesis *in vivo*. *J. Biol. Chem.* **255,** 11782–11793.
234. Grant, W., and Lennarz, W. J. (1983). Relationship between oligosaccharide-lipid synthesis and protein synthesis in mouse LM cells. *Eur. J. Biochem.* **134,** 575–583.
235. Hortin, G., and Boime, I. (1980). Inhibition of asparagine-linked glycosylation by incorporation of a threonine analog into nascent peptide chains. *J. Biol. Chem.* **255,** 8007–8010.
236. Hentze, M., Hasilik, A., and von Figura, K. (1984). Enhanced degradation of cathepsin D synthesized in the presence of the threonine analog, β-hydroxynorvaline. *Arch. Biochem. Biophys.* **230,** 375–382.
237. Docherty, P. A., and Aronson, N. N., Jr. (1985). Effect of the threonine analog β-hydroxynorvaline on the glycosylation and secretion of α-1-acid glycoprotein by rat hepatocytes. *J. Biol. Chem.* **260,** 10847–10855.
238. Rathod, P. K., Tashjian, A. H., Jr., and Abeles, R. H. (1986). Incorporation of beta-fluoroasparagine into peptides prevents N-linked glycosylation: *In vitro* studies with synthetic fluoropeptides. *J. Biol. Chem.* **261,** 6461–6469.
239. Mills, J. T., and Adamany, A. M. (1978). Impairment of dolichyl saccharide synthesis and dolichol-mediated glycoprotein assembly in aortic smooth muscle cells in culture by inhibitors of cholesterol biosynthesis. *J. Biol. Chem.* **253,** 5270–5273.
240. James, M. J., and Kandutsch, A. A. (1979). Inter-relationships between dolichol and sterol synthesis in mammalian cell cultures. *J. Biol. Chem.* **254,** 8442–8446.
241. Keller, R. K., Adair, W. L., Jr., and Ness, G. C. (1979). Studies on the regulation of glycoprotein biosynthesis: An investigation of the rate-limiting steps of dolichyl phosphate biosynthesis. *J. Biol. Chem.* **254,** 9966–9969.
242. White, D. A., Middleton, B., Pawson, S., Bradshaw, J. P., Clegg, R. J., and Hemming, F. W. (1981). Regulation of dolichol and of cholesterol biosynthesis in cholesterol-fed rabbits. *Arch. Biochem. Biophys.* **208,** 30–36.

243. Goldstein, J. L., Brown, M. S., Anderson, R. G., Russell, D. W., and Schneider, W. J. (1985). Receptor-mediated endocytosis: Concepts emerging from the LDL receptor system. *Annu. Rev. Cell Biol.* **1,** 1–39.
244. Astrand, I. M., Fries, E., Chojnacki, T., and Dallner, G. (1986). Inhibition of dolichyl phosphate biosynthesis by compactin in cultured rat hepatocytes. *Eur. J. Biochem.* **155,** 447–452.
245. Carson, D. D., and Lennarz, W. J. (1979). Inhibition of polyisoprenoid and glycoprotein biosynthesis causes abnormal embryonic development. *Proc. Natl. Acad. Sci. USA* **76,** 5709–5713.
246. Surani, M. A., Kimber, S. J., and Osborn, J. C. (1983). Mevalonate reverses the developmental arrest of preimplantation mouse embryos by compactin, an inhibitor of HMG CoA reductase. *J. Embryol. Exp. Morphol.* **75,** 205–223.
247. Kornfeld, S., and Mellman, I. (1989). The biogenesis of lysosomes. *Annu. Rev. Cell Biol.* **5,** 483–525.
248. Tartakoff, A. M., and Vassalli, P. (1977). Plasma cell immunoglobulin secretion: Arrest is accompanied by alterations of the Golgi complex. *J. Exp. Med.* **146,** 1332–1345.
249. Uchida, N., Smilowitz, H., and Tanzer, M. L. (1979). Monovalent ionophores inhibit secretion of procollagen and fibronectin from cultued human fibroblasts. *Proc. Natl. Acad. Sci. USA* **76,** 1868–1872.
250. Tartakoff, A. M. (1983). Perturbation of vesicular traffic with the carboxylic ionophore monesin. *Cell* **32,** 1026–1028.
251. Johnson, D. C., and Schlesinger, M. J. (1980). Vesicular stomatitis virus and sindbis virus glycoprotein transport to the cell surface is inhibited by ionophores. *Virology* **103,** 407–424.
252. Quinn, P., Griffiths, G., and Warren, G. (1983). Dissection of the Golgi complex. II. Density separation of specific Golgi functions in virally infected cells treated with monensin. *J. Cell Biol.* **96,** 851–856.
253. Anderson, R. G., and Pathak, R. K. (1985). Vesicles and cisternae in the trans Golgi apparatus of human fibroblasts are acidic compartments. *Cell* **40,** 635–643.
254. Misumi, Y., Misumi, Y., Miki, K., Takatsuki, A., Tamura, G., and Ikehara, Y. (1986). Novel blockade by brefeldin A of interacellular transport of secretory proteins in cultured rat hepatocytes. *J. Biol. Chem.* **261,** 11398–11403.
255. Perkel, V. S., Liu, A. Y., Miura, Y., and Magner, J. A. (1988). The effects of brefeldin-A on the high mannose oligosaccharides of mouse thyrotropin, free α-subunits and total glycoproteins. *Endocrinology* **123,** 310–318.
256. Ulmer, J. B., and Palade, G. E. (1991). Effects of brefeldin A on the Golgi complex endoplasmic reticulum and viral envelope glycoproteins in murine erythroleukemia cells. *Eur. J. Cell Biol.* **54,** 38–54.
257. Orci, L., Tagaya, M., Amkerdt, M., Perrelet, A., Donaldson, J. G., Lippincott-Schwartz, J., Klausner, R. D., and Rothman, J. E. (1991). Brefeldin A, a drug that blocks secretion, prevents the assembly of non-clathrin coated buds on Golgi cisternae. *Cell* **64,** 1183–1195.
258. Allan, V. J., and Kreis, T. E. (1986). A microtubule binding protein associated with membranes of the Golgi apparatus. *J. Cell Biol.* **103,** 2229–2239.
259. Donaldson, J. G., Lippincott-Schwartz, J., Bloom, G. S., Kreis, T. E., and Klausner, R. D. (1990). Dissection of a 110 kDa peripheral membrane protein from the Golgi apparatus is an early event in brefeldin A action. *J. Cell Biol.* **111,** 2295–3306.
260. Serafini, T., Stenbeck, G., Brecht, A., Lottsperch, F., Orci, L., Rothman, J. E., and Wieland, F. T. (1991). A coat subunit of Golgi-derived non-clathrin coated vesicles with homology to the clathrin-coated vesicle coat protein β-adaptin. *Nature* **349,** 215–220.

261. Schwarz, R. T., Mayor, S., Menon, A. K., and Cross, G. A. M. (1989). Biosynthesis of the glycolipid membrane anchor of *Trypanosoma brucei* variant surface glycoproteins: Involvement of Dol-P-man. *Biochem. Soc. Trans.* **17,** 746–748.
262. Takami, R., Oda, K., and Ikehara, Y. (1992). Aberrant processing of alkaline phosphatase precursor caused by blocking the synthesis of glycosylphosphatidylinositol. *J. Biol. Chem.* **267,** 1042–1047.
263. Lisanti, M. P., Field, M. C., Caras, I. W., Menon, A. K., and Rodriguez-Boulan, E. (1991). Mannosamine, a novel inhibitor of glycosylphosphatidylinositol incorporation into proteins. *EMBO J.* **10,** 1969–1977.
264. Pan, Y. T., Kamitani, T., Bhuvaneswaran, C., Hallaq, Y., Warren, C. D., Yeh, E. T. H., and Elbein, A. D. (1992). Inhibition of glycosylphosphatidylinositol anchor formation by mannosamine. *J. Biol. Chem.* **267,** 21250–21255.

Use of Mammalian Cell Mutants to Study the Functions of N- and O-Linked Glycosylation

Pamela Stanley

Department of Cell Biology, Albert Einstein College of Medicine, New York, New York 10461

I. Introduction

The need to understand the functions of the carbohydrates covalently bound to glycoproteins has been increasingly recognized. Driven by the fact that most recombinant therapeutics are glycoproteins, research

Cell Surface and Extracellular Glycoconjugates

on the functional roles of carbohydrates has blossomed (reviewed by Olden *et al.*, 1982; West, 1986; Dennis and Laferte, 1987; Paulson, 1989; Cumming, 1991; Stanley, 1992). It is now clear that N- and O-linked carbohydrates can be important for correct folding, solubility, stability, protease resistance, antigenicity, and biological activity of a glycoprotein. Carbohydrates also play important roles in recognition by carbohydrate-binding proteins and are key to the targeting of lysosomal enzymes to lysosomes, to certain cell–pathogen interactions, and to cell–cell interactions such as occur in the recruitment of platelets and leukocytes to sites of inflammation. The broad range of functional roles that has been described for carbohydrates means, however, that identifying a specific role for a particular structure can be difficult. With respect to the biochemical and biological properties of individual glycoproteins, there are no general rules that can be applied to predicting the role(s) of their carbohydrate complement. However, the identification of structure–function relationships is less difficult when carbohydrates function as recognition molecules for carbohydrate-binding proteins (lectins), because the latter show specificity for particular sugars or sequences of sugars independent of the protein backbone.

There are several experimental approaches that can be used to manipulate carbohydrates in order to investigate carbohydrate functions. Site-directed mutagenesis to change amino acids known to be glycosylated represents a precise approach to eliminating one or more glycosylation sites from a cloned glycoprotein. However, complete elimination of a sugar moiety can have profound effects on molecular structure. More subtle changes are often desirable to determine structure–function relationships of particular sugar residues. Removal of a single sugar or sugar sequence can be achieved by treatment with glycosidases or by production of a glycoprotein in the presence of inhibitors of carbohydrate biosynthesis (Elbein, 1987). The easiest way to vary carbohydrate complement is to produce a glycoprotein in a cell type that has the desired battery of glycosylation enzymes. Cultured mammalian cell lines that are derived from different tissues and are often obtained following mutagen or carcinogen treatment, usually express a subset of the possible range of glycosyltransferases. For example, Chinese hamster ovary (CHO) cells do not express α(1,3)fucosyltransferase activity (Campbell and Stanley, 1983), the *N*-acetylglucosaminyltransferase that adds the bisecting GlcNAc termed GlcNAc-TIII (Campbell and Stanley, 1984), α(1,3)galactosyltransferase activity (Smith *et al.*, 1990), or α(2,6)sialyltransferase activity (Lee *et al.*, 1989). In contrast, some mouse cell lines possess α(1,3)galactosyltransferase activity and express carbohydrates with terminal α(1,3)Gal residues (Stanley *et al.*, 1979; Cummings and Mattox, 1988) and others express α(1,3)fucosyltransferase activity (Cummings and Mattox, 1988). Thus a judicious choice of cell line allows variation in the possibilities for terminal glycosylation of carbohydrates.

More dramatic changes in carbohydrate complement can be achieved by using a range of glycosylation mutant cell lines that synthesize specifically truncated carbohydrate structures. Mutants that are missing a glycosylation activity synthesize carbohydrate moieties typical of intermediates in the biosynthetic pathway. There are also mutants that have acquired a glycosylating activity, such as a previously unexpressed glycosyltransferase. These mutants can be used to produce glycoproteins with more elaborate carbohydrate structures that might be particularly useful for tissue targeting applications.

Provided the biochemical and/or structural basis of a glycosylation mutation is known, quite accurate predictions of the altered carbohydrates on glycoproteins synthesized by the mutant can be made. The aim of this chapter is to describe the types of mammalian cell glycosylation mutants available and how they have been used to investigate the functions of N- and O-linked glycosylation. Several reviews describing the origins of the mutants and their properties have appeared (Stanley, 1980, 1984, 1985a; Briles, 1982; Dennis and Laferte, 1987; Krieger *et al.*, 1989; Stoll *et al.*, 1990). Reviews that have focused more on the use of glycosylation mutants in functional studies have also appeared (Stanley, 1987, 1992; Krieger *et al.*, 1989; Brandli, 1991). The well-characterized glycosylation mutants have significant potential for use in glycosylation engineering and possibilities for their exploitation in this area are also discussed.

II. Carbohydrates of Glycoproteins

In order to understand the properties and potential uses of glycosylation mutants, it is necessary to appreciate the types of carbohydrate structures that are found covalently attached to protein and the pathways of their biosynthesis. Carbohydrates are found in amide linkage to asparagine in the sequon Asn–*X*–Ser/Thr; in O-linkage to serine or threonine residues; and in amide linkage to the C terminus as part of a glycosyl phosphatidylinositol anchor. The carbohydrates that are N-linked to asparagine usually contain a minimum of seven residues and may theoretically include as many as 40 or 50 sugars in the most complex type. They are classified into oligomannosyl, hybrid, and complex (or lactosamine-containing) types, each of which includes a heterogeneous array of structures (see Section II.A). The carbohydrates that are O-linked may contain only a single sugar or a few sugars or be as complex as a glycosaminoglycan chain. Some of the simplest O-linked sugars are transferred to proteins in the cytoplasm and nucleus, whereas the more complicated structures are synthesized in the compartments of the secretory pathway, as are the N-linked carbohydrates (see Section II.B). The third type of carbohydrate covalently attached to protein is termed a glycosyl phosphatidylinositol (GPI) anchor and consists of a complex glycan attached at one end to protein via ethanolamine phosphate and at the other end to phosphatidylinositol (see Section II.C).

A. N-Linked Carbohydrates of Glycoproteins

The biosynthesis of N-linked carbohydrates is complicated as it involves synthesis of a lipid-linked oligosaccharide, transfer to protein, removal of specific sugars, and addition of other sugars to generate a mature structure. Except for the initial steps in oligosaccharide-lipid synthesis, all reactions occur within the lumen of the secretory pathway, beginning in the endoplasmic reticulum (ER) and passing through each compartment of the Golgi apparatus. The biochemical reactions involved at each step have been reviewed by Hubbard and Ivatt (1981) and Kornfeld and Kornfeld (1985). Topographical considerations and the compartmentalization of the reactions within the cell have been reviewed by Hirschberg and Snider (1987) and Pfeffer and Rothman (1987), respectively. A summary of the major features of N-linked carbohydrate biosynthesis including the origins of oligomannosyl, hybrid, and complex carbohydrates is presented in Fig. 1. The modification of oligomannosyl carbohydrates of lysosomal enzymes en route to lysosomes is also noted (reviewed by von Figura and Hasilik, 1986; Kornfeld and Mellman, 1989).

The complex class of N-linked carbohydrates exhibits the greatest range of structural variation due to the possibility of (a) up to seven different branching GlcNAc residues being added to core mannose residues (Schachter, 1991); (b) two or more lactosamine units being added to up to six of the GlcNAc residues (Saitoh *et al.* 1992); (c) enormous variation in terminal sugar addition of differentially linked sialic acids, galactose, fucose, or GalNAc residues; and (d) sulfate modification of Gal, GalNAc, or GlcNAc residues (Roux *et al.*, 1988; Hoogewerf and Bensadoun, 1991). The biosynthesis of these structures requires the sequential action of a multitude of glycosyltransferases and other enzymes such as nucleotide-sugar synthases, translocases, and sulfotransferases, all of which must be appropriately compartmentalized in particular cellular locations. Most of the glycosyltransferases are specific for the transfer of a single sugar in a specific linkage to a particular acceptor. The cell therefore expends an enormous amount of energy in

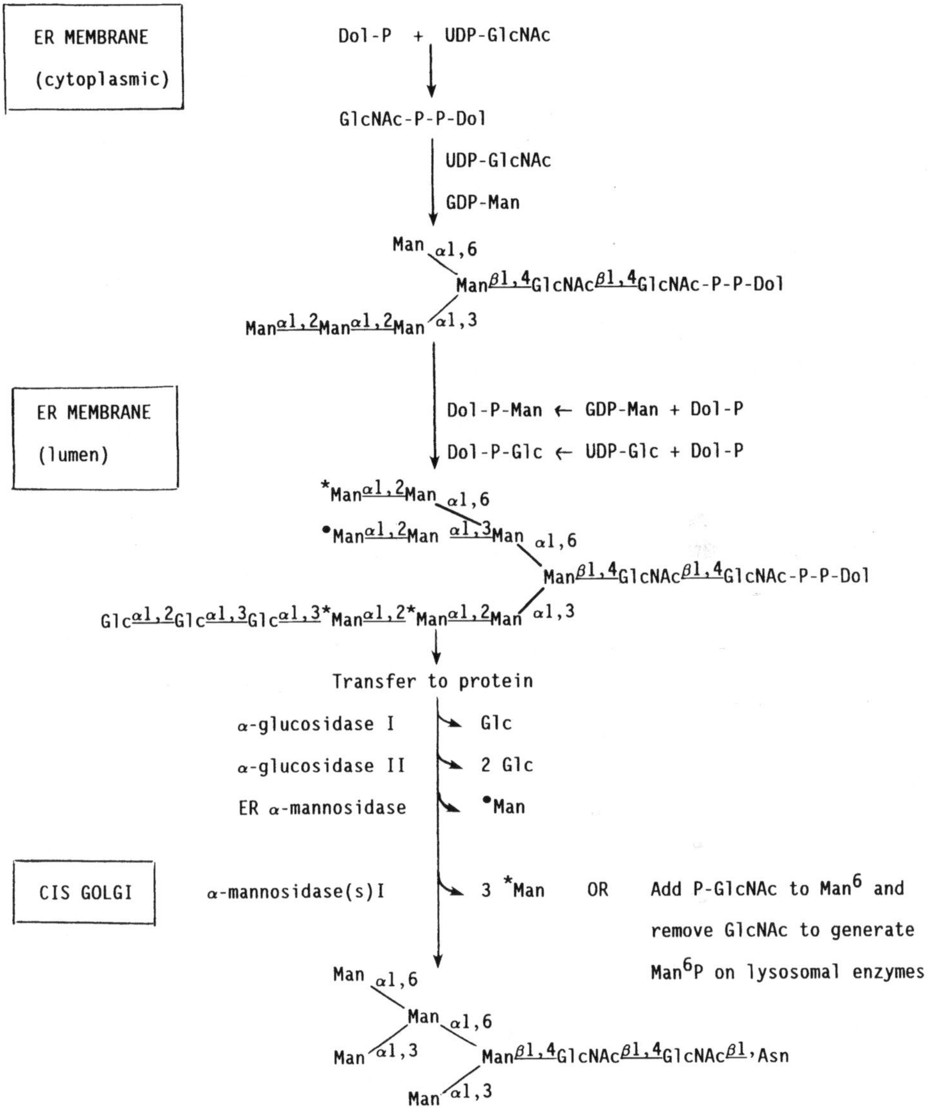

FIG. 1. The major features of N-linked carbohydrate biosynthesis are presented to show key intermediates and the theoretical range of mature N-linked complex structures that could be expressed on a glycoprotein. The final structure shown is not from an actual glycoprotein but is intended to portray many of the large number of ways in which branch GlcNAc residues can be extended to give different antennae.

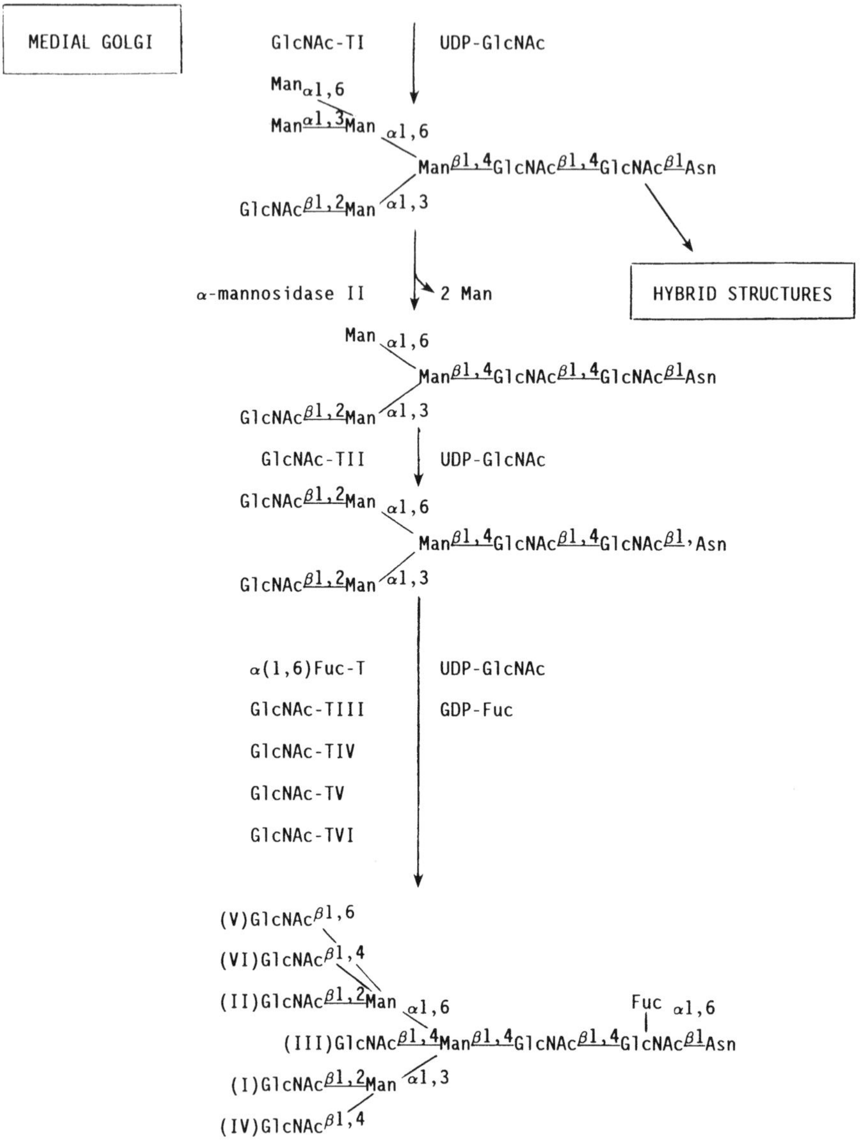

FIG. 1 *Continued*

maintaining the possibility of expressing a large variety of different N-linked structures. Recent insights into the functions of carbohydrates (see Section IV) have revealed the importance of particular sugar residues for certain functions and it seems certain that the heterogeneity

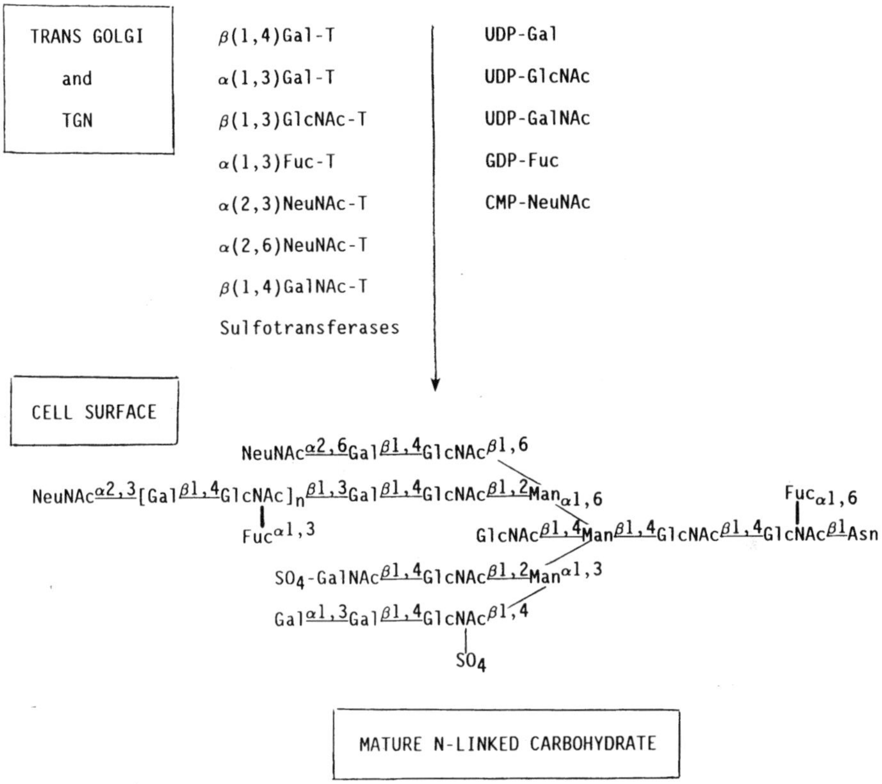

FIG. 1 *Continued*

inherent in N-linked carbohydrate biosynthesis will be found to have increasing biological significance.

B. O-Linked Carbohydrates of Glycoproteins

Carbohydrates linked to the hydroxyl group of serine or threonine and typically classified as O-linked are of the mucin type that initiate with GalNAc (Schachter and Roseman, 1980). These moieties are often small tri- and tetrasaccharides that occur in Ser–Thr-rich stretches such as those observed in leukosialin or CD43 (Carlsson *et al.*, 1986). However, O-linked structures may also be branched and the degree of branching may vary with differentiation (Carlsson *et al.*, 1986; Piller *et al.*, 1988) or disease (Higgens *et al.*, 1991). In secreted mucins, the carbohydrates can be extensively branched and very complex (Lamblin *et al.*, 1991). The synthesis of O-linked carbohydrates (Fig. 2) occurs mainly in Golgi compartments, although several reports have

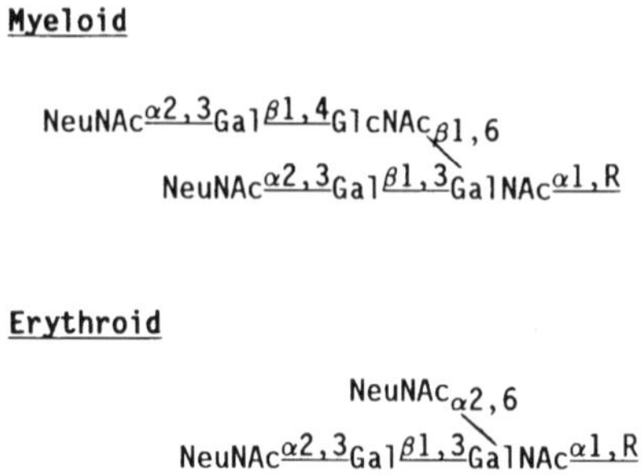

FIG. 2. O-linked oligosaccharides found on glycoproteins of cells of the myeloid or erythroid lineages (adapted from Carlsson *et al.*, 1986, *J. Biol. Chem.* **261,** 12787–12795.

suggested that the initiating GalNAc is added very quickly after protein synthesis, possibly in a late compartment of the endoplasmic reticulum. The large number of conflicting reports on the site of O-linked initiation (Piller *et al.,* 1990; Perez-Vilar *et al.,* 1991, and references therein) indicate that the addition of GalNAc to Ser or Thr may occur in the endoplasmic reticulum, a transitional compartment, or the cis-Golgi, depending on cell type or even on the state of differentiation of a cell.

A much more simple type of O-linked glycosylation that is now known to be present on many cytoplasmic and nuclear proteins was discovered. (Torres and Hart, 1984; Holt and Hart, 1986). It is the presence of a single GlcNAc residue at Ser or Thr residues (reviewed by Hart *et al.,* 1989). The peptide sequence that is recognized by the cytoplasmic enzyme complex responsible for transferring the single GlcNAc residue is very similar to the sequence recognized for phosphorylation by Ser/Thr kinases (Haltiwanger *et al.,* 1990). Thus it has been postulated that the O-GlcNAc modification may somehow be involved in protecting Ser or Thr residues from phosphorylation as part of the regulation of phosphorylation of proteins involved in signal transduction (Kearse and Hart, 1991).

Other types of simple O-linked glycosylation include the addition of Man (which is subsequently phosphorylated) to a specific cytoplasmic liver protein thought to be involved in exocytosis (Satir *et al.* 1990), the addition of Fuc to specific epidermal growth factor domains in glycoproteins (Harris *et al.,* 1992, and references therein), and the addition of Xyl–Glc and Xyl_2–Glc to certain Ser residues (Nishimura *et al.,* 1989). Although structure–function relationships of these O-linked sugars are not yet identified, it seems clear that specific glycosylation is a highly regulated feature of nuclear, cytoplasmic, secreted, and membrane proteins and such modifications must be considered in studies of the functions of carbohydrates.

C. Glycosaminoglycan Chains in Glycoproteins

Most glycoproteins carry the major types of N- and O-linked carbohydrates described in Sections II.A and II.B. However, some glycoproteins have recently been shown to also carry a glycosaminoglycan chain. For example, the growth factor colony stimulating factor 1 (CSF-1) exists predominantly in a high-molecular weight form that has a single chondroitin-sulfate chain (Price *et al.*, 1992; Suzu *et al.*, 1992). Insight into the functional consequences of producing CSF-1 without the glycosaminoglycan chain awaits studies on molecules with a truncated or modified chain. Such molecules could be produced by synthesizing CSF-1 in CHO mutants with defects in glycosaminoglycan biosynthesis (Esko, 1991).

D. Glycosyl Phosphatidylinositol-Anchored Glycoproteins

A modification of membrane proteins that is receiving increasing attention is the presence of covalently associated carbohydrate attached to phosphotidylinositol at the C-terminal amino acid of the mature protein (reviewed by Ferguson and Williams, 1988). The glycosyl phosphatidylinositol (GPI) anchor causes a protein to localize to one leaflet in a lipid bilayer and thereby to have properties quite distinct from proteins that span the membrane or are membrane-associated by other mechanisms.

The structure of the GPI anchor from rat brain Thy 1 glycoprotein has been determined (Homans *et al.*, 1988) and is shown in Fig. 3. Sugars typical of both N- and O-linked carbohydrates are found in the glycan portion of the anchor but their linkages to each other are quite unique and the presence of glucosamine rather than GlcNAc is novel. The pathway of biosynthesis of the core mannose residues has been determined (reviewed by Doering *et al.*, 1990) and reveals a connection to the synthesis of N-linked carbohydrates through the common requirement for Dol-P-Man as a mannose donor (Menon *et al.*, 1990). Therefore glycosylation mutations affecting the level of Dol-P-Man inhibit the biosynthesis of both N-linked oligomannosyl carbohydrates and GPI anchors.

The attachment of a GPI anchor occurs in the endoplasmic reticulum immediately following cleavage of a C-terminal hydrophobic signal sequence that occurs with completion of translation (reviewed in Doering *et al.*, 1990). Although no specific function for the glycan moiety of the GPI anchor has been identified, the presence of the complete anchor is important for localizing proteins to the apical membrane in

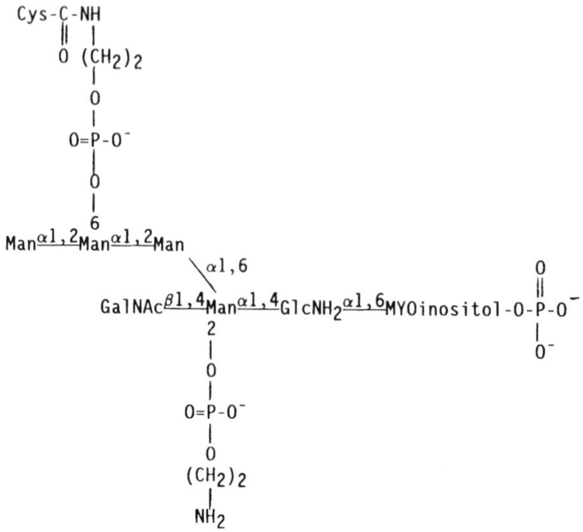

FIG. 3. Structure of the GPI anchor of rat brain Thy1 glycoprotein (taken from Homans *et al.*, 1988).

certain polarized cells (Lisanti *et al.* 1988). In addition, since the glycolipid anchor can be cleaved by specific phospholipases, GPI-anchored proteins have the potential of being rapidly released from the membrane (Low and Saltiel, 1988).

III. GLYCOSYLATION MUTANTS OF MAMMALIAN CELLS

The first somatic cell mutants shown to be defective in a specific glycosylation reaction were selected for resistance to plant lectins from Chinese hamster ovary (CHO) cell populations (Gottlieb *et al.*, 1975; Stanley *et al.*, 1975b). Lectins are carbohydrate binding proteins that interact with cells via the sugar moieties of cell surface glycoproteins (Liener *et al.*, 1986), and therefore selection for lectin resistance was expected to identify mutants expressing altered carbohydrates due to a mutation in a glycosylation pathway. Lectins are, in fact, quite specific reagents for selecting glycosylation mutants but it is important to remember that a given cytotoxic lectin will select for a range of different mutations (i.e., any mutation that causes resistance to the selective dose of the lectin; Stanley *et al.*, 1975a). Other methods designed to give rise to a range of glycosylation mutants include ^{3}H-sugar suicide (Hirschberg *et al.*, 1981), screening for reduced incorporation of labeled sugar (Stoll and Krag, 1988), and selection for resistance to N-linked carbohydrate processing inhibitors in conjunction with lectins (Lehrman and Zeng, 1989). However, glycosylation mutants are also often obtained from selections that were not intended to select for glycosylation-defective phenotypes. In fact, any selection against the cell surface expression of a particular glycoprotein is likely to give rise to mutants with a primary defect in glycosylation. Examples abound in the literature and include glycosylation mutants that arose from selections against the plasma membrane expression of Thy 1 glycoprotein (Trowbridge and Hyman, 1979), the Man-6-P receptor (Stoll *et al.*, 1982), the LDL receptor (Kingsley *et al.*, 1986b), and the influenza hemagglutinin (Hearing *et al.*, 1989b). In fact, glycosylation mutants are often the most frequent survivors of selections against cell-surface-expressed molecules. Fortunately, they are easily identified by their characteristic resistance to or binding of plant lectins that differ in sugar recognition properties (Stanley, 1985b). Such assays are also very helpful in rapidly distinguishing different glycosylation mutants from each other, although complementation analysis is necessary to identify mutants with a related phenotype that arise from mutations in different genes (Hyman and Stallings, 1974; Stanley *et al.*, 1975a).

Over the last 20 years, glycosylation mutants have been isolated from a large number of cultured mammalian cell lines. However, not all have been characterized in sufficient detail to be useful in investigations of the functions of glycosylation. Those mutants for which a biochemical basis has been identified by either structural studies of their altered carbohydrates or characterization of an altered glycosylation activity are presented in Table I and discussed in the following sections.

A. Mutants Affected in a Glycosyltransferase Activity

Mutants that are completely lacking a glycosyltransferase activity synthesize a predictable range of carbohydrate structures. The most thoroughly characterized mutants of this type are the CHO mutants lacking GlcNAc-TI (Gottlieb *et al.*, 1975; Stanley *et al.*, 1975b). They have no detectable GlcNAc-TI activity *in vitro* under saturating conditions (Chaney and Stanley, 1986) and produce the $Man_5GlcNAc_2Asn$ intermediate (Fig. 1) at complex, N-linked sites in glycoproteins (Robertson *et al.*, 1978; Li and Kornfeld, 1978). Similarly, mutants lacking GlcNAc-TV activity (Fig. 1) (Cummings *et al.*, 1982; Chaney *et al.*, 1989) or with a miscompartmentalized GlcNAc-TV enzyme (Chaney, *et al.*, 1989) synthesize N-linked carbohydrates lacking the $\beta(1,6)$GlcNAc to Man$\alpha(1,6)$ branch (Cummings and Kornfeld, 1984;

TABLE I

GLYCOSYLATION MUTANTS THAT PRODUCE GLYCOPROTEINS WITH A PREDICTABLY ALTERED CARBOHYDRATE COMPLEMENT

		Carbohydrates expected			
Mutant cell line	Biochemical change	N-linked complex	N-linked oligomannosyl	O-linked	References
I. Mutants with an altered glycosyltransferase activity					
Lecl (CHO) Clone 15B (CHO) Ric^{R}14 (BHK)	↓ GlcNAc-TI	$Man_5GlcNAc_2Asn$	+	+	Stanley *et al.* (1975b) Gottlieb *et al.* (1975) Meager *et al.* (1975)
Lec1A (CHO)	↓ GlcNAc-TI (K_m defect)	Mixture complex and $Man_5GlcNAc_2Asn$	+	+	Chaney and Stanley (1989)
Ric^{R}21 (BHK)	↓ GlcNAc-TII	$Man_3GlcNAc_2Asn$ hybrid structures	+	+	Gleeson *et al.* (1985)
PHA^{R}2.1 (BW5147) KBL1 (MDAY-D2) Lec4 (CHO) LP 3.3 (BHK)	↓ GlcNAc-TV	Complex lack $\beta(1,6)$GlcNAc branch from Man$\alpha(1,6)$	+	+	Cummings and Kornfeld (1982) Dennis *et al.* (1987) Chaney and Stanley (1989) Palcic *et al.* (1990)
Sil (KB)	↓ $\alpha(2,3)$sialyl-T	+	+	Reduced NeuNAc, terminate in Gal	Toyama *et al.* (1983)
VV6 (T cell)	↓ $\beta(1,4)$GalNAc-T	+	+	Reduced terminal GalNAc, terminate in Gal	Conzelman and Kornfeld (1984a,b)
Wa-4 (B16)	↑ $\alpha(1,3)$Fuc-T	Terminal Le^X	+	?	Finne *et al.* (1982)
LEC11 (CHO)	↑ $\alpha(1,3)$Fuc-TI	Terminal Le^X, SA-Le^X, and V1M-2	+	?	Campbell and Stanley (1983) Howard *et al.* (1987) Potvin and Stanley (1991)

LEC12 (CHO)	↑ α(1,3)Fuc-TII	Terminal Le^X and V1M-2		+	?	Campbell and Stanley (1983) Howard *et al.* (1987) Potvin and Stanley (1991)
LEC29 (CHO)	↑ α(1,3)Fuc-T	Terminal Le^X		+	?	Potvin and Stanley (1991)
LEC30 (CHO)	↑ α(1,3)Fuc-T	Terminal Le^X and V1M-2		+	?	Potvin and Stanley (1991)
LEC10 (CHO)	↑ GlcNAc-TIII	Complex have the bisected GlcNAc residue		+	+	Campbell and Stanley (1984)
II. Mutants affected in processing of N-linked carbohydrates						
Ric^R15 (BHK) Ric^R19 (BHK)	↓ α-Mannosidase II	$Man_5GlcNAc_2Asn$ hybrid structures		+	+	Hughes and Feeney (1986)
CL6 (L)	↓ α-Mannosidase I	$Man_8GlcNAc_2Asn$	$Man_{8-9}GlcNAc_2Asn$		+	Tabas and Kornfeld (1978)
PHA^R2.7 (BW5147)	↓ α-Glucosidase II	Glc_2Man_{7-9}-$GlcNAc_2Asn$	Glc_2Man_{7-9}-$GlcNAc_2Asn$		+	Reitman *et al.* (1982)
Lec23 (CHO)	↓ α-Glucosidase I	Glc_3Man_{7-9}-$GlcNAc_2Asn$	Glc_3Man_{7-9}-$GlcNAc_2Asn$		+	Ray *et al.* (1991)
III. Mutants affected in nucleotide-sugar synthesis or transport						
Lec2 (CHO) Clone 1021 (CHO)	↓ CMP–NeuNAc translocation	Reduced in NeuNAc, terminate in Gal		+	Reduced in NeuNAc, terminate in Gal	Deutscher *et al.* (1984)
Lec8 (CHO) Clone 13 (CHO)	↓ UDP–Gal translocation	Reduced in Gal, terminate in GlcNAc		+	Reduced in Gal, terminate in GalNAc	Deutscher and Hirschberg (1986)

(*continues*)

TABLE I—*Continued*

Mutant cell line	Biochemical change	Carbohydrates expected: N-linked complex	Carbohydrates expected: N-linked oligomannosyl	Carbohydrates expected: O-linked	References
MDW4 (MDAY-D2) PHA^{R}1.8 (BW5147) Ric^R(MDCK) Ric^R(PC12)					Dennis *et al.* (1986) Trowbridge *et al.* (1978) Brandli *et al.* (1988) Green and Kelly (1990)
PHA^{R}1.3 (BW5147) Lec13 (CHO)	↓ GDP–Man-2,4-dehydratase	Reduced in Fuc residues	+	Reduced in Fuc residues	Reitman *et al.* (1980) Ripka *et al.* (1986a)
ldlD (CHO)	↓ UDP–Gal/UDP–GalNAc-4-epimerase	Reduced in Gal, terminate in GlcNAc	+	Not present	Kingsley *et al.* (1986a)
D33W25-1 (MDAY-D2)	↓ CMP–NeuNAc hydroxylase	Terminate in NeuGly	+	Terminate in NeuGly	Dennis (1986)
IV. Mutants affected in oligosaccharide (OS)-dolichol biosynthesis					
B211 (CHO) (Lec5)	↓ Glucosylation OS-dolichol	Reduced transfer OS, empty N-linked sites	+	+	Krag (1979)
Thy1E (BW5147) B4-2-1 (CHO) (Lec15)	↓ Dol-P-Man synthase, reduced synthesis GPI anchors	+	$Man_{3-5}GlcNAc_2Asn$	+	Chapman *et al.* (1979) Stoll *et al.*, (1982) Fatemi and Tartakoff (1988)
F2A8 (CHO) Lec9 (CHO) 4J (CHO)	↓ Dolichol ↑ Polyprenol	Reduced transfer OS, empty N-linked sites	+	+	Stoll *et al.* (1988) Rosenwald *et al.* (1989) Beck *et al.* (1990)

PIR (CHO) (Lec35)	↑ $Man_5GlcNAc_2$-P-P-Dol	+	$Man_{3-5}GlcNAc_2Asn$	+	Lehrman and Zeng (1989)
V. Mutants with more than one glycosylation mutation					
Lec1.6 (CHO)	↓ GlcNAc-TI $Glc_{1-3}Man_7GlcNAc_2$ transferred to protein	$Man_4GlcNAc_2Asn$	$Man_{5-7}GlcNAc_2Asn$	+	Robertson *et al.* (1978) Hunt (1980)
Lec4.8 (CHO)	↓ GlcNAc-TV ↓ UDP–Gal translocase	Terminate in GlcNAc, no $\beta(1,6)$GlcNAc branch from Man$\alpha(1,6)$.	+	Terminate in GalNAc	Chaney *et al* (1989)
LEC10.8 (CHO)	↑ GlcNAc-TIII ↓ UDP–Gal translocase	Terminate in GlcNAc, have bisected GlcNAc	+	Terminate in GalNAc	Stanley *et al.* (1991)
ldlD.Lec1 (CHO)	↓ UDP–Gal-4-epimerase ↓ GlcNAc-TI	$Man_5GlcNAc_2Asn$	+	Not present	Ravdin *et al.* (1989)
Lec3.2 (CHO)	↓ CMP-NeuNAc translocase	Severe reduction in NeuNAc	+	Severe reduction in NeuNAc	Stanley (1989)
Lec3.2.1 (CHO)	↓ CMP–NeuNAc translocase ↓ GlcNAc-TI	$Man_5GlcNAc_2Asn$	+	Terminate in Gal	Stanley (1989)
Lec3.2.8 (CHO)	↓ CMP–NeuNAc and UDP–Gal translocases	Terminate in GlcNAc	+	Terminate in GalNAc	Stanley (1989)
Lec3.2.8.1 (CHO)	↓ CMP–NeuNAc and UDP–Gal translocases. ↓ GlcNAc-TI	$Man_5GlcNAc_2Asn$	+	Terminate in GalNAc	Stanley (1989)
PIR.Lec1 (CHO)	↑ $Man_5GlcNAc_2$-P-P-Dol ↓ GlcNAc-TI	$Man_3GlcNAc_2Asn$	$Man_{3-5}GlcNAc_2Asn$	+	Zeng and Lehrman (1990)

Stanley *et al.*, 1984). However, mutants with an incomplete defect in a glycosyltransferase activity are more common. Such mutants may reflect a mutation in a glycosyltransferase gene that produces partially inactive transferase or that inactivates only one of a diploid pair of genes encoding a glycosyltransferase. An example of the former type of mutant is found in Lec1A CHO cells that possess a defective GlcNAc-TI activity with a markedly reduced apparent K_m for both UDP–GlcNAc and $Man_5GlcNAc_2Asn$ substrates (Chaney and Stanley, 1986). This mutant synthesizes a mixture of complex carbohydrates and $Man_5GlcNAc_2Asn$ structures at N-linked, complex glycosylation sites (Chaney and Stanley, 1986). Other mutants that exhibit partial transferase activity include Ric^R14 BHK cells (Meager *et al.*, 1975), Ric^R21 BHK cells (Vischer and Hughes, 1981), and KBL4 MDAY-D2 cells (Dennis *et al.*, 1987). The residual activity in each mutant is $<50\%$ of parental levels, suggesting it is due to partial inactivation of a gene that is present in a hemizygous state. Whatever the genetic basis, the consequence in terms of using the mutant for functional studies is that caution must be exercised. Although the array of carbohydrates at the cell surface may be drastically altered in such a mutant, the spectrum of carbohydrates attached to a particular glycoprotein will depend on the complex set of factors that determine maturation at each glycosylation site when a specific glycosyltransferase activity is limiting. It is conceivable that some glycoproteins synthesized in this type of mutant will carry a predominantly normal rather than an altered carbohydrate complement.

Two mutants with tranferase defects specific for O-linked carbohydrates have been described. The first type was isolated from KB cells for resistance to fusion by Sendai virus and is deficient in sialytransferase activity for substrates containing O-linked Galβ1,3GalNAc moieties (Toyama *et al.*, 1983). The second type was isolated for resistance to the lectin *Vicia villosa* and is deficient in a β(1,4)GalNAc-transferase that adds a terminal GalNAc residue to an O-linked pentasaccharide (Conzelmann and Kornfeld, 1984a,b).

A third class of mutants with an altered glycosyltransferase activity has been identified. These mutants behave dominantly and are the result of *activation* of a glycosyltransferase activity expressed either at low levels or not at all in parental cells. The first example of this type of mutant was a wheat germ agglutinin-resistant B16 melanoma mutant that expressed increased α(1,3)fucosyltransferase (α(1,3)Fuc-T) activity (Tao and Burger, 1977; Finne *et al.*, 1982). Related mutants were obtained from CHO cells (Campbell and Stanley, 1983) but in these isolates an α(1,3)Fuc-T activity appeared to have been induced *de novo*. At least four α(1,3)Fuc-T activities that differ in certain biochemical properties, as well as in the fucosylated structures they produce, have been described in independent CHO mutants (Potvin and Stanley, 1991). Another glycosyltransferase activated in rare CHO mutants is GlcNAc-TIII (Fig. 1), the enzyme that transfers the bisecting GlcNAc to N-linked structures (Campbell and Stanley, 1984; Sallustio and Stanley, 1989). The best characterized mutant with this phenotype is LEC10, but there may be additional members of this class (Sallustio and Stanley, 1989). A dominant phenotype isolated from MDAY-D2 cells was found not to express a new transferase but rather a CMP–sialic acid hydroxylase that converts *N*-acetylneuraminic acid to *N*-glycolylneuraminic acid, thereby making the cells resistant to wheat germ agglutinin (Dennis, 1986).

The ability to isolate somatic cell mutants or variants that express normally silent glycosylation genes means that cultured cells and glycoproteins expressing developmentally regulated carbohydrates are available for studies of the functions of these specialized carbohydrate structures. Most importantly, each mutant expresses only one new activity, allowing the effects of the addition of a specific sugar residue to a complex structure to be studied in isolation.

B. Mutants Affected in the Processing of N-Linked Carbohydrates

Glycosylation mutants that carry N-linked carbohydrates typical of processing intermediates (Fig. 1) have also been selected with cytotoxic lectins (Table 1). Each mutant of this type possesses a block in the pathway of N-linked carbohydrate maturation due to the loss of a processing glycosidase. In the case of CL6 cells that lack α-mannosidase I activity (Fig. 1), glycoproteins express the expected intermediate $Man_8GlcNAc_2Asn$ at N-linked glycosylation sites that would normally carry complex moieties (Tabas and Kornfeld, 1978). However, in α-glucosidase mutants, a heterogeneous collection of glucosylated structures will be found at N-linked sites due to the action of the ER α-mannosidase and α-mannosidase I on terminal Man residues (Reitman *et al.*, 1982; Ray *et al.*, 1991). Further heterogeneity can be expected in these mutants due to a bypass of the α-glucosidase block by a Golgi endo-mannosidase that cleaves $Glc_{1\text{-}3}Man$ from glucosylated oligomannosyl carbohydrates (Lubas and Spiro, 1988). Similarly, a range of heterogeneous structures is expected on glycoproteins of the BHK mutants Ric^R15 and Ric^R19 with reduced α-mannosidase II activity (Hughes and Feeney, 1986). The hybrid moiety (Fig. 1), which cannot be further trimmed in these cells, can, nevertheless, be extended to contain Gal, sialic acid, and fucose residues. In addition, the α-mannosidase II defect can be bypassed if the $Man_5GlcNAc_2$-P-P-Dol intermediate (Fig. 1) is transferred to protein, since the former structure is not a substrate for α-mannosidase II but, once processed to $Man_3GlcNAc_2Asn$, can accept GlcNAc from GlcNAc-TI and go on to make complex carbohydrates (Chapman *et al.*, 1979).

C. Mutants Affected in Nucleotide-Sugar Synthesis or Transport

The first mutant shown to be defective in the synthesis of a nucleotide sugar was a pea lectin resistant line of BW5147 lymphoma cells (Reitman *et al.*, 1980). The defect was localized to the first enzyme in the pathway of conversion of GDP–Man to GDP–Fuc. A similar defect in GDP–Man-2,4-dehydratase activity was also exhibited by the CHO mutant Lec13, selected for resistance to pea lectin (Ripka *et al.*, 1986a). A second CHO mutant, Lec13A, appeared to have a different mutation affecting the same enzyme. All glycoconjugates have a low fucose content in these mutants. However, the defect can be reversed by adding fucose, which is converted directly to GDP–Fuc by a salvage pathway.

A more pleiotropic effect on carbohydrate synthesis was observed in a CHO mutant termed ldlD, which lacks UDP–Gal/UDP-GalNAc-4-epimerase (Kingsley *et al.*, 1986a). This mutation prohibits the synthesis of UDP–Gal and UDP–GalNAc and therefore stops the addition of Gal and GalNAc to glycoconjugates. This mutant phenotype can also be reversed by exogenously adding Gal or $GalNH_2$, which are taken up by the cell, phosphorylated, and converted directly to the appropriate nucleotide-sugar. Since the addition of Gal corrects only the N-linked carbohydrate lesion, whereas the addition of $GalNH_2$ corrects only the O-linked defect, the ldlD mutant has been very useful in analyzing the relative functional roles of N-linked versus O-linked carbohydrates on glycoproteins (Krieger *et al.*, 1989).

Another class of mutants that exhibits reduced addition of a sugar to all glycoconjugates is defective in nucleotide-sugar translocation into the appropriate cellular compartment. Two examples of this type have been described: Lec2 or Clone 1021 cells are defective in translocating CMP–sialic acid into the Golgi (Deutscher *et al.*, 1984) and Lec8 or Clone 13 CHO cells are defective in translocating UDP–Gal into the Golgi (Deutscher and Hirschberg, 1986). The latter mutant phenotype has been isolated from many different cell types (Table I). As might be predicted, correction of these translocation mutant phenotypes by the addition of sugar precursors is not possible.

D. Mutants Affected in Dolichol-Oligosaccharide Biosynthesis

Mutants affected in the biosynthesis of the $Glc_3Man_9GlcNAc_2$-P-P-Dol precursor that is transferred to Asn exhibit a wide variety of glycosylation phenotypes. At least two of them express glycoproteins that are underglycosylated, meaning that certain N-linked sequons do not receive a carbohydrate moiety. The consequent reduction in carbohydrate content of glycoproteins at the cell surface leads to a generalized lectin resistance phenotype. The first mutant of this type (B211 or Lec5) was selected for resistance to Con A and had temperature-sensitive growth properties (Cifone *et al.*, 1979). It was found to synthesize very low levels of $Glc_3Man_9GlcNAc_2$-P-P-Dol due to a defect in glucosylation (Krag, 1979). Two CHO mutants with similar temperature-sensitive phenotypes, but belonging to a different complementation group, are F2A8 and Lec9. Both synthesize markedly reduced levels of dilichol and accumulate its polyprenol precursor (Stoll *et al.*, 1988; Rosenwald *et al.*, 1989). Related CHO mutants, which also belong to the Lec9 complementation group (Beck *et al.*, 1990), were found to exhibit aberrant glycosylation and reduced surface expression of the influenze hemagglutinin (Hearing *et al.*, 1989a,b).

Another Con A-resistant CHO mutant with a defect in oligosaccharide-dolichol maturation was termed Lec6 and synthesized predominantly $Man_7GlcNAc_2$-P-P-Dol rather than $Man_9GlcNAc_2$-P-P-Dol (Hunt, 1980). Since this mutation exists only on a Lec1 background (Stanley *et al.*, 1975a), glycoproteins produced by the Lec1.Lec6 mutant have $Man_4GlcNAc_2Asn$ at N-linked complex glycosylation sites (Robertson *et al.*, 1978) and oligomannosyl structures that contain up to seven mannoses (Hunt, 1980). Yet another oligosaccharide-dolichol mutant that has been isolated from both BW5147 cells and CHO cells is defective in the synthesis of Dol-P-Man synthase (Chapman *et al.*, 1979; Stoll *et al.*, 1982). This mutant transfers glucosylated $Man_5GlcNAc_2$-P-P-Dol to protein and can therefore give rise to a full range of N-linked complex carbohydrates, but no hybrid moieties and only $Man_{3\text{-}5}GlcNAc_2Asn$ oligomannosyl moieties. Interestingly, however, the most dramatic effects of the loss of Dol-P-Man synthase are reflected in the inability of GPI-anchored glycoproteins to be expressed at the cell surface (Fatemi and Tartakoff, 1988) because Dol-P-Man is required as a donor of Man in the synthesis of the GPI anchor (see Section II.D). Several other mutants affected at different steps of GPI anchor biosynthesis are described in Section III.B.

Many other mutant cell lines that have altered lipid-linked oligosaccharide synthesis have been described (reviewed in Stanley 1984;

Stoll *et al.*, 1990; Nishikawa, 1991). They are generally characterized by temperature sensitivity for growth and the accumulation of the $Man_5GlcNAc_2$-P-P-Dol intermediate (Fig. 1). One of these mutant types that does not appear to be temperature-sensitive, however, is the CHO mutant selected for resistance to processing inhibitors and Con A (Lehrman and Zheng, 1989). This isolate is genetically distinct from Lec9 and Lec15 (Beck *et al.*, 1990) and may be affected in the orientation or localization of Dol-P-Man or $Man_5GlcNAc_2$-P-P-Dol (Zeng and Lehrman, 1990).

E. Mutants Affected in the Biosynthesis of GPI Anchors

As discussed earlier, selections against cell surface expression of glycoproteins often give rise to glycosylation mutants. This is particularly true for glycoproteins that are attached to the membrane by a GPI anchor (Fig. 3). Of nine distinct Thy 1^- complementation groups selected from lymphoma cells, seven are defective in GPI anchor synthesis or attachment (reviewed in Hyman, 1988). Recent analyses of the GPI anchor intermediates found in some of these mutants have identified three mutations that inhibit the synthesis of GlcNAc-PI, and two mutations that inhibit the addition of the Man added subsequent to the three Mans transferred from Dol-P-Man (Stevens and Raetz, 1991; Sugiyama *et al.*, 1991). Two classes of related mutants affected in the synthesis of GlcNAc-PI and its de-N-acetylation have been isolated from K562 cells sorted for lack of surface expression of decay accelerating factor (Hirose *et al.*, 1992).

F. Mutants That Express Multiple Glycosylation Mutations

Mutants with more than one glycosylation mutation have been selected from CHO cell populations to obtain lines that synthesize severely truncated carbohydrates affecting, in many cases, both O- and N-linked carbohydrate structures. Several mutants with two glycosylation mutations (Stanley *et al.*, 1975a, 1991; Stanley, 1989; Chaney *et al.*, 1989; Ravdin *et al.*, 1989; Zeng and Lehrman, 1991), two mutants with three mutations, and one mutant with four glycosylation mutations have been described (Stanley, 1989). The most minimal structure that has been described at N-linked complex sites is $Man_3GlcNAc_2Asn$ in the PIR.Lec1 mutant (Zeng and Lehrman, 1991). The complete absence of glycosylation at O-linked sites can be achieved with the ldlD mutant grown under appropriate culture conditions (Krieger *et al.*, 1989). If truncated carbohydrates are desired at both N- and O-linked sites, Lec3.2.8.1 cells should synthesize $Man_5GlcNAc_2Asn$ at complex N-linked sites and only GalNAc at O-linked sites (Stanley, 1989) while ldlD.Lec1 cells should synthesize $Man_5GlcNAc_2Asn$ at N-linked complex sites and add no sugar to O-linked sites. It should be possible to obtain a viable mutant that would synthesize truncated oligomannosyl carbohydrates ($Man_{3-5}GlcNAc_2Asn$), no N-linked complex moieties, and no O-linked moieties by combining the PIR, Lec1, and ldlD mutations.

IV. Use of Mutants to Analyze Carbohydrate Function

Glycosylation mutants may be used to investigate the functions of carbohydrates on glycoproteins endogenous to the mutant cell and on glycoproteins introduced into the cell by viral infection or by transfection of a cloned gene. The effects of altered carbohydrates on the properties of a particular glycoprotein are usually monitored using specific antibodies and biological assays. It is important to remember that altered properties expressed in this experimental situation might be due to a direct effect of the carbohydrates attached to the glycoprotein of interest or to an indirect effect of the altered properties of a glycoprotein required for the biosynthesis, transport, or biological activity of the glycoprotein of interest.

In addition to their functions on individual glycoproteins, carbohydrates also function as a general class of molecules expressed at the cell surface on a multitude of glycoconjugates, providing the molecular frontier of the cell. Cell surface carbohydrates present on either glycoproteins or glycolipids provide specific receptors for certain viruses, mycoplasmas, amoebae, and bacteria. Carbohydrates have also been implicated as important mediators of cell–cell interactions that occur during development, immune responses, inflammation, and the establishment of metastases. Glycosylation mutants, because they express novel arrays of carbohydrate structures at the cell surface, have been useful in investigations of the specificity of many of these recognition phenomena. In the following sections, specific examples of the use of glycosylation mutants in identifying roles for carbohydrates will be described and potential uses for glycosylation mutants in future studies will be discussed.

A. *Glycosylation and Viral Infectivity*

One of the first hypotheses concerning carbohydrate function to be readily tested using a glycosylation mutant was the notion that vesicular stomatitis virus (VSV) needed to possess sialic acid in order to infect mammalian cells (Schloemer and Wagner, 1975). If this were the case, VSV produced in Lec1 or Clone 15B CHO mutants (Table I) should not be infectious. However, VSV from both these cell lines is normally infectious, despite the complete lack of sialylated, complex N-linked carbohydrates (Schlesinger *et al.* 1976; Robertson *et al.*, 1978). Similar results with respect to infectivity were obtained with Sindbis virus (Schlesinger *et al.* 1976). However, the altered carbohydrates of Sindbis produced in Clone 15B did correlate with a reduced hemagglutination titer (Schlesinger *et al.* 1976).

Although viruses lacking complex N-linked carbohydrates were infectious in the cases tested, glycosylation mutants expressing truncated carbohydrates at the cell surface were often refractory to virus-induced cell fusion, which may be an important method of virus spreading *in vivo*. For example, Clone 15B cells were 4- to 10-fold resistant to both fusion from within and fusion from without by Newcastle disease virus (Polos and Gallagher, 1979). The BHK mutants $Ric^{R}14$ (GlcNAc-TI deficient) and $Ric^{R}21$ (GlcNAc-TII deficient) were refractory to fusion by herpes simplex virus (HSV), although the infectivity of the HSV produced was normal (Campadelli-Fiume *et al.*, 1982; Serafini-Cessi, *et al.*, 1983). Two other examples of glycosylation mutants being resistant to virus-induced fusion are the Lec2 CHO mutant (Stanley, *et al.*, 1975a), which has a broad deficiency in surface sialic acid residues, and the Sil KB mutants, which are deficient in a sialyltransferase that seems specific for O-linked carbohydrates (Toyama *et al.*, 1983). The latter mutants were actually selected for resistance to virus-induced cell fusion. Thus, enveloped viruses that themselves express truncated carbohydrates lacking sialic acid appear to have essentially unchanged infectivity and fusion abilities with wild-type cells. However, glycosylation-defective cells that express truncated cell surface carbohydrates are refractory to virus-induced cell fusion. These results demonstrate a correlation between virus-induced cell fusion and normal cell surface carbohydrate expression that has also been observed in the ability of human immunodeficiency virus (HIV) to form syncytia in cultured cells (Walker *et al.*, 1987; Montefiori *et al.*, 1988). Because the minimal truncation that leads to altered fusion ability of a glycosylation mutant is loss of terminal sialic acid residues (Stanley *et al.*, 1975a; Toyama *et al.* 1983) and because all other truncations in carbohydrate structure due to mutation or treatment with glycosylation inhibitors also involve loss of sialic acid residues, it would seem that the presence of sialic acid on cell surface carbohydrates is necessary for a cell to be fused by an enveloped virus. This, of course, may reflect a requirement for sialic acid in the initial binding of the virus rather than a specific involvement of sialic acid in the fusion process.

B. Glycosylation and Cell–Cell Fusion

Fusion between cells *in vivo* is a particular feature of muscle development as myoblasts fuse to form myotubes. This fusion event can be achieved in cell culture using a myoblast cell line. Mutant myoblasts deficient in adding sialic acid and galactose residues to membrane glycoproteins are not altered in their fusion abilities (Gilfix and Sanwal, 1982). However, L6 mutants defective in the synthesis of Dol-P-Man are unable to fuse (Parfett *et al.* 1983; Cates *et al.*, 1984). Since it is now clear that a deficiency in Dol-P-Man affects N-linked carbohydrates as well as the biosynthesis of GPI anchors (see Section II.E), the reduced fusion ability of the myoblast glycosylation mutants may reflect a role for carbohydrates in fusion or an inability to anchor glycoprotein(s) required for fusion in the plasma membrane. Determining if carbohydrates have a direct functional role in cell–cell fusion will, therefore, require much more experimentation.

C. Glycosylation and Intracellular Compartmentalization

Roles for carbohydrates in the ultimate localization of glycoproteins within the cell may be direct or indirect and glycosylation mutants have provided insight into both these functions. A direct role for carbohydrates is in targeting lysosomal enzymes to lysosomes via phosphosylated Man residues on N-linked oligomannosyl carbohydrates (von Figura and Hasilik, 1986; Kornfeld and Mellman, 1989). Human fibroblasts from patients with I cell disease lack the transferase activity that adds P-GlcNAc to oligomannosyl carbohydrates. However, I cells are very difficult to culture and no somatic cell mutant with a lesion in the phosphotransferase has been isolated from lines that are adapted to culture. Glycosylation mutants affected in other reactions, however, do exhibit changes in lysosomal targeting. For example, B211 CHO cells that have a defect in glucosylation of oligosaccharride-dolichols (Table I) exhibit decreased activity of several acid hydrolases that are not phosphosylated or internalized by the Man-6-P receptor, in contrast to hydrolases secreted from parental cells (Krag and Robbins, 1982). Lymphoma mutants deficient in glucosidase II exhibit decreased phosphosylation of oligomannosyl carbohydrates and all phosphates are present as monoesters rather than diesters (Gabel and Kornfeld, 1982). However, Thy 1E mutants deficient in Dol-P mannose have normal levels of phosphosylation of lysosomal enzymes and phosphate is present as both the mono- and diester. The spectrum of phosphosylated oligomannosyl structures synthesized by these mutants is, however, well recognized by the Man-6-P receptor (Gabel and Kornfeld, 1982).

Another important role for carbohydrates in compartmentalization is their presence in the GPI anchor that targets proteins to the cell surface. Glycosylation mutants with an inability to synthesize GPI anchors (Table I) were instrumental in revealing the importance of the glycan portion of the anchor. In its absence, proteins are either degraded rapidly or secreted from the cell. It seems likely that the specificity for membrane targeting of GPI-anchored proteins will be found to reside in the phosphatidylinositol region of the anchor (Brown and Rose, 1992). However, since different anchors carry different sugar extensions of the core glycan (Ferguson and Williams, 1988), it is possible that a role for the carbohydrates of the anchor may exist. However, it is clear from studies with MDCK mutants affected in terminal glycosylation of N-linked carbohydrates that alterations in the complex structures associated with the typically GPI-anchored proteins targeted to the apical membrane do not affect their localization (Green *et al.* 1981; Brandli *et al.* 1988).

D. Glycosylation and Adhesion of Pathogens

Because glycosylation mutants express novel arrays of cell surface carbohydrates, they have been very useful in studies of specific carbohydrate recognition by pathogens. One of the first studies using glycosylation mutants for this purpose compared the attachment of pili from *Neisseria gonorrhoeae* with CHO mutants expressing truncated carbohydrates (Gubish *et al.*, 1982). The pili showed significantly reduced attachment to clone 13 CHO cells and even more reduced attachment (~50% compared with wild type) to clone 15B CHO cells, indicating that cells expressing sialylated lactosamines on surface glycoproteins would be preferentially infected. In contrast, pathogenic *Escherichia coli* bind preferentially to BHK glycosylation mutants expressing an increased proportion of oligomannosyl carbohydrates (Firon *et al.*, 1985) whereas *Entamoeba histolytica* bind to carbohydrate with terminal Gal or GalNAc residues. CHO mutants lacking Gal residues on N-linked carbohydrates are very deficient in binding the amoebae, whereas mutants lacking only terminal sialic acid residues exhibit increased binding, consistent with the presence of a Gal-binding lectin on the amoebae (Li *et al.*, 1988, 1989; Ravdin *et al.*, 1989). The absence of Gal-containing N-linked carbohydrates correlates with markedly reduced binding at 4°C (Li *et al.*, 1988, 1989; Ravdin *et al.*, 1989), but the difference is only marginal at 37°C (Ravdin *et al.*, 1989). Thus it seems that O-linked, Gal-containing carbohydrates can act as receptors at a physiological temperature. Evidence consistent with this was obtained from the ldlD.Lecl mutant, which lacks both N- and O-linked Gal-containing carbohydrates and which exhibits essentially no binding of amoebae at 37°C (Ravdin *et al.*, 1989). The ldlD.Lecl mutant was also highly resistant to amoebically induced cytolysis (Ravdin *et al.*, 1989). In contrast, the 15B or Lec1 CHO mutant, which expresses a normal spectrum of O-linked carbohydrates, was only partially resistant to cytolysis (Li *et al.*, 1989; Ravdin *et al.*, 1989). While these studies demonstrate the usefulness of glycosylation mutants for determining the specificity of pathogen–cell interactions, they also reveal the complexities involved in interpreting the results. It is clearly important to compare mutants with alterations in both N- and O-linked carbohydrates before reaching conclusions regarding those carbohydrates that may act as receptors.

E. Glycosylation in Tumorigenicity and Metastasis

One of the most well-studied biological correlations is the change in expression of cell surface carbohydrates that occurs following transformation of a mammalian cell (reviewed by Olden *et al.*, 1982). The first clear demonstration that expression of a glycosylation mutation could result in a change in cancerous properties was reported by Tao and Burger (1979). These authors showed that wheat germ agglutinin-resistant B16 melanoma cells, although still able to form tumors, were drastically reduced in metastatic ability. The glycosylation change expressed by the mutant was an increase in $\alpha(1,3)$fucosylation of lactosamine units (Finne *et al.*, 1982). Revertants selected for loss of $\alpha(1,3)$fucose expression were also reverted for their metastatic properties (Finne *et al.*, 1989). Similar studies were performed by Kerbel and colleagues, who selected lectin-resistant MDAY-D2 mutants that possessed reduced metastatic properties (Kerbel, 1979). In these mutants, the correlation with altered metastatic abilities appears to lie in the loss of formation or the lack of extension of the $\beta(1,6)$GlcNAc to $\alpha(1,6)$Man antenna of N-linked carbohydrates (Dennis *et al.*, 1987). Although they do not metastasize, glycosylation mutants lacking GlcNAc-TV or expressing $\alpha(1,3)$fucosylated lactosamines are able to form tumors (Tao and Burger, 1977: Kerbel, 1979; Ripka *et al.*, 1986b; Dennis *et al.*, 1987). In fact, most glycosylation-defective cells are able to form tumors. Among 15 CHO glycosylation mutants, the Lec9 mutant was the only one found not to give rise to tumors (Ripka *et al.*, 1986b). Independent Lec9 mutants were nontumorigenic, and independent revertants or double mutants such as Lec9.Lec1 expressing the Lec9 mutation (causes an increase in the proportion of $\beta(1,6)$GlcNAc-branched N-linked carbohydrates) and a Lec1 mutation (truncates all complex structures to $Man_5GlcNAc_2Asn$) had regained their tumorigenicity. The Lec9 mutants were, however, tumorigenic in irradiated mice, suggesting that their inability to form tumors in nude mice was due to their inactivation by a radiation-sensitive cell. Whether the carbohydrates expressed at the surface of Lec9 cells or the other glycosylation mutants with altered cancerous properties were directly or indirectly responsible for their reduced tumorigenicity or metastasis remains to be addressed.

Several other mutant lines expressing altered cell surface carbohydrates have been shown to have altered tumorigenic or metastatic properties (Dennis and Laferte, 1987). Since terminal sugars or whole antennae are missing in these glycosylation mutants, a general conclusion of these studies is that mature, branched, sialylated, N-linked carbohydrates are necessary for cells to metastasize. However, this is clearly not the case for solid tumor formation because many mutants expressing severely truncated carbohydrates are normally tumorigenic. Since sialic acid is the most common sugar missing due to truncation, it is tempting to speculate that sialic acid on fully branched N-linked carbohydrates is important for metastasis. In addition, the presence of an $\alpha(1,3)$fucose on sialylated lactosamine units would generate the SA-Le^X determinant, which is recognized by at least two members of the selectin family of molecules (Polley *et al.*, 1991, and references therein). Interactions with this class of molecules on endothelial cells might be one mechanism by which metastatic cells that express the SA-Le^X determinant leave the bloodstream to localize in tissues.

F. Glycosylation Mutants and Cell Adhesion

Mammalian cells interact both with extracellular matrix molecules and with other cells during development and differentiation. Glycosylation mutants provide model cell lines that possess an altered array of cell surface carbohydrates for investigating possible roles for carbohydrates in cell adhesion phenomena. Although not exploited recently in studies of cell adhesion to matrix, several reports in the late 1970s described differences in the ability of glycosylation mutants to adhere to a plastic surface (Hughes, 1978; Juliano, 1978, and references therein). However, opposite results were obtained with glycosylation mutants expressing the same carbohydrate defect on a different cellular background. Hughes and colleagues (1978) found that BHK glycosylation mutant cells with a GlcNAc-TI defect were less adherent to plastic than parental cells, whereas Juliano (1978) showed that Lec1 CHO mutants (termed C2P1 in that report) with a GlcNAc-TI defect adhered with the same properties as wild-type CHO cells. However, Lec1 cells were more readily released by trypsin than wild-type CHO cells. Similar properties have been reported for other CHO glycosylation mutants, which have generally been found to be more rounded rather than fibroblastic on plastic, more easily detached from plastic, and more clumpy in suspension culture (Stanley and Sudo, 1981; Stanley, 1983).

Studies of an MDAY-D2 glycosylation mutant indicated a role for carbohydrates in cellular adhesion to collagen and fibronectin but not to laminin. Adherence to collagen type IV and fibronectin was increased in MDAY-D2 mutants lacking surface sialic acid and Gal residues (Dennis *et al.*, 1982). Although these studies were provocative, they have not been pursued and too few data are available to come to general conclusions. Now that much more is known about the domain structure of molecules of the extracellular matrix and it has become clear that molecules with lectin domains abound in animal tissues (Drickamer, 1988), a reexamination of the adhesion properties of well-characterized glycosylation mutants may provide valuable insights into the molecular mechanisms of cell adhesion. The mutants may be used directly or after modification by a glycosyltransferase that adds a specific sugar to available carbohydrate acceptors on membrane glycoproteins. Using exogenously added UDP–Gal and β(1,4)galactosyltransferase, Cornil *et al.*, (1990) showed that terminal Gal residues mediated binding of MDAY-D2 cells to microvascular endothelial cells by controlling the addition of Gal to a mutant of the MDW4 type (Table I).

The comparatively recent demonstration that specific carbohydrates are recognized by the selectin group of molecules represents another area in which glycosylation mutants can provide molecular insights. In cell–cell binding assays performed with endothelial cell cultures or platelets, the carbohydrate ligand recognized with highest affinity by E-selectin (formerly ELAM-1) and P-selectin (formerly PADGEM or GMP-140) is the sialylated Lewis^{X} determinant. This terminal tetrasaccharide is expressed by the LEC11 CHO cell mutant, which has been shown to bind both E- and P-selectins expressed by activated endothelial cells or platelets, respectively (Phillips *et al.*, 1990; Kumar *et al.*, 1991; Polley *et al.*, 1991). The closely related CHO mutant LEC12 that expresses the Le^{X} (but not the SA-Le^{X}) determinant is not bound by activated endothelial cells nor activated platelets. Both CHO mutant lines express the VIM-2 determinant (Potvin and Stanley, 1991), which is therefore not behaving as a ligand for E-selectin or P-selectin in cell–cell binding assays, despite the fact that a glycolipid with the

VIM-2 carbohydrate structure bound COS cell transfectants expressing E-selectin (Tiemeyer *et al.*, 1991). The ligand for L-selectin is a mucin in which the critical carbohydrates recognized in the binding of lymphocytes to high-endothelial venules are presented on a specific protein backbone (Lasky *et al.*, 1992). Evidence that the ligand for P-selectin involves a specific protein that presents the required carbohydrates for recognition has also been reported (Larsen *et al.*, 1992; Moore *et al.*, 1992). Thus, in attempting to establish assays that reflect these recognition events, it will be important to express the appropriate protein in a cell that synthesizes the correct carbohydrates in order to generate the relevant ligand. Glycosylation mutants that express developmentally regulated structures such as those carrying α(1,3)fucose or the bisecting GlcNAc should continue to be useful in such assays either directly, as in the case of LEC11, or after transfection with a gene encoding the appropriate ligand protein.

A summary of the selectin binding and carbohydrate-specific monoclonal antibody binding properties of LEC11 and LEC12 cells is given in Stanley (1992; Table III) and Table I. Similar results were obtained for E-selectin with CHO cells transfected with cDNAs encoding α(1,3)-Fuc-T enzymes that generate the Le^X or SA-Le^X determinant (Lowe *et al.*, 1990, 1991). Despite these apparently straightforward results, the physiological ligands for E-selectin or P-selectin are not yet known. This becomes clear from the fact that a carbohydrate not expressed by LEC11 cells binds to E-selectin (Berg *et al.*, 1991) and from experiments in which different assay methods for the same molecular interactions have given different results. For example, when recombinant P-selectin was expressed at the surface of transfected CHO cells, it was not as well recognized by LEC11 CHO cells (Larsen *et al.*, 1992) as when it was expressed on activated platelets (Polley *et al.*, 1991). Second, a cloned α(1,3)fucosyltransferase cDNA expressed in CHO cells caused them to bind recombinant E-selectin expressed on mouse cells (Goelz *et al.*, 1990), whereas the same α(1,3)fucosyltransferase expressed to even higher specific activity from genomic DNA in CHO transfectants did *not* cause the latter to bind to activated endothelial cells (Lowe *et al.*, 1991; Kumar *et al.*, 1991). Clearly, the method of presentation of the selectin and its ligand is of critical importance in determining if and at what level binding may occur. Assays using activated cells that express endogenous selectins would seem the most likely to approximate physiological conditions. Assays involving molecules expressed from transfected genes will depend on the level of expression of the gene product and the molecular density achieved at the cell surface (Larkin *et al.*, 1992). Although not exactly physiological, such assays should be more likely to detect relevant interactions than those involving recombinant selectin molecules bound to plastic or carbohydrate ligands separated on thin-layer plates (Tiemeyer *et al.*, 1991; Larkin *et al.*, 1992). Ultimately, more sophisticated assays performed *in vivo* will be required to identify the carbohydrate-containing ligands of E-selectin and P-selectin. Meanwhile glycosylation mutants may be useful in directing the search for biologically relevant carbohydrate ligands.

G. *Glycosylation Mutants and Glycoprotein Function*

Because glycosylation mutants express altered carbohydrates on endogenous as well as introduced glycoproteins, they are ideal hosts for investigating the role of carbohydrates in the functions of glycoproteins. As mentioned in Section III, endogenous glycoproteins affected by altered glycosylation include the LDL receptor, the Man-6-P receptor in CHO cells, and Thy 1 in BW5147 lymphoma cells. The LDL receptor of CHO cells was found to function normally with highly truncated N-linked carbohydrates (Kingsley *et al.*, 1986b) but, in the absence of O-linked carbohydrates, it was unstable at the cell surface (Kozarsky *et al.*, 1988b). Thus, one function of the O-linked carbohydrates of the LDL receptor seems to be protection from cleavage at the plasma membrane by a protease. The mechanism by which Man-6-P receptor function is reduced in mutants such as B4-2-1 lacking Dol-P-Man synthase activity (Stoll, *et al.*, 1982) is not known, but is presumably due to an indirect effect of carbohydrate deficiency. Reduced Dol-P-Man synthase activity in BW5147 mutants causes Thy 1 molecules to be secreted rather than localized to the plasma membrane due to the lack of a glycophosphatidylinositol anchor (Conzelmann *et al.*, 1988; Fatemi and Tartakoff, 1988).

In contrast to the dramatic effects of mutations affecting GPI anchor formation, severe truncation of N-linked carbohydrates has no detectable effect on the localization, distribution, and turnover of many glycoproteins of the CHO cell membrane (Fitzgerald *et al.*, 1984). Lec1 CHO mutants that do not synthesize complex or hybrid N-linked carbohydrates, but express only oligomannosyl N-linked carbohydrates, exhibit no change in the localization or functional properties of the LDL receptor (Kingsley *et al.*, 1986b) or the hexose transporter (Haspel *et al.*, 1988). Similarly, no change in the expression of glycophorin A was observed in Lec2 or Lec8 CHO cells that synthesize N-linked carbohydrates lacking sialic acid or sialic acid and Gal residues, respectively (Remaley *et al.*, 1991). However, several plasma membrane proteins are severely affected when produced in the ldlD CHO mutant under conditions that prohibit the synthesis of O-linked carbohydrates. In the absence of O-linked sugars, the LDL receptor (Kozarsky *et al.*, 1988b), the Epstein–Barr virus envelope glycoprotein (Kozarsky *et al.*, 1988a), and decay accelerating factor (Reddy *et al.*, 1989) are transported with apparently normal kinetics to the cell surface but, once there, are rapidly cleaved to a soluble form that cannot function at the membrane. Glycophorin A (Remaley *et al.*, 1991) and the interleukin 2 receptor (Kozarsky *et al.*, 1988a) are also not found at the cell surface when synthesized without O-linked carbohydrates. The IL-2 receptor lacking O-linked carbohydrates appears to be missorted intracellularly but it is not known why the glycophorin A molecule lacking O-linked carbohydrates is not expressed at the cell surface. The absence of O-linked carbohydrates has not so far been correlated with altered folding or oligomerization of a glycoprotein. However, conformational effects that inhibit transit through the secretory pathway are a common consequence of complete removal of N-linked carbohydrates from membrane glycoproteins (Hurtley and Helenius, 1989). Experiments to investigate the properties of a glycoprotein lacking N-linked carbohydrates must be performed using tunicamycin or molecules with mutated glycosylation sites because no mutant lacking N-linked carbohydrates has been isolated. Such a mutant would need to be conditional-lethal because it is clear that the presence of at least truncated N-linked carbohydrates is required for cell viability.

V. Use of Glycosylation Mutants to Optimize Carbohydrate Function: Glycosylation Engineering

Once the function of the carbohydrates associated with a glycoprotein have been determined, glycosylation mutants can be used to synthesize that glycoprotein with the optimal complement of carbohydrate structures. Thus to reduce heterogeneity, it might be desirable to use a mutant that synthesizes the most truncated carbohydrates known to be compatible with optimum biological activity. Producing a glycoprotein with minimal carbohydrate heterogeneity should be advantageous for purification, crystallization, and quality control. It may also be useful in tissue targeting. For example, if it were desirable to target a therapeutic glycoprotein to the liver, it could be synthesized in Lec2 or Lec3.2 CHO cells (Table I) so that it would lack sialic acid residues on N- and O-linked carbohydrates and be efficiently targeted to hepatocytes by the binding of its terminal Gal residues to the asialoglycoprotein receptor.

A specific instance in which truncated carbohydrates are needed and have been used for appropriate tissue targeting is in glucocerebrosidase used to treat patients with Gaucher's disease (Barton *et al.*, 1990). The enzyme used in therapy is purified from placenta and subsequently treated with three exoglycosidases to remove sialic acid, Gal, and GlcNAc and to generate terminal Man residues, which cause targeting of the enzyme to cells of the reticulo-endothelial system. A much simpler way of obtaining glucocerebrosidase with the desired carbohydrates would be to synthesize the recombinant glycoprotein in Lec1 CHO cells (Table I). This product would carry $Man_5GlcNAc_2Asn$ at each N-linked glycosylation site and should be even more avidly recognized by the mannose receptor of reticulo-endothelial cells than the exposed Man_3Gn_2Asn core of the exoglycosidase-treated enzyme.

Other glycosylation mutants that may be useful for tissue targeting purposes are those dominant mutants that express a new glycosyltransferase (Table I). Analogous cell lines may also be engineered by transfection of a cloned glycosyltransferase gene. In this manner, stable CHO lines expressing an $\alpha(2,6)$ sialyltransferase (Lee *et al.*, 1989), an $\alpha(1,3)$fucosyltransferase (Lowe *et al.*, 1990, 1991; Goelz *et al.*, 1990, Kumar *et al.*, 1991), or an $\alpha(1,3)$galactosyltransferase (Smith *et al.*, 1990) have been established. Cells expressing a transferase that modifies terminal glycosylation might be helpful for tissue targeting applications as well as for causing glycosylation to be more complete. An example of the former case would be the production of a glycoprotein in a LEC11 CHO cell in order to target it to activated endothelial cells due to the presence of the SA-LE^X determinant on its carbohydrates (Table I). An example of the latter case, in which more complete glycosylation would be desirable, is erythropoietin. Recombinant erythropoietin has optimal *in vivo* biological activity when it carries tetraantennary, N-linked carbohydrates that are fully sialylated to prevent clearance to the liver by the asialoglycoprotein receptor (reviewed by Takeuchi and Kobata, 1991). More complete sialylation might be achieved by producing erythropoietin in a CHO cell expressing high levels of a cloned sialyltransferase. Increased branching of N-linked carbohydrates may also be induced by transfection of appropriate GlcNAc-transferases. However, a variant CHO line that synthesizes a high proportion of tetraantennary carbohydrates on erythropoietin has already been identified (Takeuchi *et al.*, 1989).

Clearly glycosylation mutants and cell lines engineered to express particular glycosyltransferases have a lot of potential for use in glycosylation engineering. The possibilities outlined above are but a few theoretical examples that illustrate the potential power of this approach. Since most therapeutic products of biotechnology companies are glycoproteins, the optimization of glycosylation pattern should be an important consideration from the very beginning of product development. Pilot experiments in which products synthesized by an appropriate group of glycosylation mutants are functionally compared might greatly facilitate the rapid development of a therapeutic with properties close to those of the native molecule.

VI. Conclusions

Although the range of carbohydrate structures synthesized by a mammalian cell is rather large and the functions of carbohydrates often vary depending on the molecule with which they are associated, there are now several approaches to determining structure–function relationships directly. One of these approaches is the use of glycosylation mutants that synthesize altered carbohydrates. By judicious choice of a mutant host, endogenous or introduced glycoproteins can be obtained with N- and/or O-linked carbohydrates that are either specifically truncated or specifically embellished by the addition of a particular sugar. The biological and biochemical properties of molecules with such altered carbohydrates can be compared to identify specific functions for particular carbohydrate structures. Once functional roles are clear, the mutant cells can be used for glycosylation engineering to optimize the carbohydrate complement of a recombinant glycoprotein. Such optimization might improve the homogeneity of a preparation, the ability to purify the glycoprotein or to crystallize it, the biological activity, the intracellular localization, or the tissue targeting of the glycoprotein. Glycosylation engineering using cell lines altered in carbohydrate biosynthesis has already provided a powerful approach for investigating basic properties of glycoproteins and the functions of carbohydrates. In the future, glycosylation engineering should have an application in optimizing the properties of glycoproteins.

References

Barton, N. W., Furbish, F. S., Murray, G. J., Garfield, M., and Brady, R. O. (1990). Therapeutic response to intravenous infusions of glucocerebrosidase in a patient with Gaucher disease. *Proc. Natl. Acad. Sci. USA* **87,** 1913–1916.

Beck, P. J., Gething, M-J., Sambrook, J., and Lehrman, M. A. (1990). Complementing mutant alleles define three loci involved in mannosylation of Man_5-$GlcNAc_2$-P-P-Dol in Chinese hamster ovary cells. *Som. Cell. Mol. Genet.* **16,** 539–548.

Berg, E. L., Robinson, M. K., Mansson, O., Butcher, E. C., and Magnani, J. L. (1991). A carbohydrate domain common to both sialyl Le^a and sialyl Le^x is recognized by the endothelial cell leukocyte adhesion molecule ELAM-1. *J. Biol. Chem.* **266,** 14869–14872.

Brändli, A. W. (1991). Mammalian glycosylation mutants as tools for the analysis and reconstitution of protein transport. *Biochem. J.* **276,** 1–12.

Brändli, A. W., Hansson, G. C., Rodriquez-Boulan, E., and Simons, K. (1988). A polarized epithelial cell mutant deficient in translocation of UDP–galactose into the Golgi complex. *J. Biol. Chem.* **263,** 16283–16290.

Briles, E. B. (1982). Lectin-resistant cell surface variants of eukaryotic cells. *Int. Rev. Cytol.* **75,** 101–165.

Brown, D. A., and Rose J. K. (1992). Sorting of GPI-anchored proteins to glycolipid-enriched membrane subdomains during transport to the apical cell surface. *Cell* **68,** 533–544.

Campadelli-Fiume, G., Poletti, L., Dall-Olio, F., and Serafini-Cessi, F. (1982). Infectivity and glycoprotein processing of herpes simplex virus type I grown in a ricin-resistant cell line deficient in *N*-acetylglucosaminylytransferase. *J. Virol.* **43,** 1061–1071.

Campbell, C., and Stanley, P. (1983). Regulatory mutations in CHO cells induce expression of the mouse embryonic antigen SSEA-1. *Cell* **35,** 303–309.

Campbell, C., and Stanley, P. (1984). A dominant mutation to ricin-resistance in CHO cells induces UDP–GlcNAc: Glycopeptide β-4-*N*-acetyllglucosaminyltransferase III activity. *J. Biol. Chem.* **261,** 13370–13378.

Carlsson, S. R., Sasaki, H., and Fukuda, M. (1986). Structural variations of O-linked oligosaccharides present in leukosialin isolated from erythroid, myeloid, and T-lymphoid cell lines. *J. Biol. Chem.* **261,** 12787–12795.

Cates, G. A., Brickenden, A. M., and Sanwal, B. D. (1984). Possible involvement of a cell surface glycoprotein in the differentiation of skeletal myoblasts. *J. Biol. Chem.* **259,** 2646–2650.

Chaney, W., and Stanley, P. (1986). Lec1A Chinese hamster ovary cell mutants appear to arise from a structural alteration in *N*-acetylglucosaminyltransferase I. *J. Biol. Chem.* **261,** 10551–10557.

Chaney, W., Sundaram, S., Friedman, N., and Stanley, P. (1989). The Lec4A CHO glycosylation mutant arises from miscompartmentalization of a Golgi glycosyltransferase. *J. Cell Biol.* **109,** 2089–2096.

Chapman, A., Trowbridge, I. S., Hyman, R., and Kornfeld, S. (1979). Structure of the lipid-linked oligosaccharides that accumulate in class E Thy-1-negative mutant lymphomas. *Cell* **17,** 509–515.

Cifone, M. A., Hynes, R. O., and Baker, R. M. (1979). Characteristics of concanavalin A-resistant Chinese hamster ovary cells and certain revertants. *J. Cell. Physiol.* **100,** 39–54.

Conzelmann, A., and Kornfeld, S. (1984a). β-linked *N*-acetylgalactosamine residues present at the nonreducing termini of O-linked oligosaccacharides of a cloned murine cytotoxic T lymphocyte line are absent in a *Vicia villosa* lectin-resistant mutant cell line. *J. Biol. Chem.* **259,** 12528–12535.

Conzelmann, A., and Kornfeld, S. (1984b). A murine cytotoxic T lymphocyte cell line resistant to *Vicia villosa* lectin is deficient in UDP–GalNAc: β-Galactose β1,4-*N*-acetylgalactosaminyltransferase. *J. Biol. Chem.* **259,** 12536–12542.

Conzelmann, A., Spiazzi, A., Bron, C., and Hyman, R. (1988). No glycolipid anchors are added to Thy-1 glycoprotein in Thy-1-negative mutant thymoma cells of four different complementation classes. *Mol. Cell. Biol.* **8,** 674–678.

Cornil, I. Kerbel, R. S., and Dennis, J. W. (1990). Tumor cell surface β-1,4-linked galactose binds to lectin(s) on microvascular endothelial cells and contributes to organ colonization. *J. Cell Biol.* **111,** 773–781.

Cumming, D. A. (1991). Glycosylation of recombinant protein therapeutics: Control and functional implications. *Glycobiology* **1,** 115–130.

Cummings, R. D., and Kornfeld, S. (1984). The distribution of repeating [Galβ1,4,-GlcNAcβ1,3] sequences in asparagine-linked oligosaccharides of the mouse lymphoma cell lines BW5147 and PHA^{R}2.1. *J. Biol. Chem.* **259,** 6253–6260.

Cummings, R. D., and Mattox, S. A. (1988). Retinoic acid-induced differentiation of the mouse teratocarcinoma cell line F9 is accompanied by an increase in the activity of UDP–galactose: β-D-Galactosyl-α1,3-galactosyltransferase. *J. Biol. Chem.* **263,** 511–519.

Cummings, R. D., Trowbridge, I. S., and Kornfeld, S. (1982). A mouse lymphoma cell line resistant to the leukoagglutinating lectin from *Phaseolus vulgaris* is deficient in UDP–GlcNAc: α-D-Mannoside β1,6-*N*-acetylglucosaminyltransferase. *J. Biol. Chem.* **257,** 13421–13427.

Dennis, J. W. (1986). Different metastatic phenotypes in two genetic classes of wheat germ agglutinin-resistant tumor cell mutants. *Cancer Res.* **46,** 4594–4600.

Dennis, J. W., and Laferté, S. (1987). Tumor cell surface carbohydrate and the metastatic phenotype. *Cancer Metastasis Rev.* **5,** 185–204.

Dennis, J. W., Laferté, S., Fukuda, M., Dell, A., and Carver, J. P. (1986). Asn-linked oligosaccharides in lectin-resistant tumor-cell mutants with varying metastatic potential. *Eur. J. Biochem.* **161,** 359–373.

Dennis, J. W., Laferté, S., Waghorne, C., Breitman, M. L., and Kerbel, R. S. (1987). β1,6 branching of Asn-linked oligosaccharides is directly associated with metastasis. *Science* **236,** 582–585.

Dennis, J., Waller, C., Timpl, R., and Schirrmacher, V. (1982). Surface sialic acid reduces attachment of metastatic tumor cells to collagen type IV and fibronectin. *Nature* **300,** 274–276.

Deutscher, S. L., and Hirschberg, C. B. (1986). Mechanism of galactosylation in the Golgi apparatus: A Chinese hamster ovary cell mutant deficient in translocation of UDP–galactose across Golgi vesicle membranes. *J. Biol. Chem.* **261,** 96–100.

Deutscher, S. L., Nuwayhid, N., Stanley, P., Briles, E. B., and Hirschberg, C. B. (1984). Translocation across Golgi membranes: A CHO glycosylation mutant deficient in CMP-sialic acid transport. *Cell* **39,** 295–299.

Doering, T. L., Masterson, W. J., Hart, G. W., and Englund, P. T. (1990). Biosynthesis of glycosylphosphatidylinositol membrane anchors. *J. Biol. Chem.* **265,** 611–614.

Drickamer, K. (1988). Two distinct classes of carbohydrate-recognition domains in animal lectins. *J. Biol. Chem.* **263,** 9557–9560.

Elbein, A. D. (1987). Inhibitors of the biosynthesis and processing of N-linked oligosaccharide chains. *Annu. Rev. Biochem.* **56,** 497–534.

Esko, J. D. (1991). Genetic analysis of proteoglycan structure, function and metabolism. *Curr. Opin. Cell Biol.* **3,** 805–816.

Fatemi, S. H., and Tartakoff, A. M. (1988). The phenotype of five classes of T-lymphoma mutants: Defective glycophospholipid anchoring, rapid degradation, and secretion of Thy-1 glycoprotein. *J. Biol. Chem.* **263,** 1288–1294.

Ferguson, M. A. J., and Williams, A. F. (1988). Cell-surface anchoring of proteins via glycosylphosphatidylinositol structures. *Annu. Rev. Biochem.* **57,** 285–320.

Finne, J., Burger, M. M., and Prieels, J-P. (1982). Enzymatic basis for a lectin-resistant phenotype: Increase in a fucosyltransferase in mouse melanoma cells. *J. Cell Biol.* **92,** 277–282.

Finne, J., Castori, S., Feizi, T., and Burger, M. M. (1989). Lectin-resistant variants and revertants of mouse melanoma cells: Differential expression of a fucosylated cell-surface antigen and altered metastasizing capacity. *Int. J. Cancer* **43,** 300–304.

Firon, N., Duksin, D., and Sharon, N. (1985). Mannose-specific adherence of *Escherichia coli* to BHK cells that differ in their glycosylation patterns. *FEMS Microbiol. Lett.* **27,** 161–165.

Fitzgerald, L. A., Denny, J. B., Baumbach, G. A., Ketcham, C. M., and Roberts, R. M. (1984). Effect of altered oligosaccharide structure on the cell surface number, distribution and turnover of the high molecular weight acidic glycoproteins of CHO cells. *J. Cell Sci.* **67,** 1–23.

Gabel, C. A., and Kornfeld, S. (1982). Lysosomal enzyme phosphorylation in mouse lymphoma cell lines with altered asparagine-linked oligosaccharides. *J. Biol. Chem.* **257,** 10605–10612.

Gilfix, B. M., and Sanwal, B. D. (1982). Lectin resistant myoblasts. *In* "Muscle Development: Molecular and Cellular Control" (M. L. Pearson and H. F. Epstein, eds.), pp. 329–336. Cold Spring Harbor Laboratory, Cold Spring Harbor, New York.

Gleeson, P. A., Feeney, J., and Hughes, R. C. (1985). Structures of N-glycans of a ricin-resistant mutant of baby hamster kidney cells: Synthesis of high-mannose and hybrid N-glycans. *Biochemistry* **24,** 493–503.

Goelz, S. E., Hession, C., Goff, D., Griffiths, B., Tizard, R., Newman, B., Chi-Rosso, G., and Lobb, R. (1990). ELFT: A gene that directs the expression of an ELAM-1 ligand. *Cell* **63,** 1349–1356.

Gottlieb, C., Baenziger, J., and Kornfeld, S. (1975). Deficient uridine diphosphate-*N*-acetylglucosamine: Glycoprotein *N*-acetylglucosaminyltransferase activity in a clone of Chinese hamster ovary cells with altered surface glycoproteins. *J. Biol. Chem.* **250,** 3303–3309.

Green, S. A., and Kelly, R. B. (1990). Endocytic membrane traffic to the Golgi apparatus in a regulated secretory cell line. *J. Biol. Chem.* **265,** 21269–21278.

Green, R. F., Meiss, H. K., and Rodriguez-Boulan, E. (1981). Glycosylation does not determine segregation of viral envelope proteins in the plasma membrane of epithelial cells. *J. Cell Biol.* **89,** 230–239.

Gubish, E. R., Chen, K. C. S., and Buchanan, T. M. (1982). Attachment of gonococcal pili to lectin-resistant clones of Chinese hamster ovary cells. *Infect. Immun.* **37,** 189–194.

Haltiwanger, R. S., Holt, G. D., and Hart, G. W. (1990). Enzymatic addition of *O*-GlcNAc to nuclear and cytoplasmic proteins: Identification of a uridine diphospho-*N*-acetylglucosamine:peptide β-*N*-acetylglucosaminyltransferase. *J. Biol. Chem.* **265,** 2563–2568.

Harris, R. J., Ling, V. T., and Spellman, M. W. (1992). O-linked fucose is present in the first epidermal growth factor domain of factor XII but not protein C. *J. Biol. Chem.* **267,** 5102–5107.

Hart, G. W., Haltiwanger, R. S., Holt, G. D., and Kelly, W. G. (1989). Glycosylation in the nucleus and cytoplasm. *Annu. Rev. Biochem.* **58,** 841–874.

Haspel, H. C., Revillame, J., and Rosen, O. M. (1988). Structure, biosynthesis, and function of the hexose transporter in Chinese hamster ovary cells deficient in *N*-acetylglucosaminyltransferase I activity. *J. Cell. Physiol.* **136,** 361–366.

Hearing, J., Gething, M-J., and Sambrook, J. (1989a). Addition of truncated oligosaccharides to influenza virus hemagglutinin results in its temperature-conditional cell-surface expression. *J. Cell Biol.* **108,** 355–365.

Hearing, J., Hunter, E., Rodgers, L., Gething, M-J., and Sambrook, J. (1989b). Isolation of Chinese hamster ovary cell lines temperature conditional for the cell-surface expression of integral membrane proteins. *J. Cell Biol.* **108,** 339–353.

Higgins, E. A., Siminovitch, K. A., Zhuang, D., Brockhausen, I., and Dennis, J. W. (1991). Aberrant O-linked oligosaccharide biosynthesis in lymphocytes and platelets from patients with the Wiskott–Aldrich syndrome. *J. Biol. Chem.* **266,** 6280–6290.

Hirose, S., Mohney, R. P., Mutka, S. C., Ravi, L., Singleton, D. R., Perry, G., Tartakoff, A. M., and Medof, M. E. (1992). Derivation and characterization of glycoinositol-phospholipid anchor-defective human K562 cell clones. *J. Biol. Chem.* **267,** 5272–5278.

Hirschberg, C. B., and Snider, M. D. (1987). Topography of glycosylation in the rough endoplasmic reticulum and Golgi apparatus. *Annu. Rev. Biochem.* **56,** 63–87.

Hirschberg, C. B., Baker, R. M., Perez, M., Spencer, L. A., and Watson, D. (1981). Selection of mutant Chinese hamster ovary cells altered in glycoproteins by means of tritiated fucose suicide. *Mol. Cell. Biol.* **1,** 902–909.

Holt, G. D., and Hart, G. W. (1986). The subcellular distribution of terminal *N*-acetylglucosamine moieties. *J. Biol. Chem.* **261,** 8049–8057.

Homans, S. W., Ferguson, M. A. J., Dwek, R. A., Rademacher, T. W., Anand, R., and Williams, A. F. (1988). Complete structure of the glycosyl phosphatidylinositol membrane anchor of rat brain Thy-l glycoprotein. *Nature* **333,** 269–272.

Hoogewerf, A. J., and Bensadoun, A. (1991). Occurrence of sulfate in an asparagine-linked complex oligosaccharide of chicken adipose lipoprotein lipase. *J. Biol. Chem.* **266,** 1048–1057.

Howard, D. R., Fukuda, M., Fukuda, M. N., and Stanley, P. (1987). The GDP-fucose: *N*-Acetylglucosaminide 3-α-L-fucosyltransferases of LEC11 and LEC12 Chinese hamster ovary mutants exhibit novel specificities for glycolipid substrates. *J. Biol. Chem.* **262,** 16830–16837.

Hubbard, S. C., and Ivatt, R. J. (1981). Synthesis and processing of asparagine-linked oligosaccharides. *Annu. Rev. Biochem.* **50,** 555–583.

Hughes, R. C. (1978). Lectin receptors and cell surface recognition. *In* "Cell Surface Carbohydrates and Biological Recognition," pp. 657–668. Alan R. Liss, New York.

Hughes, R. C., and Feeney, J. (1986). Ricin-resistant mutants of baby-hamster-kidney cells deficient in α-mannosidase-II-catalyzed processing of asparagine-linked oligosaccharides. *Eur. J. Biochem.* **158,** 227–237.

Hunt, L. A. (1980). CHO cells selected for phytohemagglutinin and Con A resistance are defective in both early and late stages of protein glycosylation. *Cell* **21,** 407–415.

Hurtley, S. M., and Helenius, A. (1989). Protein oligomerization in the endoplasmic reticulum. *Annu. Rev. Cell Biol.* **5,** 277–307.

Hyman, R. (1988). Somatic genetic analysis of the expression of cell surface molecules. *Trends Genet.* **4,** 5–8.

Hyman, R., and Stallings, V. (1974). Complementation patterns of Thy-1 variants and evidence that antigen loss variants "pre-exist" in the parental population. *J. Natl. Cancer Inst.* **52,** 429–436.

Juliano, R. L. (1978). Adhesion and detachment characteristics of Chinese hamster ovary cell membrane mutants. *J. Cell Biol.* **76,** 43–49.

Kearse, K. P., and Hart, G. W. (1991). Lymphocyte activation induces rapid changes in nuclear and cytoplasmic glycoproteins. *Proc. Natl. Acad. Sci. USA* **88,** 1701–1705.

Kerbel, R. S. (1979). Immunologic studies of membrane mutants of a highly metastatic murine tumor. *Am. J. Pathol.* **97,** 609–622.

Kingsley, D. M., Kozarsky, R. F., Hobbie, L., and Krieger, M. (1986a). Reversible defects in O-linked glycosylation and LDL receptor expression in a UDP–Gal/UDP–GalNAc 4-epimerase mutant. *Cell* **44,** 749–759.

Kingsley, D. M., Kozarsky, K. F., Sege, R. D., Segal, M., and Krieger, M. (1986b). Three types of LDL receptor-deficient mutant have pleiotropic defects in the synthesis of N-linked, O-linked and lipid-linked carbohydrate chains. *J. Cell Biol.* **102,** 1576–1585.

Kornfeld, R., and Kornfeld, S. (1985). Assembly of asparagine-linked oligosaccharides. *Annu. Rev. Biochem.* **54,** 631–664.

Kornfeld, S., and Mellman, I. (1989). The biogenesis of lysosomes. *Annu. Rev. Cell Biol.* **5,** 483–525.

Kozarsky, K. F., Call, S. M., Dower, S. K., and Krieger, M. (1988a). Abnormal intracellular sorting of O-linked carbohydrate-deficient interleukin-2 receptors. *Mol. Cell. Biol.* **8,** 3357–3363.

Kozarsky, K., Kingsley, D., and Krieger, M. (1988b). Use of a mutant cell line to study the kinetics and function of O-linked glycosylation of low density lipoprotein receptors. *Proc. Natl. Acad. Sci. USA* **85**, 4335–4339.

Krag, S. S. (1979). A concanavalin-A-resistant Chinese hamster ovary cell line is deficient in the synthesis of [^{3}H]glucosyloligosaccharide-lipid. *J. Biol. Chem.* **254**, 9167–9177.

Krag, S. S., and Robbins, A. R. (1982). A Chinese hamster ovary cell mutant deficient in glucosylation of lipid-linked oligosaccharide synthesizes lysosomal enzymes of altered structure and function. *J. Biol. Chem.* **257**, 8424–8431.

Krieger, M., Reddy, P., Kozarsky, K., Kingsley, D., Hobbie, L., and Penman, M. (1989). Analysis of the synthesis, intracellular sorting and function of glycoproteins using a mammalian cell mutant with reversible glycosylation defects. *Methods Cell Biol.* **32**, 57–84.

Kumar, R., Potvin, B., Muller, W. A., and Stanley, P. (1991). Cloning of a human α(1,3)-fucosyltransferase gene that encodes ELFT but does not confer ELAM-1 binding on CHO cell transfectants. *J. Biol. Chem.* **266**, 21777–21783.

Lamblin, G., Lhermitte, M., Klein, A., Houdret, N., Scharfman, A., Ramphal, R., and Roussel, P. (1991). The carbohydrate diversity of human respiratory mucins: A protection of the underlying mucosa? *Am. Rev. Respir. Dis.* **144**, S19–S24.

Larkin, M., Ahern, T. J., Stoll, M. S., Shaffer, M., Sako, D., O'Brien, J., Yuen, C-T., Lawson, A. M., Childs, R. A., Barone, K. M., Langer-Safer, P. R., Hasegawa, A., Kiso, M., Larsen, G. R., and Feizi, T. (1992). Spectrum of sialylated and non-sialylated fuco-oligosaccharides bound by the endothelial-leukocyte adhesion molecule E-selectin. *J. Biol. Chem.* **267**, 13661–13668.

Larsen, G. R., Sako, D., Ahern, T. J., Shaffer, M., Erban, J., Sajer, S. A., Gibson, R. M., Wagner, D. D., Furie, B. C., and Furie, B. (1992). P-selectin and E-selectin: Distinct but overlapping leukocyte ligand specificities. *J. Biol. Chem.* **267**, 11104–11110.

Lasky, L. A., Singer, M. S., Dowbenko, D., Imai, Y., Henzel, W. J., Grimley, C., Fennie, C., Gillett, N., Watson, S. R., and Rosen, S. D. (1992). An endothelial ligand for L-selectin is a novel mucin-like molecule. *Cell* **69**, 927–938.

Lee, E. U., Roth, J., and Paulson, J. C. (1989). Alteration of terminal glycosylation sequences on N-linked oligosaccharides of Chinese hamster ovary cells by expression of β-galactoside α2,6-sialyltransferase. *J. Biol. Chem.* **264**, 13848–13855.

Lehrman, M. A., and Zeng, Y. (1989). Pleiotropic resistance to glycoprotein processing inhibitors in Chinese hamster ovary cells. *J. Biol. Chem.* **264**, 1584–1593.

Li, E., and Kornfeld, S. (1978). Structure of the altered oligosaccharide present in glycoproteins from a clone of Chinese hamster ovary cells deficient in *N*-acetylglucosaminyltransferase activity. *J. Biol. Chem.* **253**, 6426–6431.

Li, E., Becker, A., and Stanley, S. L., Jr. (1988). Use of Chinese hamster ovary cells with altered glycosylation patterns to define the carbohydrate specificity of *Entamoeba histolytica* adhesion. *J. Exp. Med.* **167**, 1725–1730.

Li, E., Becker, A., and Stanley, S. L., Jr. (1989). Chinese hamster ovary cells deficient in *N*-acetylglucosaminyltransferase I activity are resistant to *Entamoeba histolytica*-mediated cytotoxicity. *Infect. Immun.* **57**, 8–12.

Liener, I. E., Sharon, N., and Goldstein, I. J. (1986). "The Lectins." Academic Press, San Diego.

Lisanti, M. P., Sargiacomo, M., Graeve, L., Saltiel, A. R., and Rodriguez-Boulan, E. (1988). Polarized apical distribution of glycosylphosphatidylinositol anchored proteins in a renal epithelial cell line. *Proc. Natl. Acad. Sci. USA* **85**, 9557–9561.

Low, M. G., and Saltiel, A. R. (1988). Structural and functional roles of glycosylphosphatidylinositol in membranes. *Science* **239**, 268–275.

Lowe, J. B., Kukowska-Latallo, J. F., Nair, R. P., Larsen, R. D., Marks, R. M., Macher, B. A., Kelly, R. J., and Ernst, L. K. (1991). Molecular cloning of a human fucosyltransferase gene that determines expression of the Lewis X and V1M-2 epitopes but not ELAM-1-dependent cell adhesion. *J. Biol. Chem.* **266,** 17467–17477.

Lowe, J., Stoolman, L. M., Nair, R. P., Larsen, R. D., Berhend, T. L., and Marks, R. M. (1990). ELAM-1-dependent cell adhesion to vascular endothelium determined by a transfected human fucosyltransferase cDNA. *Cell* **63,** 475–484.

Lubas, W. A., and Spiro, R. G. (1988). Evaluation of the role of rat liver Golgi endo-α-D-mannosidase in processing N-linked oligosaccharides. *J. Biol. Chem.* **263,** 3990–3998.

Meager, A., Ungkitchanukit, A., Nairn, R., and Hughes, R. C. (1975). Ricin resistance in baby hamster kidney cells. *Nature* **257,** 137–139.

Menon, A. K., Mayor, S., and Schwarz, R. T. (1990). Biosynthesis of glycosyl-phosphatidylinositol lipids in *Trypanosoma brucei:* Involvement of mannosylphosphoryldolichol as the mannose donor. *EMBO J.* **9,** 4249–4258.

Montefiori, D. C., Robinson, W. D., Jr., and Mitchell, W. (1988). Role of protein N-glycosylation in pathogenesis of human immunodeficiency virus type 1. *Proc. Natl. Acad. Sci. USA* **85,** 9248–9252.

Moore, K. L., Stults, N. L., Diaz, S., Smith, D. F., Cummings, R. D., Varki, A. and McEver, R. P. (1992). Identification of a specific glycoprotein ligand for P-selectin (CD62) on myeloid cells. *J. Cell Biol.* **118,** 445–456.

Nishikawa, Y. (1991). An FM3A mutant, G258, with a mutation that affects both cell growth and oligosaccharide synthesis. *Biochim. Biophys. Acta.* **1091,** 135–140.

Nishimura, H., Kawabata, S., Kisiel, W., Hase, S., Ikenaka, T., Takao, T., Shimonishi, Y., and Iwanaga, S. (1989). Identification of a disaccharide (Xyl–Glc) and a Trisaccharide (Xyl_2-Glc) O-glycosidically linked to a serine residue in the first epidermal growth factor-like domain of human factors VII and IX and protein 2 and bovine protein 2. *J. Biol. Chem.* **264,** 20320–20325.

Olden, K., Parent, J. B., and White, S. L. (1982). Carbohydrate moieties of glycoproteins: A re-evaluation of their function. *Biochim. Biophys. Acta* **650,** 209–232.

Palcic, M. M., Ripka, J., Kaur, K. J., Shoreibah, M., Hindsgaul, O., and Pierce, M. (1990). Regulation of N-acetylglucosaminyltransferase V activity: Kinetic comparisons of parental, rous sarcoma virus-transformed BHK and L-phytohemagglutinin-resistant BHK cells using synthetic analogues and an inhibitory substrate analogue. *J. Biol. Chem.* **265,** 6759–6769.

Parfett, C. L. J., Jamieson, J. C., and Wright, J. A. (1983). Changes in cell surface glycoproteins on non-differentiating L6 rat myoblasts selected for resistance to concanavalin A. *Exp. Cell Res.* **144,** 405–425.

Paulson, J. C. (1989). Glycoproteins: What are the sugar chains for? *Trends Biochem. Sci.* **14,** 272–276.

Perez-Villar, J., Hidalgo, J., and Velasco, A. (1991). Presence of terminal *N*-acetylgalactosamine residues in subregions of the endoplasmic reticulum is influenced by cell differentiation in culture. *J. Biol. Chem.* **266,** 23967–23976.

Pfeffer, S. R., and Rothman, J. E. (1987). Biosynthetic protein transport and sorting by the endoplasmic reticulum and Golgi. *Annu. Rev. Biochem.* **56,** 829–852.

Phillips, M. L., Nudelman, E., Gaeta, F. C. A., Perez, M., Singhal, A. K., Hakomori, S., and Paulson, J. C. (1990). ELAM-1 mediates cell adhesion by recognition of a carbohydrate ligand, sialosyl-Le^x. *Science* **250,** 1130–1132.

Piller, F., Piller, V., Fox, R. I., and Fukuda, M. (1988). Human T-lymphocyte activation is associated with changes in O-glycan biosynthesis. *J. Biol. Chem.* **263,** 15146–15150.

Piller, V., Piller, F., and Fukuda, M. (1990). Biosynthesis of truncated O-glycans in the T cell line Jurkat. *J. Biol. Chem.* **265,** 9264–9271.

Polley, M. J., Phillips, M. L., Wayner, E., Nudelman, E., Singhal, A. K., Hakomori, S., and Paulson, J. C. (1991). CD62 and endothelial cell–leukocyte adhesion molecule 1 (ELAM-1) recognize the same carbohydrate ligand, sialyl-Lewis X. *Proc. Natl. Acad. Sci. USA* **88,** 6224–6228.

Polos, P. G., and Gallagher, W. R. (1979). Insensitivity of a ricin-resistant mutant of Chinese hamster cells to fusion induced by Newcastle disease virus. *J. Virol.* **30,** 69–75.

Potvin, B., and Stanley, P. (1991). 5-Azacytidine induces expression of two apparently novel α(1,3)fucosyltransferase activities in CHO cells. *Cell Regul.* **2,** 989–1000.

Price, L. K. H., Choi, H. U., Rosenberg, L., and Stanley, E. R. (1992). The predominant form of secreted colony stimulating factor-1 is a proteoglycan. *J. Biol. Chem.* **267,** 2190–2199.

Ravdin, J. I., Stanley, P., Murphy, C. F., and Petri, W. A., Jr. (1989). Characterization of cell surface carbohydrate receptors for *Entamoeba histolytica* adherence lectin. *Infect. Immun.* **57,** 2179–2186.

Ray, M., Yang, J., Sundaram, S., and Stanley, P. (1991). A novel glycosylation phenotype expressed by Lec23, a CHO mutant deficient in α-glucosidase I. *J. Biol. Chem.* **266,** 22818–22825.

Reddy, P., Caras, I., and Krieger, M. (1989). Effects of O-linked glycosylation on the cell surface expression and stability of decay-accelerating factor, a glycophospholipid-anchored membrane protein. *J. Biol. Chem.* **264,** 17329–17336.

Reitman, M. L., Trowbridge, I. S., and Kornfeld, S. (1980). Mouse lymphoma cell lines resistant to pea lectin are defective in fucose metabolism. *J. Biol. Chem.* **255,** 9900–9906.

Reitman, M. L., Trowbridge, I. S., and Kornfeld, S. (1982). A lectin-resistant mouse lymphoma cell line is deficient in glucosidase II, a glycoprotein-processing enzyme. *J. Biol. Chem.* **257,** 10357–10363.

Remaley, A. T., Ugorski, M., Wu, N., Litzky, L., Burger, S. R., Moore, J. S., Fukuda, M., and Spitalnik, S. L. (1991) Expression of human glycophorin A in wild-type and glycosylation-deficient Chinese hamster ovary cells. *J. Biol. Chem.* **266,** 24176–24183.

Ripka, J., Adamany, A., and Stanley, P. (1986a). Two Chinese hamster ovary glycosylation mutants affected in the conversion of GDP–mannose to GDP–fucose. *Arch. Biochem. Biophys.* **249,** 533–545.

Ripka, J., Shin, S.-I., and Stanley, P. (1986b). Decreased tumorigenicity correlates with expression of altered cell surface carbohydrates in Lec9 CHO cells. *Mol. Cell. Biol.* **6,** 1268–1275.

Robertson, M. A., Etchison, J. R., Robertson, J. S., Summers, D. F., and Stanley, P. (1978). Specific changes in the oligosaccharide moieties of VSV grown in different lectin-resistant CHO cells. *Cell* **13,** 515–526.

Rosenwald, A., Stanley, P., and Krag, S. S. (1989). Control of carbohydrate processing: Increased β1,6-branching in the N-linked carbohydrates of Lec9 CHO mutants appears to arise from a defect in oligosaccharide-dolichol biosynthesis. *Mol. Cell. Biol.* **9,** 914–924.

Roux, L., Holojda, S., Sundblad, G., Freeze, H. H., and Varki, A. (1988). Sulfated N-linked oligosaccharides in mammalian cells. *J. Biol. Chem.* **263,** 8879–8889.

Saitoh, O., Wang, W-C., Lotan, R., and Fukuda, M. (1992). Differential glycosylation and cell surface expression of lysosomal membrane glycoproteins in sublines of a human colon cancer exhibiting distinct metastatic potentials. *J. Biol. Chem.* **267,** 5700–5711.

Sallustio, S., and Stanley, P. (1989). Novel genetic instability associated with a developmentally regulated glycosyltransferase locus in Chinese hamster ovary cells. *Som. Cell Mol. Genet.* **15,** 387–400.

Satir, B. H., Srisomsap, C., Reichman, M., and Marchase, R. B. (1990). Parafusin, an exocytic-sensitive phosphoprotein, is the primary acceptor for the glucosylphosphotransferase in *Paramecium* and rat liver. *J. Cell Biol.* **111,** 901–907.

Schachter, H. (1991). The "yellow brick road" to branched complex N-glycans. *Glycobiology* **1,** 453–461.

Schachter, H., and Roseman, S. (1980). Mammalian glycosyltransferases: Their role in the synthesis and function of complex carbohydrates and glycolipids. *In* "The Biochemistry of Glycoproteins and Proteoglycans" (W. J. Lennarz, ed.), pp. 85–160. Plenum, New York.

Schlesinger, S., Gottlieb, C., Feil, P., Gelb, N., and Kornfeld, S. (1976). Growth of enveloped RNA viruses in a line of Chinese hamster ovary cells with deficient *N*-acetylglucosaminyltransferase activity. *J. Virol.* **17,** 239–246.

Schloemer, R., and Wagner, R. (1975). Mosquito cells infected with vesicular stomatitis virus yield unsialylated virions of low infectivity. *J. Virol.* **15,** 1029–1032.

Serafini-Cessi, F., Dall'Olio, F., Scannavini, M., and Campadelli-Fiume, G. (1983). Processing of herpes simplex virus-l glycans in cells defective in glycosyl transferases of the Golgi system: Relationship to cell fusion and virion egress. *Virology* **131,** 59–70.

Smith, D. F., Larsen, R. D., Mattox, S., Lowe, J. B., and Cummings, R. D. (1990). Transfer and expression of a murine UDP–gal:β-D-gal-α1,3-galactosyltransferase gene in transfected Chinese hamster ovary cells. *J. Biol. Chem.* **265,** 6225–6234.

Stanley, P. (1980). Surface carbohydrate alterations of mutant mammalian cells selected for resistance to plant lectins. *In* "Biochemistry of Glycoproteins and Proteoglycans" (W. J. Lennarz, ed.), pp. 161–189. Plenum, New York.

Stanley, P. (1983). Selection of lectin-resistant mutants of animal cells. *In* "Methods in Enzymology" (S. Fleischer and B. Fleischer, eds.), Vol. 96, pp. 157–184. Academic Press, San Diego.

Stanley, P. (1984). Glycosylation mutants of animal cells. *Annu. Rev. Genet.* **18,** 525–552.

Stanley, P. (1985a). Lectin-resistant glycosylation mutants. *In* "Molecular Cell Genetics: The Chinese Hamster Cell" (M. M. Gottesman, ed.), pp. 745–772. Wiley, New York.

Stanley, P. (1985b). Membrane mutants of animal cells: Rapid identification of those with a primary defect in glycosylation. *Mol. Cell. Biol.* **5,** 923–929.

Stanley, P. (1987). Glycosylation mutants and the functions of mammalian carbohydrates. *Trends Genet.* **3,** 77–81.

Stanley, P. (1989). Chinese hamster ovary mutants with multiple glycosylation defects for production of glycoproteins with minimal carbohydrate heterogeneity. *Mol. Cell. Biol.* **9,** 377–383.

Stanley, P. (1992). Glycosylation engineering. *Glycobiology* **2,** 99–107.

Stanley, P., and Sudo, T. (1981). Microheterogeneity among carbohydrate structures at the cell surface may be important in recognition phenomena. *Cell* **23,** 763–769.

Stanley, P., Caillibot, V., and Siminovitch, L. (1975a). Selection and characterization of eight phenotypically distinct lines of lectin-resistant Chinese hamster ovary cells. *Cell* **6,** 121–128.

Stanley, P., Narasimhan, S., Siminovitch, L., and Schachter, H. (1975b). Chinese hamster ovary cells selected for resistance to the cytotoxity of phytohemagglutinin are deficient in a UDP–*N*-acetylglucosamine-glycoprotein *N*-acetylglucosaminyltransferase activity. *Proc. Natl. Acad. Sci. USA* **72,** 3323–3327.

Stanley, P., Sundaram, S., and Sallustio, S. (1991). A subclass of cell surface carbohydrates revealed by a CHO mutant with two glycosylation mutations. *Glycobiology* **1,** 307–314.

Stanley, P., Vivona, G., and Atkinson, P. H. (1984). ^{1}H NMR spectroscopy of carbohydrates from the G glycoprotein of vesicular stomatitis virus grown in parental and Lec4 Chinese hamster ovary cells. *Arch. Biochem. Biophys.* **230,** 363–374.

Stanley, W. S., Peters, B. P., Blake, D. A., Yep, D., Chu, E. H. Y., and Goldstein, I. J. (1979). Interaction of wild-type and variant mouse 3T3 cells with lectins from *Bandeiraea simplicifolia* seeds. *Proc. Natl. Acad. Sci. USA* **76,** 303–307.

Stevens, V. L., and Raetz, C. R. H. (1991). Defective glycosyl phosphatidylinositol biosynthesis in extracts of three Thy-1 negative lymphoma cell mutants. *J. Biol. Chem.* **266,** 10039–10042.

Stoll, J., and Krag, S. S. (1988). A mutant of Chinese hamster ovary cells with a reduction in levels of dolichyl phosphate available for glycosylation. *J. Biol. Chem.* **263,** 10766–10773.

Stoll, J., Robbins, A. R., and Krag, S. S. (1982). Mutant of Chinese hamster ovary cells with altered mannose-6-phosphate receptor activity is unable to synthesize mannosylphosphoryldolichol. *Proc. Natl. Acad. Sci. USA* **79,** 2296–2300.

Stoll, J., Rosenwald, A. G., and Krag, S. S. (1988). A Chinese hamster overy cell mutant F2A8 utilizes polyprenol rather than dolichol for its lipid-dependent asparagine-linked glycosylation reactions. *J. Biol. Chem.* **263,** 10774–10782.

Stoll, J., Rosenwald, A., and Krag, S. S. (1990). Mutants: A fruitful approach to the understanding of asparagine-linked glycosylation. *In* "Nato ASI Series" (J. A. F. Op den Kamp, ed.), Vol. H40, pp. 151–166. Springer-Verlag, Berlin/Heidelberg.

Sugiyama, E., DeGasperi, R., Urakaze, M., Chang, H-M., Thomas, L. J., Hyman, R., Warren, C. D., and Yeh, E. T. H. (1991). Identification of defects in glycophosphatidylinositol anchor biosynthesis in the Thy-1 expression mutants. *J. Biol. Chem.* **266,** 12119–12122.

Suzu, S., Ohtsuki, T., Yanai, N., Takatsu, Z., Kawashima, T., Takaku, F., Nagata, N., and Motoyoshi, K. (1992). Identification of a high molecular weight macrophage colony-stimulating factor as a glycosaminoglycan-containing species. *J. Biol. Chem.* **267,** 4345–4385.

Tabas, I., and Kornfeld, S. (1978). The synthesis of complex-type oligosaccharides. III. Identification of an α-mannosidase activity involved in a late stage of processing of complex-type oligosaccharides. *J. Biol. Chem.* **253,** 7779–7786.

Takeuchi, M., and Kobata, A. (1991). Structures and functional roles of the sugar chains of human erythropoietins. *Glycobiology* **1,** 337–446.

Takeuchi, M., Inoue, N., Strickland, T. W., Kubota, M., Wada, M., Shimizu, R., Hoshi, S., Kozutsumi, H., Takasaki, S., and Kobata, A. (1989). Relationship between sugar chain structure and biological activity of recombinant human erythropoietin produced in Chinese hamster ovary cells. *Proc. Natl. Acad. Sci. USA* **86,** 7819–7822.

Tao, T.-W., and Burger, M. M. (1977). Non-metastasizing variants selected from metastasizing melanoma cells. *Nature* **270,** 437–438.

Tiemeyer, M., Swiedler, S. J., Ishihara, M., Moreland, M., Schweingruber, H., Hirtzer, P., and Brandley, B. K. (1991). Carbohydrate ligands for endothelial-leukocyte adhesion molecule 1. *Proc. Natl. Acad. Sci. USA* **88,** 1138–1142.

Torres, C.-R., and Hart, G. W. (1984). Topography and polypeptide distribution of terminal *N*-acetylglucosamine residues on the surfaces of intact lymphocytes: Evidence for O-linked GlcNAc. *J. Biol. Chem.* **259,** 3308–3317.

Toyama, S., Koyama, A. H., and Toyama, S. (1983). Deficient cytidine monophospho-*N*-acetyl neuraminic acid:glycoprotein sialyltransferase activity in a clone of KB cells with altered cell fusion ability. *J. Biol. Chem.* **258,** 9147–9152.

Trowbridge, I. S., and Hyman, R. (1979). Abnormal lipid-linked oligosaccharides in class E Thy-1-negative mutant lymphomas. *Cell* **17,** 503–508.

Trowbridge, I. S., Hyman, R., Ferson, T., and Mazauskas, S. (1978). Expression of Thy-1 glycoprotein on lectin-resistant lymphoma cell lines. *Eur. J. Immunol.* **8,** 716–723.

Vischer, P., and Hughes, R. C. (1981). Glycosyltransferases of baby-hamster-kidney (BHK) cells and ricin-resistant mutants. *Eur. J. Biochem.* **117,** 275–284.

Von Figura, K., and Hasilik, A. (1986). Lysosomal enzymes and their receptors. *Annu. Rev. Biochem.* **55,** 167–193.

Walker, B. D., Kowalski, M., Goff, W. C., Kozarsky, K., Krieger, M., Rosen, C., Rohrschneider, L., Haseltine, W. A., and Sadroski, J. (1987). Inhibition of human immunodeficiency virus syncytium formation and virus replication by castanospermine. *Proc. Natl. Acad. Sci. USA* **84,** 8120–8124.

West, C. M. (1986). Current ideas on the significance of protein glycosylation. *Mol. Cell. Biochem.* **72,** 3–20.

Zeng, Y., and Lehrman, M. A. (1990). A block at $Man_5GlcNAc_2$-pyrphosphoryldolichol in intact but not disrupted castanospermine and swainsonine-resistant Chinese hamster ovary cells. *J. Biol. Chem.* **265,** 2296–2305.

Zeng, Y., and Lehrman, M. A. (1991). Isolation of Chinese hamster ovary cell lines producing $Man_3GlcNAc_2$ asparagine-linked glycans. *Anal. Biochem.* **193,** 266–271.

The Structure and Function of Glycosphingolipid Receptors in Inflammation and Cancer

Bruce A. Macher and Cheryl L. M. Stults

Department of Chemistry and Biochemistry, San Francisco State University, San Francisco, California 94132

I. Introduction

Two topics addressed in this chapter concern the interaction of cell surface glycosphingolipids and proteins. The first section deals with the interaction of glycosphingolipids and E-selectin (previously known as ELAM-1, or endothelial leukocyte adhesion molecule 1). Questions concerning the natural ligand for this protein will be addressed and structure–activity studies of the carbohydrate ligand for E-selectin will be reviewed. The second section reviews the available literature on the potential function of glycosphingolipids in mediating cell–substratum interactions and their possible involvement in melanoma cell adhesion. Several methods have been developed to characterize protein-glycosphingolipid binding and these will be considered throughout the chapter.

Cell Surface and Extracellular Glycoconjugates

II. Glycosphingolipids and E-Selectin

The selectin field has developed quickly since these molecules were cloned and expressed in mammalian cells (for reviews, see Brandley *et al.*, 1990; Lasky, 1991; Osborn, 1990; Stoolman, 1989). A substantial amount of evidence has been accumulated by Rosen and co-workers (see Rosen and Yednock, 1986, for example) to indicate that at least one member (i.e., L-selectin) of this class of cell adhesion molecules had lectinlike properties before the sequence of these proteins was known and shown to contain a C-type lectin domain. This information caused an intense search to begin to identify the carbohydrate structure recognized by these proteins. Much of the early work focused on the discovery of the carbohydrate ligand for E-selectin. This protein is expressed on the surface of endothelial cells that have become activated early in the inflammatory process (Bevilacqua *et al.*, 1987). Expression of E-selectin begins about 4 hr after the initial activation response (e.g., interleukin 1) by an upregulation of the message for this protein. Expression of E-selectin initiates (neutrophil rolling on the surface of activated endothelium) a series of complex adhesion processes that ultimately results in leukocyte extravasation (Butcher, 1991). The observation that anti-E-selectin antibodies, under normal shear force conditions, can block the subsequent steps of the inflammation response argues that E-selectin binding is primarily responsible for the initial binding (rolling) of neutrophils to endothelial cells (von Andrian *et al.*, 1991).

The fact that the ligand for E-selectin occurs on a limited number of cells (primarily myeloid cells) led to a search for a compound that had restricted expression among hemopoietic cells. The work of Michiko and Minuro Fukuda (reviewed in Fukuda, 1985) and our studies (Macher and Beckstead, 1990) demonstrated that myeloid cells were unique among hemopoietic cells in expressing carbohydrates that were both fucosylated and sialylated. Immunocytochemical analysis of human marrow with antibodies directed to fucosylated and sialyl-fucosylated antigens revealed that these antigens are expressed throughout the differentiation process of myeloid cells, but not by other hemopoietic cell lines. Such a restricted expression of glycoconjugates led several groups to evaluate the ability of E-selectin to bind to glycosphingolipids containing both sialic acid and fucose.

A. Fucosylated Gangliosides as Ligands for E-Selectin

Phillips *et al.* (1990) reported that E-selectin bound to a difucosylated Sialyl–Lewis X (sLe^x) glycosphingolipid (see Table I for structures of

TABLE I
GLYCOSPHINGOLIPID STRUCTURES

NeuAcα2–3Galβ1–4GlcNAcβ1–3Galβ1–4GlcNAcβ1–3Galβ1–4Glcβ1–1Cer
3 | α1Fuc (on first GlcNAc); 3 | α1Fuc (on second GlcNAc)

Difucosylated Sialyl–Lewis X

NeuAcα2–3Galβ1–4GlcNAcβ1–3Galβ1–4GlcNAcβ1–3Galβ1–4Glcβ1–1Cer
3 | α1Fuc (on second GlcNAc)

VIM2

NeuAcα2–3Galβ1–4GlcNAcβ1–3Galβ1–4GlcNAcβ1–3Galβ1–4Glcβ1–1Cer
3 | α1Fuc (on first GlcNAc)

Sialyl–Lewis X

glycosphingolipids referred to in this chapter) and that an antibody to the sLe^x tetrasaccharide epitope could block the interaction between HL-60 cells and E-selectin. Furthermore, a CHO lectin-resistant mutant cell line (LEC11) known to express sLe^x glycoconjugates bound to E-selectin. What was not clear from this initial study was if a monofucosylated sLe^x compound could be recognized by E-selectin. Walz *et al.* (1990) also reported that an anti-sLe^x antibody could block E-selectin interaction with myeloid cells, whereas an antibody directed against a structural analog of sLe^x, VIM2, or CD62 could not. This finding contradicted the results of a subsequent study by Tiemeyer *et al.* (1991), who reported the isolation of the E-selectin ligand from human chronic myelogenous leukemia (CML) cells. The ligand identified in this latter study was a sialylated, fucosylated decasaccharide (VIM2) that had been previously identified as a myeloid-specific ganglioside by Macher *et al.* (1988). Thus, among the three initial reports on the carbohydrate epitope that is recognized by E-selectin, the report by Tiemeyer *et al.* (1991) was the only one that reported an attempt to isolate a natural ligand from a myeloid cell.

Lowe and co-workers (1990) used another novel approach to provide evidence that sLe^x acts as a ligand for E-selectin. They showed that transiently transfecting COS and CHO cells with an α1–3 fucosyltransferase resulted in cells that expressed sLe^x and that were capable of binding to activated human endothelial cells in an E-selectin-dependent manner. Similar results were reported by Goelz *et al.* (1990) and Stanley and co-workers (1992). Taken together these results indicated that a sialylated–fucosylated glycoconjugate contains the ligand for E-selectin. However, two questions remain to be answered. First, is VIM2 a ligand for E-selectin? Second, what is the natural ligand for E-selectin?

B. Is VIM2 a Ligand for E-Selectin?

In attempting to answer the first question, it would be worthwhile to evaluate the methods that were used to identify VIM2 as a ligand and those that were used to argue against it. One approach that has already been mentioned is the use of monoclonal antibodies to block the interaction of E-selectin with its ligand on myeloid cells. As applied by Walz *et al.* (1990), this approach demonstrated that anti-sLe^x, but not the VIM2 antibody, could block the binding of E-selectin to myeloid cells. The ability of an antibody to block or not block this binding can at best be used to infer an interaction or lack of one. For example, blocking could result from a steric blocking process, rather than from direct binding. A lack of blocking could result from a difference in the affinity of the antibody vs that of the selectin for the carbohydrate antigen or from other factors. The fact that one antibody blocks this binding, whereas another does not, should be used only in conjunction with other approaches. Two other approaches were utilized to evaluate the role of VIM2 as a ligand for E-selectin.

The results reported by Tiemeyer *et al.* (1991) were obtained by a classical purification approach that relied on a TLC overlay assay using radiolabeled COS cells that had been transfected with a plasmid coding for E-selectin. The assay was a modification of an approach originally reported by Magnani *et al.* (1980) to identify glycosphingolipids that bound cholera toxin and later modified by Magnani *et al.* (1981) to identify tumor-associated antigens recognized by monoclonal antibodies. The assay allowed Tiemeyer and co-workers (1991) to detect several E-selectin binding gangliosides extracted from CML cells and purify a fraction containing a ganglioside that had properties (carbohydrate composition, methylation analysis, and FAB–MS) consistent with a VIM2 structure. As shown in Table I, sLe^x and VIM2 gangliosides are structural isomers of one another and therefore carbohydrate composition and methylation analysis would not distinguish between these two gangliosides. However, FAB–MS gives quite different fragmentation patterns due to the difference in location of the branching fucose

(Macher *et al.*, 1988). The authors (Tiemeyer *et al.*, 1991) also reported that a desialylated form of the purified ganglioside could be hydrolyzed by β-galactosidase and β-hexosaminidase. This would be possible only with a VIM2 and not an sLe^x ganglioside because the latter desialylated compound would be resistant to β-galactosidase hydrolysis due to the presence of an α1–3 fucose on the adjacent GlcNAc residue.

An important experiment that was not reported in the article by Tiemeyer *et al.* (1991) would be to analyze the desialylated ganglioside(s) for immunoreactivity to anti-Le^x antibodies. If small quantities of sLe^x gangliosides were present in the VIM2 ganglioside fractions, they would probably not be detected by FAB–MS; however, they would be detected by the TLC immunostaining procedure. Of course the presence of a small quantity of sLe^x gangliosides would be significant only if there is a substantial difference in the binding affinity of E-selectin for sLe^x vs that for VIM2 gangliosides. To evaluate this possibility we (P. James and B. A. Macher, unpublished results) isolated VIM2 gangliosides from CML cells with the goal of obtaining fractions that contained this ganglioside structure, which were free of sLe^x gangliosides as judged by TLC immunostaining assays using anti-sLe^x and VIM2 antibodies. When these gangliosides were tested in the TLC overlay assay using radiolabeled COS cells expressing E-selectin, we found no binding compared with positive binding with chemically synthesized sLe^x gangliosides. This material was subsequently tested in a solid-phase, microtiter plate binding assay for E-selectin binding. In this assay, gangliosides were adsorbed to the plastic wells and binding of radiolabeled COS cells expressing E-selectin was monitored. As reported by Tyrrell *et al.* (1991), there was minimal binding of the COS cells to these VIM2 gangliosides (1.5 to 2 orders of magnitude lower) when compared with that of sLe^x gangliosides.

Another approach to evaluate the possibility that VIM2 functions as a ligand for E-selectin was reported by Lowe and co-workers (1991). These workers found that transfection of CHO cells with a myeloid form of α1–3 fucosyltransferase resulted in the expression of VIM2, but not sLe^x glycoconjugates. These cells were unable to bind to E-selectin. In contrast, CHO cells transfected with a plasmid construct encoding the Lewis α1–3 fucosyltransferase, which allows cells to express both sLe^x and VIM2, bound E-selectin. Taken together these results make it unlikely that VIM2 is a ligand recognized by E-selectin.

C. The Natural Ligand for E-Selectin

Little information is available to answer the question: What is the natural ligand for E-selectin? In the context of this chapter the point of this question is to ascertain if sLe^x gangliosides mediate E-selectin binding to myeloid cells *in vitro* and to neutrophils *in vivo*. Picker *et al.* (1991a) have presented information that supports the idea that E-selectin binds to a glycoprotein that can be released by protease treatment. Their results specifically indicate that L-selectin, expressed on neutrophils, contains sLe^x and constitutes a natural ligand for E-selectin. In contrast, Larsen *et al.* (1992) reported that treatment of either HL-60 cells or neutrophils with a variety of proteases does not eliminate their binding to E-selectin. Interestingly, these scientists found that the ligand for another selectin (P-selectin) on these same cells is destroyed by protease treatment. Further work is clearly needed to provide a definitive answer to the question of what glycoconjugate(s) is the natural ligand for E-selectin.

D. Functional Groups of Sialyl–Lewis X Recognized by E-Selectin

Both acute and chronic inflammation are serious disease states. It appears that blocking E-selectin–ligand interaction can effectively block the initial stages of the inflammatory response. Furthermore, it is also possible that this lectin–carbohydrate interaction may mediate tumor cell metastasis and that drugs that can block this interaction may be useful in inhibiting the metastasis. Therefore, it is no surprise that a significant effort has been applied toward a better understanding of the structure–activity relationship of E-selectin and sLe^x in hope of finding effective anti-inflammatory drugs.

As an example, we initially evaluated if E-selectin would bind to sLe^x ganglioside that contained an *N*-glycolylneuraminic acid (NeuGc) rather than an *N*-acetylneuraminic acid (NeuAc). Since human gangliosides generally contain NeuAc, we isolated gangliosides from an animal source (rabbit thymus) that we knew contained sialylparagloboside (IV^3-NeuGc-nLc$_4$Cer) that contained nearly 100% NeuGc (P. He and B. A. Macher, unpublished results). This material was checked for its ability to bind E-selectin and as expected did not. This material was fucosylated using a purified preparation of human milk α1–3 fucosyltransferase (Prieels *et al.*, 1981). The product of this reaction was characterized and found to contain fucose linked to the 3-OH of *N*-acetylglucosamine and to the 3-OH of both *N*-acetylglucosamine and the glucose of some molecules. The FAB–MS spectra of the ganglioside acceptor and product are shown in Fig. 1. As illustrated in the upper panel of Fig. 1, the starting material has a major molecular ion at *m/z* 1532.7 for the parent molecule containing NeuGc and a $C_{16:0}$ fatty acid. Following fucosylation (lower panel of Fig. 1) the molecular ions are *m/z* 1678.3

for a monofucosylated ganglioside containing NeuGc and 1825.2 for a difucosylated ganglioside containing NeuGc. The fragmentation pattern supports the assigned structure as shown in the insets of Fig. 1. TLC immunostaining with E-selectin (Fig. 2) clearly demonstrates that the selectin binds to the NeuGc-Le^x ganglioside that was biosynthesized *in vitro*.

A series of structure–activity studies were done to more clearly define the functional groups of the sLe^x epitope that are recognized by E-selectin. These results have been published by Tyrrell *et al.* (1991) and demonstrate the following important features about the functional groups on sLe^x for E-selectin binding: (a) the sialic acid group must be linked α2–3 to galactose (i.e., NeuAcα2–6Gal(Fucα1–3)β1–4GlcNAc-R is not recognized by E-selectin); (b) the glycerol side chain (C8–9) of sialic acid is not necessary in the sLe^x structure; and (c) Sialyl–$Lewis^a$ (sLe^a) is recognized by E-selectin. This latter finding has also been demonstrated by other investigators (Berg *et al.*, 1991). This observation is rationalized by the fact that sLe^x and sLe^a share a minimum-energy conformation as first proposed for Le^x and Le^a by Lemieux and co-workers (Spohr *et al.*, 1985).

In summary, E-selectin recognizes the tetrasaccharide epitope sLe^x, which has a restricted expression on human myeloid cells. The chemical nature (gangliosides and/or other glycoconjugates) of the natural ligand on human myeloid cells for E-selectin is still not settled. Interestingly, it has been reported that a subclass of T cells involved in skin inflammation binds to E-selectin, but these cells apparently do not express sLe^x (Picker *et al.*, 1991b; Shimizu *et al.*, 1991). Further work will be required to determine the structural range of natural ligands for E-selectin. *In vitro* assays have demonstrated that fucose and certain portions of the sialic acid group are critical for E-selectin binding, but further work is necessary to establish the minimal recognition domain of sLe^x, a critical issue for the development of optimal anti-inflammatory drugs.

III. Function of Glycosphingolipids in Mediating Cell–Substratum Interactions

The previous section dealt primarily with the nature of the carbohydrate ligand involved in the initial neutrophil–endothelial cell interaction, which ultimately results in leukocyte extravasation. It is apparent that glycosphingolipids may be involved in such cell–cell interactions. There have also been several studies that have provided strong evidence for glycosphingolipid involvement in cell–substratum interactions. The

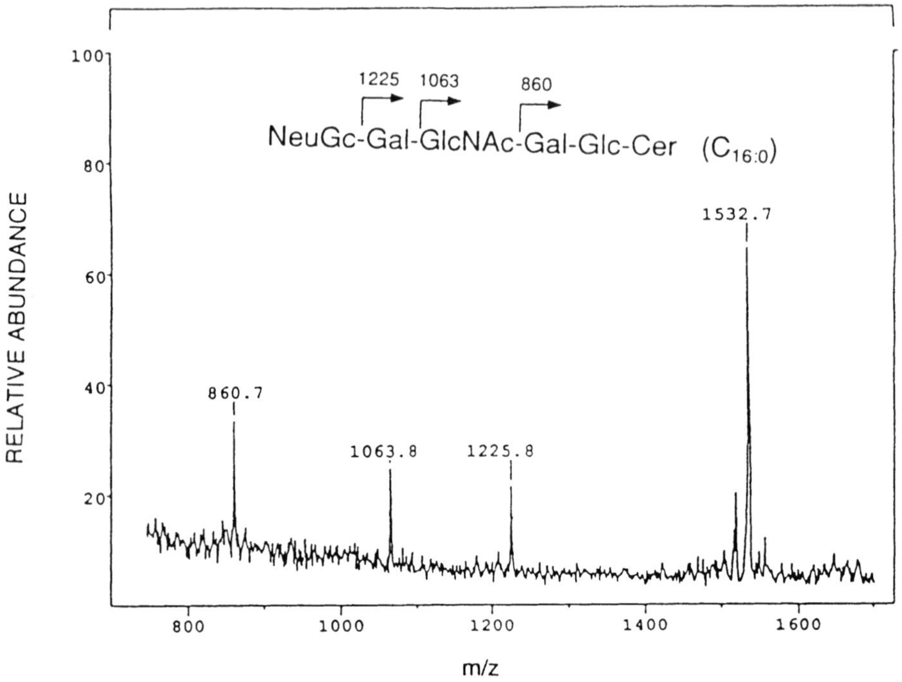

1225
1063
860
NeuGc-Gal-GlcNAc-Gal-Glc-Cer (C16:0)
1532.7
860.7
1063.8
1225.8
RELATIVE ABUNDANCE
m/z

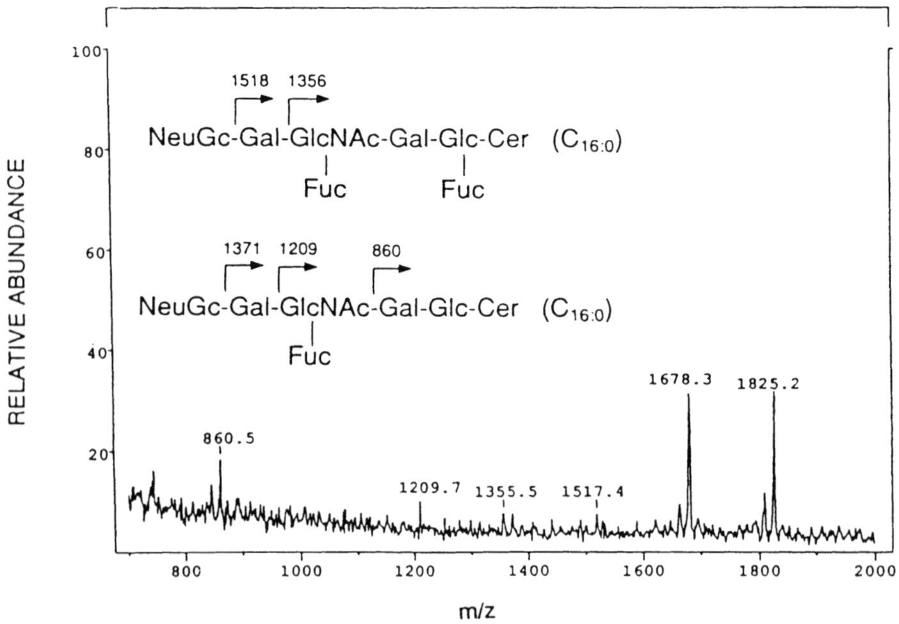

1518
1356
NeuGc-Gal-GlcNAc-Gal-Glc-Cer (C16:0)
Fuc
Fuc
1371
1209
860
NeuGc-Gal-GlcNAc-Gal-Glc-Cer (C16:0)
Fuc
1678.3
1825.2
860.5
1209.7
1355.5
1517.4
RELATIVE ABUNDANCE
m/z

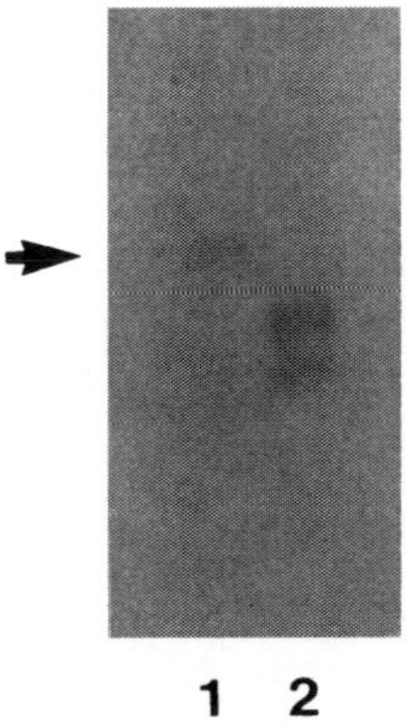

FIG. 2. Binding of E-selectin to NeuGc-Le^x ganglioside. IV$^3\alpha$NeuGc-nLc$_4$Cer was fucosylated enzymatically to generate III$^3\alpha$Fuc,IV$^3\alpha$NeuGc-nLc$_4$Cer and I^3,III$^3\alpha$Fuc$_2$,IV$^3\alpha$-NeuGc-nLc$_4$Cer, and analyzed for E-selectin binding using a TLC binding assay. The binding of E-selectin (an IgG chimera was used) was detected using biotinylated anti-human IgG, avidin–alkaline phosphatase and a commercial alkaline phosphatase substrate (Vector Labs, Burlingame, CA). The solvent system used was chloroform/methanol/0.2% aqueous $CaCl_2$, 60/40/9. Lane 1, IV$^3\alpha$NeuAc-nLc$_4$Cer standard (designated by the arrow), Lane 2, III$^3\alpha$Fuc,IV$^3\alpha$NeuGc-nLc$_4$Cer (upper band), and I^3,III$^3\alpha$Fuc$_2$,IV$^3\alpha$Neu-Gc-nLc$_4$Cer (lower band). (Experimental result courtesy of Dr. Paul James, Glycomed, Inc.)

attachment process is prerequisite to cell spreading and motility. These events are necessary for normal growth and development in multicellular organisms. Attachment is also a critical event for wound healing and the formation of tumor metastases.

The initial step in cell–substratum interactions involves the attachment of a ligand on the cell surface to an extracellular matrix protein such as fibronectin, vitronectin, laminin, von Willebrand factor, and thrombospondin. There are four families of compounds on cell surfaces that are involved in binding these extracellular matrix proteins (Albelda and Buck, 1990): integrins (made of unique combinations of α and β subunits), proteins (integral membrane and peripheral), proteoglycans (heparan sulfate and syndecan), and glycosphingolipids (sul-

---

FIG. 1. Fast atom bombardment mass spectra (FAB-MS) of nonfucosylated (upper panel) and enzymatically fucosylated (lower panel) ganglioside analyzed for E-selectin binding. The ganglioside was isolated from rabbit thymus tissue (upper panel) and fucosylated enzymatically (lower panel). Negative ion FAB-MS were acquired on a Finnigan (San Jose, CA) MAT 900 spectrometer with a triethanolamine matrix. Glycolipids (1–5 nmol) were suspended in $CHCl_3/MeOH/H_2O$, 1/1/0.1, and mixed with the matrix for analysis. (FAB-MS analysis provided by Mr. Hans Schweingruber, Glycomed, Inc.)

fated and sialylated). The first three families have been implicated in direct binding, whereas there is some debate about the role of glycosphingolipids in cell–substratum interactions. Therefore we have chosen to present a review of what has been demonstrated concerning the role of glycosphingolipids in the cell attachment process. More specifically, the role of glycosphingolipids in melanoma cell–substratum interactions will be discussed in detail.

A. Melanoma-Associated Gangliosides

A melanoma is a tumor of neural crest origin. Normal melanocytes are derived from nevus cells, which have only two major gangliosides (sialylated glycosphingolipid)—GM_3 and GD_3 (Hamanaka, 1984) (see Table II for structures). According to Dippold *et al.* (1985) the intensity and uniformity of GD_3 immunostaining on melanoma cells was much greater than that on normal melanocytes. This indicates that the level of GD_3 is probably higher on melanoma cells than on normal melanocytes. Carubia *et al.* (1984) compared melanoma cells with normal melanocytes and found that the GD_3 level was higher and the GM_3 level was lower than normal. It is not clear if the total ganglioside content in melanoma cells is greater than that found in normal nevi. However, it is known that the level of ganglioside expression is higher in a melanoma metastatic tumor than in the normal tissue to which it has metastasized (Portoukalian *et al.*, 1979); it is also higher than that found in tumors of other origin (Tsuchida *et al.*, 1987b).

In addition to GM_3 and GD_3, other melanoma-associated gangliosides have been identified (see Table II for structures). The gangliosides associated with human melanoma may be classified into three groups based on tissue distribution (see Table III): gangliosides that are widely distributed in normal tissues but have elevated levels in melanoma (GM_3); gangliosides that are found in only trace amounts or have a narrow tissue distribution but are expressed in large quantities in melanoma (GD_3, GM_2, and GD_2); and ganglioside derivatives that are unique to human melanoma (9-OAc-GD_3, NeuGc-GM_3, NeuGc-GD_3, and NeuGc-GM_2).

The structures of these gangliosides were proved by conventional methods (Protoukalian *et al.*, 1979; Cheresh and Reisfeld, 1984; Hirabayashi *et al.*, 1987; Pukel, 1982). Each compound was isolated and characterized by TLC mobility compared with standards, glycosidase (including neuraminidase) treatment, chemical degradation by alkali or periodate, and binding of monoclonal antibodies specific for each ganglioside. In some cases fast atom bombardment–mass spectrometry and NMR (9-OAc-GD_3; Thurin, 1985) were used.

TABLE II

STRUCTURE OF MELANOMA-ASSOCIATED GANGLIOSIDES AND MONOCLONAL ANTIBODIES THAT BIND TO THEM

Name	Structure	Monoclonal antibody[a]	Reference[b]
GM_3[c]	NeuAcα2–3Galβ1–4Glcβ1–1Cer	**M2590,** FCM1*, DH2 (lactone)	Yamaguchi *et al.* (1987), Dohi *et al.* (1988)
GD_3	NeuAcα2–8NeuAcα2–3Galβ1–4Glcβ1–1Cer	**R-24,** MB3.6, **4.2, 2B2, MG-21,** ME24, HJM1*, ME36.1, A1-267, 11C64, IF4	Cheresh *et al.* (1984), Koprowski *et al.* (1978), Yamaguchi *et al.* (1987), Thurin *et al.* (1987), Kawashima *et al.* (1990), Cheresh *et al.* (1986), Hellstrom *et al.* (1985)
GM_2	GalNAcβ1–4Galβ1–4Glcβ1–1Cer 3 | NeuAcα2	3-207*, **5-3**	Yamaguchi *et al.* (1990)
GD_2	GalNAcβ1–4Galβ1–4Glcβ1–1Cer 3 | NeuAcα2–8NeuAcα2	**Leo Mel 3,** 126, 14.18, ME36.1, L72, A1-425, A1-267, 3-207*, **3F8**	Cheresh *et al.* (1984), Thurin *et al.* (1987), Cahan *et al.* (1982), Kawashima *et al.* (1990), Yamaguchi *et al.* (1990)
NeuGc–GM_3	NeuGcα2–3Galβ1–4Glcβ1–1Cer	**5-3** (GM2), 3-207 (GM2), anti-GM3 (NeuGc)	Yamaguchi *et al.* (1990), Hirabayashi *et al.* (1983)
9-*O*-Ac GD_3	9-OAc-Neu5Acα2–8NeuAcα2–3Galβ1–4Glcβ1–1Cer	**D1.1,** ME311, JONES	Thurin *et al.* (1985), Blum and Barnstable (1987)

[a] Monoclonal antibodies that have been derived from human lymphocytes are indicated by an asterisk.
[b] References are given for those antibodies whose characterization was not previously summarized (shown in bold) by Magnani (1986).
[c] Gangliosides are abbreviated according to Svennerholm (1963).

TABLE III

TISSUE DISTRIBUTION AND IMMUNOGENICITY OF MELANOMA-ASSOCIATED GANGLIOSIDES

Ganglioside	Tissue distribution	Immunogenicity[a]
GM3	Most tissues, elevated in melanoma	−
GD3	Melanoma, other tumors, brain	−
GM2	Neural crest tumors, brain	+
GD2	Neural crest tumors, brain	+
NeuGc–GM3 (GM2,GD3)	Melanoma, fetal tissue, colon cancer	Unknown
9-OAc-GD3	Melanoma	+

[a] The ability of a compound to induce an immune response (e.g., the production of antibodies that bind specifically to the compound of interest).

It has been shown that the ganglioside distribution pattern varies in each case of melanoma (Portoukalian *et al.*, 1979; Tsuchida *et al.*, 1987b). Also, as mentioned earlier, in some instances the total quantity of gangliosides is significantly elevated (Berthier-Vergnes, 1985). It has been proposed that these elevated levels are a result of increased levels of synthesis and/or blocked synthesis of more complex structures (Singhal and Hakomori, 1990). Furthermore it has been shown that GD_2 expression and elevated *N*-acetylgalactosaminyltransferase levels correlate with the ability of the melanoma to metastasize (Thurin *et al.*, 1986).

GD_3 ganglioside has received much attention as a melanoma-associated ganglioside, especially in the clinical arena (Hersy *et al.*, 1986, 1987; Ritter, 1991; Houghton *et al.*, 1985). This is probably due to the fact that GD_3 is more abundantly expressed than other gangliosides, and GD_3 is generally a minor component in the few tissues in which it is expressed. However, several researchers have shown that the other gangliosides found in melanoma cells may be equally important. For example, Tsuchida *et al.* (1987a) showed that GM_2 levels correlated positively with tumorigenicity. Similarly, Thurin *et al.* (1986) found that the presence of GD_2 correlated with melanoma tumor progression. In contrast, high GM_3 levels (relative to GD_3) have been cited as an indicator for good prognoses (e.g., low probability of metastasis) (Ravindranath *et al.*, 1991; Tsuchida *et al.*, 1989). Similarly Kono *et al.* (1990) showed that GM_3 content correlated inversely with agar plating efficiency. This finding implies that melanoma cells with high GM_3 content are less tumorigenic than those with low GM_3 content. The role of each of these gangliosides in melanoma–substratum interactions will be considered after a discussion of the assays used to evaluate this biological process.

B. Experimental Design for Assessing Cell Attachment

Several methodologies have been used to infer that gangliosides are involved in attachment events. The first evidence that gangliosides might be associated with the attachment process was obtained by incubation of cell cultures with exogenously added gangliosides (Kleinman *et al.,* 1979; Perkins *et al.,* 1982; Yamada *et al.,* 1981). In all three studies it was found that the addition of gangliosides to the culture media inhibited cell binding to fibronectin.

Another method widely used by many researchers involves the use of monoclonal antibodies (see Table II for a list of those used in the work described here) that bind specifically to the ganglioside of interest. An early report by Dippold *et al.* (1984) suggested that melanoma cell rounding up and detachment from culture plates was caused by incubation with the monoclonal antibody R-24, which binds GD_3. More recent experiments with melanoma cells have used monoclonal antibodies as preblocking agents for experiments that have tested binding of melanoma cells to various extracellular matrix proteins and other compounds. The general protocol involves the following steps:

1. Cells growing as monolayers are detached from a tissue culture flask.
2. After washing with medium, cells are incubated in medium containing a monoclonal antibody.
3. Cells are subsequently washed, resuspended in medium, and transferred to 96-well microtiter plates that have been precoated with extracellular matrix proteins and blocked with BSA.
4. After allowing cells to adhere for a specified period of time (usually a few hours), the plate is washed with PBS to remove unbound cells.
5. The wells are then treated with trypsin to remove the attached cells, which may then be counted.

As described in Section II.B, there are a few pitfalls (e.g., nonspecific interaction) that are inherent in methods that utilize monoclonal antibodies as blocking agents. Therefore, other methods have also been used in conjunction with these to substantiate involvement of gangliosides in melanoma cell attachment to extracellular matrix proteins. For example, Cheresh *et al.* (1986) used controlled periodate oxidation of gangliosides on the surface of human melanoma cells to show that GD_2 and GD_3 are required for attachment to fibronectin and laminin.

A more general approach using monoclonal antibodies has also been used. The underlying assumption of these studies was that cell growth is prevented by inhibition of cell attachment. Several investigators have found that adding monoclonal antibodies specific for GM_3 (Dohi *et al.*, 1988) or GD_3 (Dippold *et al.*, 1984; Werkmeister *et al.*, 1987) to culture medium inhibited cell growth. *In vivo* studies have also shown that growth of melanoma cells implanted into mice can be inhibited by administration of anti-GD_2,GD_3 (Illiopoulos *et al.*, 1989), anti-GM_3 (Dohi *et al.*, 1988), and anti-GD_3 (Hellstrom *et al.*, 1985) monoclonal antibodies. In the latter case (Hellstrom *et al.*, 1985), it was suggested that the lack of tumor growth was due to antibody-dependent cellular cytotoxicity. Therefore lack of cell attachment may or may not be a significant factor in this case.

The experimental approaches described here are representative of those used by most researchers that have studied melanoma cell–substratum interactions. In each methodology inferences are made on the basis of results obtained in systems that attempt to mimic the natural environment. The weaknesses of a given system are overcome by using appropriate controls and multiple methods and examining the work of many investigators. An ideal system for studying the interaction of melanoma cells with extracellular matrix components would allow one to view directly an intact sample containing all the components involved—perhaps by labeling them with fluorescent tags. Such a system was used by Cheresh *et al.* (1987) to show that the vitronectin receptor and GD_2 are colocalized on human melanoma cells. The methods described above have contributed to an accumulation of much information about the role of gangliosides and melanoma cell–substratum interactions.

C. Role of Gangliosides in Cell–Substratum Interactions

In this section two different theories regarding the role of gangliosides in cell attachment are discussed. Evidence is presented to support each. The basic question is, "Do gangliosides interact directly or indirectly with extracellular matrix components?"

1. Direct Binding of Cell Surface Gangliosides to Substratum

As mentioned earlier the first clue that gangliosides might be involved in cell attachment was obtained from experiments in which gangliosides exogenously added to the cell culture medium inhibited fibronectin binding. Of the melanoma gangliosides, GD_3 has been most extensively studied. Several reports indicate that this ganglioside is directly involved in melanoma cell–substratum interactions. Growth inhibition by incubation of melanoma cells with monoclonal antibodies to GD_3 suggested that this ganglioside is involved in attachment. Interestingly, Werkmeister *et al.* (1987) suggested that GD_3 expression on the cell surface correlated directly with cell attachment. In this study it was found that GD_3 content decreased approximately 30% while adherent human melanoma cells were maintained in suspension culture for 48 hr; a burst of GD_3 expression was recorded after the same cells were subsequently grown as adherent monolayers. A more specific investigation by Cheresh *et al.* (1984) indicated that GD_3 was found not only on the surface of human melanoma cells but also in focal-adhesion plaques. In that same study it was found that neither fibronectin nor chondoitin sulfate was present in the focal-adhesion plaques.

In addition to GD_3, other gangliosides have been implicated as having a direct role in various melanoma cell–substratum interactions. Cheresh *et al.* (1986) used monoclonal antibodies and chemical degradation experiments to demonstrate that both GD_2 and GD_3 were involved in melanoma cell attachment to fibronectin, laminin, vitronectin, and collagen IV. The effect of each ganglioside was independent of the other. As with GD_3, GD_2 was localized to the focal-adhesion plaque. Finally, periodate oxidation was used to demonstrate that the GD_3 ganglioside derivative 9-OAc-GD_3 was not involved in the cell attachment process.

Taken together, the evidence suggests that melanoma cell attachment occurs via direct binding of specific cell surface gangliosides. There is some precedence for this theory based on other studies. For example, Kojima and Hakomori (1991) have shown that B16 mouse melanoma cells attached themselves to microtiter wells coated with Gg_3Cer, LacCer, and Gb_4Cer via interaction with GM_3 on the cell surface. Furthermore, Blackburn and Schnaar (1983) reported cell attachment via immobilized glycolipids. Gangliosides have also been shown to mediate cell adhesion via selectins as discussed in the first section of this chapter.

2. Gangliosides as Modifiers of Cell–Substratum Interactions

Although there is evidence for direct binding of gangliosides to extracellular matrix components, more recent studies favor an indirect role for gangliosides in cell–substratum interactions. Because gangliosides were shown to interact with Ca^{2+} (Cheresh *et al.*, 1986), a function of the gangliosides might be that of creating the proper electrostatic environment for an extracellular matrix protein receptor and its ligand. Investigations on two different melanoma systems (human and mouse) provide support for this theory.

a. Human Melanoma: GD_2 and GD_3 Gangliosides Gangliosides have been found to be associated with protein receptors in human melanoma cells. Two groups found that GD_2 and the vitronectin receptor are associated on human melanoma cells. Cheresh *et al.* (1987) found that the GD_2 copurified with the vitronectin receptor in the presence of Ca^{2+} on affinity columns. Burns *et al.* (1988) identified vitronectin receptor as a coprecipitate with GD_2 (and GD_3). Further evidence for the association of ganglioside with receptors was obtained from two-color fluorescent staining studies by Burns *et al.* (1988). In these studies fluorescent labeling of two different monoclonal antibodies (antivitronectin receptor and anti-GD_2(GD_3)) was used to show that a stable ganglioside–receptor complex exists on the cell membrane. In the same study it was shown that treatment with monoclonal antibody Leo Mel 3 (anti-GD_2(GD_3)) resulted in capping of the melanoma cells. Subsequent analysis showed this capping resulted in colocalization of the gangliosides and the vitronectin receptor into clusters.

Several studies provide further information regarding the localization of the ganglioside and vitronectin receptor. GD_2 was localized to focal-adhesion plaques (Cheresh *et al.,* 1984), and by immunoelectron microscopy (Cheresh and Klier, 1986), to the microprocesses that extend from the melanoma cell toward the fibronectin substratum. In a recent study one form of the vitronectin receptor ($\alpha_v\beta_3$) was also localized to the focal contact (Wayner *et al.,* 1991). Taken together, these results imply that the ganglioside-associated integrin is $\alpha_v\beta_3$. This integrin is known to bind thrombospondin, vitronectin, fibronectin, von Willebrand factor, and fibrinogen. Furthermore, sulfatides on the melanoma cell surface are thought to bind to specific domains on thrombospondin and thus mediate melanoma cell spreading according to the model developed by Roberts and Ginsburg (1988).

b. Mouse Melanoma: GM_3 It is quite apparent that GD_2 and GD_3 are involved in human melanoma cell–substratum interactions. A different ganglioside, GM_3, has been implicated in mouse melanoma cell–substratum interactions. A mouse monoclonal antibody, M2590, which specifically binds GM_3, inhibits mouse melanoma cell attachment. Additionally, this monoclonal antibody binds only to GM_3 on melanoma cells and not to GM_3 on cells from normal tissue (Taniguchi and Wakabayashi, 1984). It was later discovered that the binding specificity seemed to result from high-density patches of GM_3 on melanoma cells that are not found on normal cells (Harada *et al.,* 1989). Furthermore it was also reported by Harada *et al.* (1989) that an 80-kDa glycoprotein was associated with GM_3 on the cell surface. The function of this protein was not identified; its low molecular weight would suggest that the GM_3-associated protein is not an integrin (MW $>$ 100 kDa). However, an antibody to this protein did inhibit cell attachment. Clustering of GM_3 on mouse melanoma cells may result from an association with this glycoprotein. An investigation at the level of electron microscopy could be used to elucidate the details of this interaction. It would also be interesting to determine if GM_3 functions in a similar manner in human melanoma.

The studies on human and mouse melanoma indicate that the role of gangliosides is an indirect one that may involve association with cell surface glycoproteins. The gangliosides may modulate the environment in such a way that the associated protein can move rather rapidly in the plasma membrane (attachment times are less than 15 min) to facilitate binding (or other functions) of the protein. The exact nature of the role that gangliosides have in cell–substratum interactions will require further investigation.

D. Implications for Treatment of Melanoma

Specific gangliosides that play an indirect role in melanoma cell attachment have been identified. Some of these compounds are also targets for immunotherapy, because their expression is limited in normal cells. Furthermore, these gangliosides have been found to induce an immune response in some melanoma patients (see Table III). Monoclonal antibodies to melanoma-associated gangliosides have been used to treat patients with advanced cases of melanoma. Limited success has been achieved with this approach. For example, mouse monoclonal antibody R-24 (anti-GD_3, see Table II) was administered by iv infusion and achieved tumor regression in 3 out of 12 patients (Houghton *et al.*, 1985). Greater success was achieved with a human monoclonal antibody (Irie and Morton, 1986). In this case, cutaneous metastatic melanoma was treated by intralesional injections of monoclonal antibody L72 (anti-GD_2 from EBV-transformed peripheral blood lymphocytes) and remission was observed in 8 out of 10 cases. It has been proposed that antibody-mediated cellular cytotoxicity may be the mechanism by which tumor growth is inhibited in this type of therapy (Kawashima *et al.*, 1990). However, the actual mechanism is unknown and it is important to keep cell–substratum interactions in mind when developing therapies for melanoma.

References

Albelda, S. M., and Buck, C. A. (1990). Integrins and other cell adhesion molecules. *FASEB J.* **4,** 2868–2880.

Berg, E. L., Robinson, M. K., Mansson, O., Butcher, E. C., and Magnani, J. (1991). A carbohydrate domain common to both sialyl Lea and sialyl Lex is recognized by the endothelial cell leukocyte adhesion molecule ELAM-1. *J. Biol. Chem.* **266,** 14869–14872.

Berthier-Vergnes, O., Portoukalian, J., and Dore, J.-F. (1985). Organization and neuraminidase susceptibility of sialic acid residues in human melanoma cell lines with different heterotransplantabilities in nude mice. *J. Natl. Cancer Inst.* **75,** 605–611.

Bevilacqua, M. P., Pober, J. S., Mendrick, D. L., Cotran, R. S., and Gimbrone, M. A., Jr. (1987). Identification of an inducible endothelial-leukocyte adhesion molecule. *Proc. Natl. Acad. Sci. USA* **84,** 9238–9242.

Blackburn, C. C., and Schnaar, R. L. (1983). Carbohydrate-specific cell adhesion is mediated by immobilized glycolipids. *J. Biol. Chem.* **258,** 1180–1188.

Blum, A. S., and Barnstable, C. J. (1987). O-acetylation of a cell-surface carbohydrate creates discrete molecular patterns during neural development. *Proc. Natl. Acad. Sci. USA* **84,** 8716–8720.

Brandley, B. K., Swiedler, S. J., and Robbins, P. W. (1990). Carbohydrate ligands of the LEC cell adhesion molecules. *Cell* **63,** 861–863.

Burns, G. F., Lucas, C. M., Krissansen, G. W., Werkmeister, J. A., Scanlon, D. B., Simpson, R. J., and Vadas, M. A. (1988). Synergism between membrane gangliosides and arg–gly–asp-directed glycoprotein receptors in attachment to matrix proteins by melanoma cells. *J. Cell Biol.* **107,** 1225–1230.

Butcher, E. C. (1991). Leukocyte–endothelial cell recognition: Three (or more) steps to specificity and diversity. *Cell* **67,** 1033–1036.

Cahan, L. D., Irie, R. F., Singh, R., Cassidenti, A., and Paulson, J. C. (1982). Identification of a human neuroectodermal tumor antigen (OFA-I-2) as ganglioside GD_2. *Proc. Natl. Acad. Sci. USA* **79,** 7629–7633.

Carubia, J. M., Yu, R. K., Macala, L. J., Kirkwood, J. M., and Varga, J. M. (1984). Gangliosides of normal and neoplastic human melanocytes. *Biochem. Biophys. Res. Commun.* **120,** 500–504.

Cheresh, D. A., and Klier, F. G. (1986). Disialoganglioside GD_2 distributes preferentially into substrate-associated microprocesses on human melanoma cells during their attachment to fibronectin. *J. Cell Biol.* **102,** 1887–1897.

Cheresh, D. A., and Reisfeld, R. A. (1984). O-acetylation of disialoganglioside GD_3 by human melanoma cells creates a unique antigenic determinant. *Science* **225,** 844–846.

Cheresh, D. A., Harper, J. R., Schulz, G., and Reisfeld, R. A. (1984). Localization of the gangliosides GD_2 and GD_3 in adhesion plaques and on the surface of human melanoma cells. *Proc. Natl. Acad. Sci. USA* **81,** 5767–5771.

Cheresh, D. A., Pierschbacher, M. D., Herzig, M. A., and Mujoo, K. (1986). Disialogangliosides GD_2 and GD_3 are involved in the attachment of human melanoma and neuroblastoma cells to extracellular matrix proteins. *J. Cell Biol.* **102,** 688–696.

Cheresh, D. A., Pytela, R., Pierschbacher, M. D., Klier, F. G., Rouslahti, E., and Reisfeld, R. A. (1987). An arg–gly–asp-directed receptor on the surface of human melanoma cells exists in a divalent cation-dependent functional complex with disialoganglioside GD_2. *J. Cell Biol.* **105,** 1163–1173.

Dippold W. G., Knuth, A., and zum Büschenfeld, K.-H. M. (1984). Inhibition of human melanoma cell growth *in vitro* by monoclonal anti-GD_3-ganglioside antibody. *Cancer Res.* **44,** 806–810.

Dippold, W. G., Dienes, H. P., Knuth, A., zum Büschenfeld K.-H. M. (1985). Immunohistochemical localization of ganglioside GD_3 in human malignant melanoma, epithelial tumors, and normal tissues. *Cancer Res.* **45,** 3699–3705.

Dohi, T., Nores, G., and Hakomori, S. (1988). An IgG_3 monoclonal antibody established after immunization with GM_3 lactone: Immunochemical specificity and inhibition of melanoma cell growth *in vitro* and *in vivo*. *Cancer Res.* **48,** 5680–5685.

Fukuda, M. (1985). Cell surface glycoconjugates as onco-differentiation markers in hematopoietic cells. *Biochim. Biophys. Acta* **780,** 119–150.

Goelz, S. E., Hession, C., Goff, D., Giffiths, B., Tizard, R., Newman, B., Chi-Rosso, G., and Lobb, R. (1990). ELFT: A gene that directs the expression of an ELAM-1 ligand. *Cell* **63,** 1349–1356.

Hamanaka, S., Asagami, C., and Yamaguchi, Y. (1984). Studies on glycolipid of the skin. II. Glycolipids of nevus cell: Nevus and malignant melanoma. *Nishinihon J. Dermatol.* **46,** 173–176.

Harada, Y., Sakatsume, M., Nores, G. A., Hakomori, S., and Taniguchi, M. (1989). Density of GM_3 with normal primary structure determines mouse melanoma antigenicity. A new concept of tumor antigen. *Jpn. J. Cancer Res.* **80,** 988–992.

Hellstrom, I., Brankovan, V., and Hellstrom, K. E. (1985). Strong antitumor activities of IgG3 antibodies to a human melanoma-associated ganglioside. *Proc. Natl. Acad. Sci. USA* **82,** 1499–1502.

Hersey, P., Schibeci, S. D., Townsend, P., Burns, C., and Cheresh, D. A. (1986). Potentiation of lymphocyte responses by monoclonal antibodies to the ganglioside GD_3. *Cancer Res.* **46,** 6083–6090.

Hersey, P., MacDonald, M., Burns, C., and Cheresh, D. A. (1987). Enhancement of cytotoxic and proliferative responses of lymphocytes from melanoma patients by incubation with monoclonal antibodies against ganglioside GD_3. *Cancer Immunol. Immunother.* **24,** 144–150.

Hirabayashi, Y., Suzuki, T., Suzuki, Y., Taki, T., Matsumoto, M., Higashi, H., and Kato, S. (1983). A new method for purification of anti-glycosphingolipid antibody: Avian anti-hematoside (NeuGc) antibody. *J. Biochem.* **94,** 327–330.

Hirabayashi, Y., Higashi, H., Kato, S., Taniguchi, M., and Matsumoto, M. (1987). Occurrence of tumor-associated antigens with hanganutziu-deicher antigenic activity on human melanomas. *Jpn. J. Cancer Res.* **78,** 614–620.

Houghton, A. N., Mintzer, D., Cordon-Cardo, C., Welt, S., Fliegel, B., Vadhan, S., Carswell, E., Melamed, M. R., Oettgen, H. F., Old, L. J. (1985). Mouse monoclonal IgG3 antibody detecting GD_3 ganglioside: A phase I trial in patients with malignant melanoma. *Proc. Natl. Acad. Sci. USA* **82,** 1242–1246.

Iliopoulos, D., Ernst, C., Steplewski, Z., Jambrosic, J. A., Rodeck, U., Herlyn, M., Clark, W. H., Jr., Koprowski, H., and Herlyn D. (1989). Inhibition of metastases of a human melanoma xenograft by monoclonal antibody to the GD_2/GD_3 gangliosides. *J. Natl. Cancer Inst.* **81,** 440–444.

Irie, R. F., and Morton, D. L. (1986). Regression of cutaneous metastatic melanoma by intralesional injection with human monoclonal antibody to ganglioside GD_2. *Proc. Natl. Acad. Sci. USA* **83,** 8694–8698.

Kawashima, I., Tada, N., Fujimori, T., and Tai, T. (1990). Monoclonal antibodies to disialogangliosides: Characterization of antibody-mediated cytotoxicity against human melanoma and neuroblastoma cells *in vitro. J. Biochem.* **108,** 109–115.

Kleinman, H. K., Martin, G. R., and Fishman, P. H. (1979). Ganglioside inhibition of fibronectin-mediated cell adhesion to collagen. *Proc. Natl. Acad. Sci. USA* **76,** 3367–3371.

Kojima, N., and Hakomori, S. (1991). Cell adhesion, spreading, and motility of GM_3-expressing cells based on glycolipid–glycolipid interaction. *J. Biol. Chem.* **266,** 17552–17558.

Kono, K., Tsuchida, T., Kern, D. H., and Irie, R. (1990). Ganglioside composition of human melanoma and response to antitumor treatment. *Cancer Invest.* **8,** 161–167.

Koprowski, H., Steplewski, Z., Herlyn, D., and Herlyn, M., (1978). Study of antibodies against human melanoma produced by somatic cell hybrids. *Proc. Natl. Acad. Sci. USA* **75,** 3405–3409.

Kumar, R., Potvin, B., Muller, W. A., and Stanley, P. (1991). Cloning of a human α(1,3)-fucosyltransferase gene that encodes ELFT but does not confer ELAM-1 recognition on Chinese hamster ovary cell transfectants. *J. Biol. Chem.* **266,** 21777–21783.

Larsen, G. R., Sako, D., Ahern, T. J., Shaffer, M., Erban, J., Sajer, S. A., Gibson, R. M., Wagner, D. D., Furie, B. C., and Furie, B. (1992). P-selectin and E-selectin. *J. Biol. Chem.* **267,** 11104–11110.

Lasky, L. A. (1991). Lectin cell adhesion molecules (LEC-CAMs): A new family of cell adhesion proteins involved with inflammation. *J. Cell. Biochem.* **45,** 139–146.

Lowe, J. B., Stoolman, L. M., Nair, R. P., Larsen, R. D., Berhend, T. L., and Marks, R. M. (1990). ELAM-1-dependent cell adhesion to vascular endothelium determined by a transfected human fucosyltransferase cDNA. *Cell* **63,** 475–484.

Lowe, J. B., Kukowska-Latallo, J., Nair, R. P., Larsen, R. D., Marks, R. M., Macher, B. A., Kelly, R. J., and Ernst, L. K. (1991). Molecular cloning of a human fucosyltransferase gene that determines expression of the Lewis x and VIM-2 epitopes but not ELAM-1-dependent cell adhesion. *J. Biol. Chem.* **266,** 17467–17477.

Macher, B. A., and Beckstead, J. H. (1990). Distribution of VIM-2 and SSEA-1 glycoconjugates among human leukocytes and leukemia cells. *Leukemia Res.* **13,** 119–130.

Macher, B. A., Buehler, J., Scudder, P., Knapp, W., and Feizi, T. (1988). A novel carbohydrate, differentiation antigen on fucogangliosides of human myeloid cells recognized by monoclonal antibody VIM-2. *J. Biol. Chem.* **263,** 10186–10191.

Magnani, J. L. (1986). Carbohydrate sequences detected by murine monoclonal antibodies. *Chem. Phys. Lipids* **42,** 65–74.

Magnani, J. L., Smith, D. F., and Ginsburg, V. (1980). Detection of gangliosides that bind cholera toxin: Direct binding of ^{125}I-labeled toxin to thin-layer chromatograms. *Anal. Biochem.* **109,** 399–402.

Magnani, J. L., Brockhaus, M., Smith, D. F., Ginsburg, V., Blaszczyk, M., Mitchell, K. F., Steplewski, Z., and Koprowski, H. (1981). Specific antigen in serum of patients with colon carcinoma. *Science* **212,** 53–56.

Osborn, L. (1990). Leukocyte adhesion to endothelium in inflammation. *Cell* **62,** 3–6.

Perkins, R. M., Kellie, S., Patel, B., and Critchley, D. R. (1982). Gangliosides as receptors for fibronectin. *Exp. Cell. Res.* **141,** 231–243.

Phillips, M. L., Nudelman, E., Gaeta, F. C. A., Perez, M., Singhal, A. K., Hakomori, S., and Paulson, J. C. (1990). ELAM-1 mediates cell adhesion by recognition of a carbohydrate ligand, Sialyl-Le^x. *Science* **250,** 1130–1132.

Picker, L. J., Warnock, R. A., Burns, A. R., Doerschuk, C. M., Berg, E. L., and Butcher, E. C. (1991a). The neutrophil selectin LECAM-1 presents carbohydrate ligands to the vascular selectins ELAM-1 and GMP-140. *Cell* **66,** 921–933.

Picker, L. J., Kishimoto, T. K., Smith, C. W., Warnock, R. A., and Butcher, E. C. (1991b). ELAM-1 is an adhesion molecule for skin-homing T cells. *Nature* **349,** 796–799.

Portoukalian, J., Zwingelstein, G., and Dore, J.-F. (1979). Lipid composition of human malignant melanoma tumors at various levels of malignant growth. *Eur. J. Biochem.* **94,** 19–23.

Prieels, J.-P., Monnom, D., Dolmans, M., Beyer, T. A., and Hill, R. L. (1981). Copurification of the lewis blood group *N*-acetylglucosaminide $\alpha 1 \rightarrow 4$ fucosyltransferase and an *N*-acetylglucosaminide $\alpha 1 \rightarrow 3$ fucosyltransferase from human milk. *J. Biol. Chem.* **256,** 10456–10463.

Pukel, C. S., Lloyd, K. O., Travassos, R., Dippold, W. G., Oettgen, H. F., and Old, L. J. (1982). GD_3, a prominent ganglioside of human melanoma. Detection and characterization by mouse monoclonal antibody. *J. Exp. Med.* **155,** 1133–1147.

Ravindranath, M. H., Tsuchida, T., Morton, D. L., and Irie, R. F. (1991). Ganglioside $GM_3:GD_3$ ratio as an index for the management of melanoma. *Cancer* **67,** 3029–3035.

Ritter, G., Boosfeld, E., Adluri, R., Calves, M., Oettgen, H. F., Old, L. J., and Livingston, P. (1991). Antibody response to immunization with ganglioside GD_3 and GD_3 congeners (lactones, amide and gangliosidol) in patients with malignant melanoma. *Int. J. Cancer* **48,** 379–385.

Roberts, D. D., and Ginsburg, V. (1988). Sulfated glycolipids and cell adhesion. *Arch. Biochem. Biophys.* **267,** 405–415.

Rosen, S. D., and Yednock, T. A. (1986). Lymphocyte attachment to high endothelial venules during recirculation: A possible role for carbohydrates as recognition determinants. *Mol. Cell. Biochem.* **72,** 153–164.

Shimizu, Y., Shaw, S., Graber, N., Gopal, T. V., Horgan, K. J., Van Seventer, G. A., and Newman, W. (1991). Activation-independent binding of human memory T cells to adhesion molecule ELAM-1. *Nature* **349,** 799–802.

Singhal, A., and Hakomori, S. (1990). Molecular changes in carbohydrate antigens associated with cancer. *BioEssays* **12,** 223–230.

Spohr, U., Hindsgaul, O., and Lemieux, R. V. (1985). Molecular recognition. II. The binding of the Lewis b and Y human blood group determinants by the lectin IV Griffonia simplicifolia. *Can. J. Chem.* **63,** 2644–2652.

Stoolman, L. M. (1989). Adhesion molecules controlling lymphocyte migration. *Cell* **56,** 907–910.

Svennerholm, L. (1963). Chromatographic separation of human brain gangliosides. *Neurochemistry* **10,** 613–623.

Taniguchi, M., and Wakabayashi, S. (1984). Shared antigenic determinant expressed on various mammalian melanoma cells. *Jpn. J. Cancer Res.* **75,** 418–426.

Thurin, J., Herlyn, M., Hindsgaul, O., Stromberg, N., Karlsson, K. A., Elder, D., Steplewski, Z., and Koprowski, H. (1985). Proton NMR and fast-atom bombardment mass spectrometry analysis of the melanoma-associated ganglioside 9-*O*-acetyl-GD_3. *J. Biol. Chem.* **260,** 14556–14563.

Thurin, J., Thurin, M., Herlyn, M., Elder, D. E., Steplewski, Z., Clark, W. H., Jr., and Koprowski, H. (1986). GD_2 ganglioside biosynthesis is a distinct biochemical event in human melanoma tumor progression. *FEBS Lett.* **208,** 17–22.

Thurin, J., Thurin, M., and Kimoto, Y. (1987). Monoclonal antibody-defined correlation in melanoma between levels of GD_2 and GD_3 antigens and antibody-mediated cytotoxicity. *Cancer Res.* **47,** 1229–1233.

Tiemeyer, M., Swiedler, S. J., Ishihara, M., Moreland, M., Schweingruber, H., Hirtzer, P., and Brandley, B. (1991). Carbohydrate ligands for endothelial-leukocyte adhesion molecule 1. *Proc. Natl. Acad. Sci. USA* **88,** 1138–1142.

Tsuchida, T., Saxton, R. E., and Irie, R. F., (1987a). Gangliosides of human melanoma: GM_2 and tumorigenicity. *J. Natl. Cancer Inst.* **78,** 55–60.

Tsuchida, T., Saxton, R. E., Morton, D. J., and Irie, R. F. (1987b). Gangliosides of human melanoma. *J. Natl. Cancer Inst.* **78,** 45–54.

Tsuchida, Y., Saxton, R. E., Morton, D. L., and Irie, R. F. (1989). Gangliosides of human melanoma. *Cancer* **63,** 1166–1174.

Tyrrell, D., James, P., Rao, N., Foxall, C., Abbas, A., Dasgupta, F., Nashed, M., Hasegawa, A., Kiso, M., Kidd, J., and Brandley, B. (1991). Structural requirements for the carbohydrate ligand of E-selectin. *Proc. Natl. Acad. Sci. USA* **88,** 10372–10376.

von Andrian, U. H., Chambers, J. D., McEvoy, L., Bargatze, R. F., Arfors, K. E., and Butcher, E. C. (1991). A two-step model of leukocyte–endothelial cell interaction in inflammation: Distinct roles for LECAM-1 and the leukocyte b2 intergrins *in vivo*. *Proc. Natl. Acad. Sci. USA* **88,** 7538–7542.

Walz, G., Aruffo, A., Kolanus, W., Bevilacqua, M., and Seed, B. (1990). Recognition by ELAM-1 of the Sialyl-Le^x determinant on myeloid and tumor cells. *Science* **250,** 1132–1135.

Wayner, E. A., Orlando. R. A., and Cheresh, D. A. (1991). Integrins $\alpha_v\beta_3$ and $\alpha_v\beta_5$ contribute to cell attachment to vitronectin but differentially distribute on the cell surface. *J. Cell Biol.* **113,** 919–929.

Werkmeister, J. A., Triglia, T., Mackay, I. R., Dowling, J. P., Varigos, G. A., Morstyn, G., and Burns, G. F. (1987). Fluctuations in the expression of a glycolipid antigen associated with differentiation of melanoma cells monitored by a monoclonal antibody, Leo Mel 3. *Cancer Res.* **47,** 225–230.

Yamada, K. M., Kennedy, D. W., Grotendorst, G. R., and Momoi, T. (1981). Glycolipids: Receptors for fibronectin? *J. Cell Physiol.* **109,** 343–351.

Yamaguchi, H., Furukawa, K., Fortunato, S. R., Livingston, P. O., Lloyd, K. O., Oettgen, H. F., and Old, L. J. (1987). Cell-surface antigens of melanoma recognized by human monoclonal antibodies. *Proc. Natl. Acad. Sci. USA* **84,** 2416–2420.

Yamaguchi, H., Furukawa, K., Fortunato, S. R., Livingston, P. O., Lloyd, K. O., Oettgen, H. F., and Old, L. J. (1990). Human monoclonal antibody with dual GM_2/GD_2 specificity derived from an immunized melanoma patient. *Proc. Natl. Acad. Sci. USA* **87,** 3333–3337.

Biochemical and Molecular Biological Approaches to Modify Oligosaccharides: Analysis of Glycosylation Function in Human Choriogonadotropin

O. P. Bahl,* N. R. Thotakura,† and W. Y. Chen*

*Department of Biological Sciences, State University of New York at Buffalo, Buffalo, New York 14260; and †Molecular and Cellular Endocrinology Branch, NIDDK, NIH, Bethesda, Maryland 20892

I. Introduction

Because of the importance of glycoproteins in various biological phenomena and in recognition, such as in host–pathogen interactions, cell–cell interaction, and protein targeting, and because of our ability to produce them in adequate amounts from heterologous cell systems with different carbohydrates by recombinant methodology, it has become imperative to know the role of carbohydrates in their biological function. The general approach to study the role of carbohydrates in the function of glycoproteins has been to completely or partially delete the carbohydrate chemically, enzymatically, metabolically, by site-directed mutagenesis, or by the expression of a glycoprotein in a mutant cell line and evaluate the effect of this modification on its biological properties. Although the various methods for the modification of carbohydrates described below pertain to human chorionic gonadotropin (hCG), they are of a general nature and are equally applicable to studying the significance of carbohydrates in the function of any other glycoprotein.

II. Structures of Glycoprotein hormones

Glycoprotein hormones, including lutropin (LH), thyrotropin (TSH), follitropin (FSH), and human choriogonadotropin (hCG), are heterodimers made of two noncovalently bonded subunits designated α and β. Both subunits are necessary for their biological activity. As far as we know, the individual subunits have no biological activity. The hormones and their subunits contain about 20 to 30% carbohydrate. The human hormones share a common α-subunit with an identical polypeptide chain, but varying carbohydrate structures. The number and location of the glycosylation sites in the α- and β-subunits are summarized in Table I. Since the α-subunits in various hormones are interchangeable, their β-subunits must have some sequence homology with one another. HCGβ and LHβ are the closest in amino acid sequence, having about 86% homology. One of the unique features of hCG

TABLE I
THE NUMBER AND LOCATION OF GLYCOSYLATION SITES IN HUMAN GLYCOPROTEIN HORMONES[a]

		Number		
Hormone	Subunit	N-linked	O-linked	Location
HCG	α	2		Asn52, Asn78
	β	2	4	Asn13, Asn30, Ser121, Ser127, Ser132, Ser138
HFSH	α	2		Asn52, Asn78
	β	2		Asn7, Asn24
HTSH	α	2		Asn52, Asn78
	β	1		Asn23
HLH	α	2		Asn52, Asn78
	β	1		Asn30

[a] From Pierce and Parsons (1981).

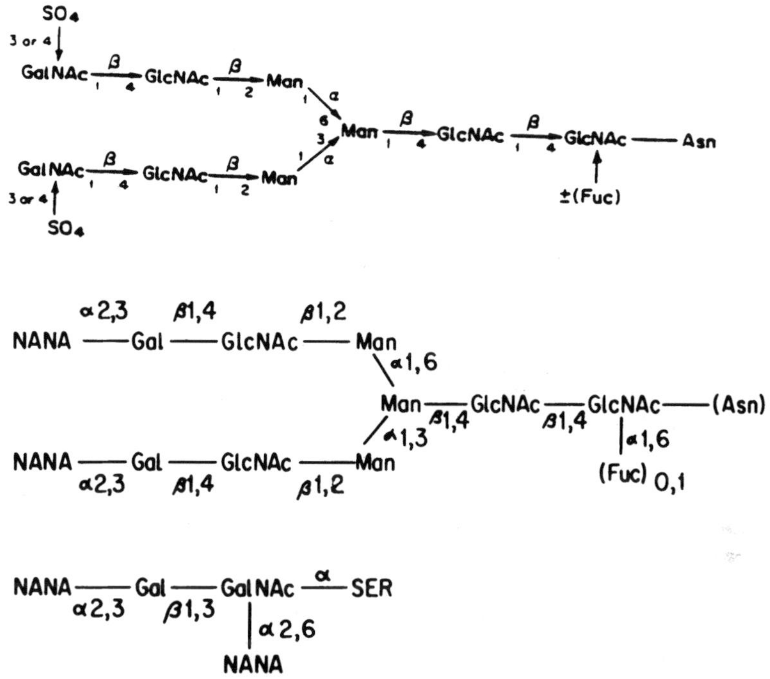

FIG. 1. Carbohydrate structures of the glycoprotein hormones. (top) N-linked oligosaccharides of bLH (Pierce and Parsons, 1981); (middle) N-linked oligosaccharides of hCG (Kessler *et al.*, 1979a), and (bottom) O-linked oligosaccharides of hCG (Kessler *et al.*, 1979b).

molecule is the presence of a 30-residue carboxy-terminal extension containing four O-linked glycosylation sites. The predominant N- and O-linked carbohydrate structures present in hCGα and hCGβ are given in Fig. 1. Obviously, like other glycoproteins, hCG also has been found to have various glycoforms differing in the N- and O-linked carbohydrate structures (Blithe, 1990; Mizuochi *et al.*, 1983; Cole *et al.*, 1985).

III. MODIFICATION OF OLIGOSACCHARIDES AND ITS EFFECTS ON FUNCTION

A. *Biochemical Methods*

1. CHEMICAL

Anhydrous trifluoromethane sulfonic acid (TFMS) and HF have been employed for the deglycosylation of hCG and other glycoprotein hormones (Sojar and Bahl, 1987). These reagents hydrolyze glycosidic bonds, those involving neutral monosaccharides more readily than those involving amino sugars. Both N- and O-linked glycans are hydrolyzed, the latter requiring harsher conditions. Anisole or dimethyl sulfide was used as a scavenger in earlier studies (Edge *et al.*, 1981), but it was found to be unnecessary if the reaction was carried out under controlled conditions. Removal of 70% or more of hexosamines from hCG was reported using TFMS and about 50% using HF by different groups (Kalyan and Bahl, 1981; Sojar and Bahl, 1987; Manjunath and Sairam, 1982; Chen *et al.*, 1982). Not only is TFMS more effective than HF, it is also more convenient to use than HF, which requires a special apparatus and handling.

Deglycosylation reactions can be carried out on intact hormones without significant dissociation of the subunits or on individual subunits, which can later be recombined to a dimer. The integrity of the polypeptide chain after deglycosylation was not altered as judged by polyacrylamide gel electrophoresis and UV and circular dichroic spectra compared with those of the native hormones (Kalyan and Bahl, 1983; Sojar and Bahl, 1987). However, altered circular dichroic spectra (Merz, 1988; Sairam, 1989), increased subunit dissociation (Merz, 1988), or increased thermal stability (Sairam, 1989) resulting from chemical deglycosylation has been reported. It appears that the integrity of the polypeptide chain is unaffected by the chemical treatment per se, excluding the effects of carbohydrate removal, such as the exposure of additional antigenic sites and subtle conformational changes due to the lack of carbohydrate interactions.

Deglycosylation by chemical methods did not reduce but rather increased significantly the receptor binding activity of hCG. On the other hand, the *in vitro* biological activities, as measured by adenylyl cyclase stimulation in ovarian membranes and cAMP accumulation and steroidogenesis in granulosa cells, were drastically impaired (Kalyan and Bahl, 1983; Bahl *et al.*, 1984; Manjunath and Sairam, 1982). These results are shown in Fig. 2. Similar results on the deglycosylation of LH, FSH, and TSH from human and other species were reported by various laboratories (Sairam, 1989; Amr *et al.*, 1987; Reichert *et al.*, 1991; Calvo *et al.*, 1986).

The exact mechanism of signal transduction by glycoprotein hormones and how the removal of the N-linked oligosaccharides affect this process are not currently known. It has been hypothesized that the presence of one or more oligosaccharides on the hormones maintains a conformation that is suitable for signal transduction by inducing the coupling of the hormone–receptor complex to the G protein. The deglycosylation prevents the coupling of the hormone–receptor complex

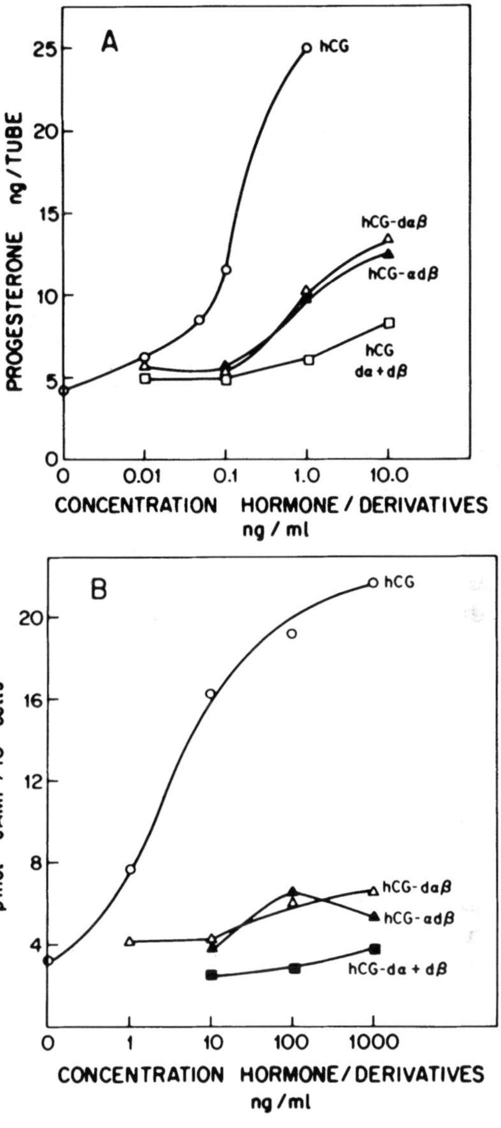

FIG. 2. Stimulation of progesterone (A) and cAMP (B) production by hCG and its reconstituted derivatives in corpus luteum cells. (A) Various amounts of hCG and its derivative were incubated with 4×10^5 cells/tube at 37°C for 90 min. (B) About 1×10^6 cells were incubated in the presence of the indicated amounts of hormones at 37°C for 90 min. The medium contained 0.2 mM methyl 1,3-isobutylxanthine. Both assays were done in triplicate and represent the mean ± SE (Kalyan and Bahl, 1982).

to the G protein–adenylyl cyclase system (Thotakura and Bahl, 1982). This may be due to the fact that the oligosaccharide chain(s) on one or both subunits may interact with a lectin-like component on the receptor to bring about receptor–G protein interaction (Thotakura *et al.*, 1990) or may indirectly impart a conformational change in the other subunit that is necessary for receptor activation, but not for hormone binding to the receptor. Supporting this hypothesis are studies demonstrating the ability of glycopeptides to inhibit the hormone–receptor binding (Thotakura *et al.*, 1991) and the conversion of deglycosylated hCG from an antagonist to an agonist by hCG antibodies (Rebois and Fishman, 1984). Although the chemical removal of oligosaccharides from hCG drastically reduces the *in vitro* bioactivities of the hormone, *in vivo* studies in monkeys indicate that the deglycosylated hormone possesses full agonistic activity to stimulate steroidogenesis (Liu *et al.*, 1989). Some of the recent *in vitro* studies also question the usefulness of deglycosylated hCG as an antagonist (Browne *et al.*, 1990). In any case, chemically deglycosylated derivatives of glycoprotein hormones represent a unique class of reagents (antagonists/partial agonists) to study the mechanism of signal transduction and the role of carbohydrates in this process.

2. Enzymatic

Methodological considerations have been previously reviewed for glycoproteins in general (Tarentino, 1989) and for hCG in particular (Thotakura and Bahl, 1987). Removal or modification of oligosaccharides by enzymatic methods is preferred over chemical methods since the oligo- or monosaccharides can be specifically removed and doubts about nonspecific effects on the polypeptide do not arise. However, this method is not suitable for large-scale or complete deglycosylation of intact native glycoproteins. The use of endoglycosidases to hydrolyze intact oligosaccharide chains or exoglycosidases, sequentially or as a mixture, has been reported by different laboratories (Goverman *et al.*, 1982; Moyle *et al.*, 1978; Lee *et al.*, 1986; Cole *et al.*, 1987; Thotakura and Bahl, 1987; Thotakura *et al.*, 1990b).

Different endoglycosidases, including endo-*N*-acetyl-β-D-glucosaminidase H (Endo H) from *Streptomyces plicatus* (for the removal of high-mannose-type chains), Endo F, and peptide-N4-(*N*-acetyl-β-D-glycosaminyl) asparagine amidase or peptide-*N*-glycosidase F (PNGase F) (for the removal of high-mannose and complex chains), both from *Flavobacterium meningosepticum,* have been used for the removal of N-linked glycan chains (Tarentino *et al.*, 1989). Different isozymes of Endo F, Endo $F_{1,2,3}$, have been isolated from the culture filtrates of

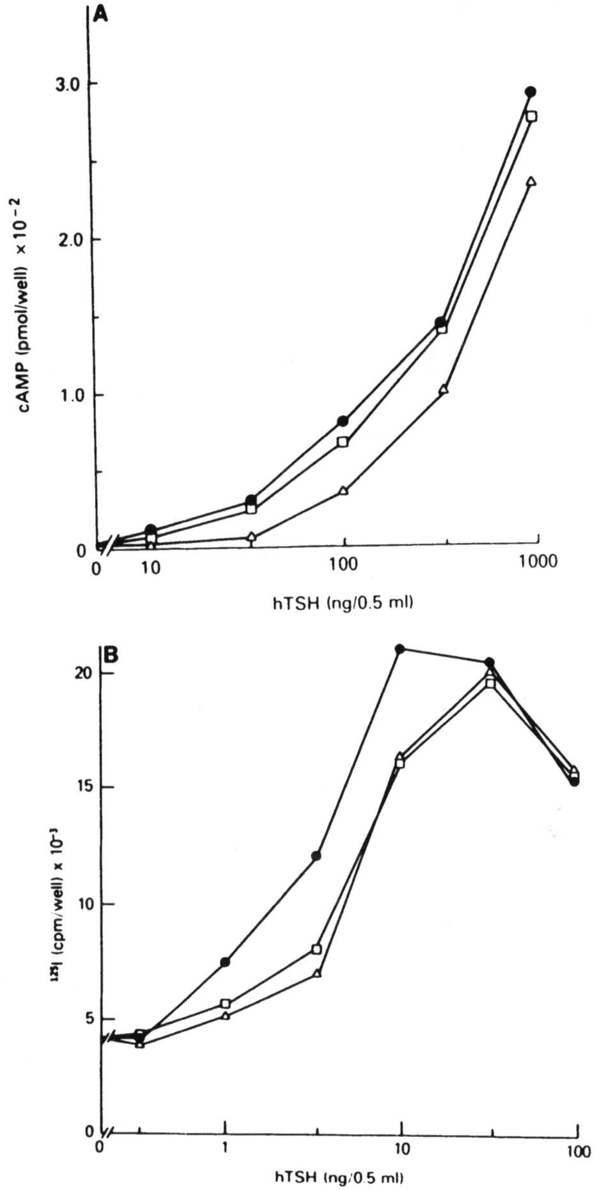

FIG. 3. Stimulation of cAMP production and iodide uptake response on enzymatically deglycosylated purified hTSH in FRTL-5 cells. Dose–response curves before (●) and after treatment with PNGase F (△) and Endo F (□) are shown (Nissim *et al.*, 1987).

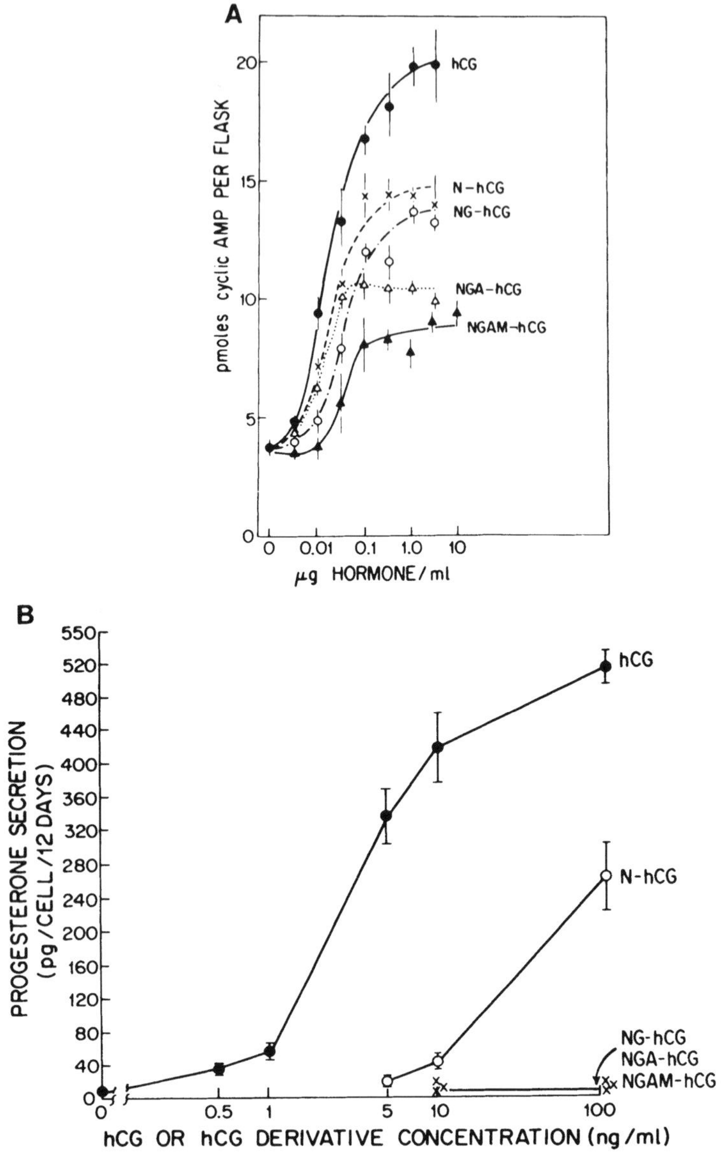

FIG. 4. (A) Cyclic AMP production in response to hCG and its derivatives. Testes cells (400,000) were incubated in triplicate with hCG and its derivatives *in vitro*. The average picomoles of cyclic AMP present in each flask after 10 min at 37°C are shown. Vertical bars extend to the limits of the SEM. This experiment is unusual inasmuch as this is the one in which NGhCG, NGAhCG, and NGAMhCG produced the largest response. In all other experiments, the response was less than this or there was no response. Although the derivatives were not assayed in the range 0.01 to 1.0 μg/ml in this experiment, in experiments where these concentrations were examined, their effect on cyclic AMP accumulation was always equal to or less than that effect elicited by 1.0, 3.0, or 10 μg/ml. The responses of hCG and NhCG in this experiment are typical (Moyle *et*

F. meningosepticum with various preferences toward complex and high-mannose N-linked glycans (Tarentino *et al.*, 1992). PNGase F is broader in its specificity compared with that of Endo F. In addition to these enzymes, endo-*N*-acetyl-α-D-galactosaminidase (*O*-glycanase) has been used for the removal of O-linked chains.

For studying the role of glycan chains in glycoproteins using endoglycosidases, the digestions have to be carried out on native proteins but the enzymes are not efficient in the complete removal of the glycans (Thotakura and Bahl, 1987). In addition, large amounts of the enzymes may be required to attain a significant amount of deglycosylation (Thotakura *et al.*, 1990b). A crude mixture of Endo F and PNGase F may be more effective in complete deglycosylation. One of the chains in the α-subunit of bovine (Thotakura *et al.*, 1990b) and human (Ronin *et al.*, 1987) TSH appears to be more accessible than the other. This property, however, can be used for selective deglycosylation after establishing which chain is more accessible. The limitation of this approach is that the glycans cannot be selectively removed from one or more sites. Furthermore, PNGase F, being an amidase, changes the Asn to Asp and might alter the polypeptide conformation.

Earlier studies on ovine and human LH and bovine and human TSH using endoglycosidases had reported very little changes in the activity of the hormones due to deglycosylation (Cole *et al.*, 1987; Nissim *et al.*, 1987). In later studies, where complete removal of the oligosaccharides was achieved, deglycosylated bovine TSH showed unaltered receptor binding activity and reduced potency in *in vitro* assay (Fig. 3), but the effects were less drastic than those seen with chemically deglycosylated hormones (Thotakura *et al.*, 1990b). This again raised doubts concerning the integrity of the polypeptide chains of glycoprotein hormones deglycosylated chemically. Studies were also conducted to define the effects of enzymatic as well as chemical modification of oligosaccharides on individual subuints and it appears that the glycan chains of both the subunits affect the hormone activity (Thotakura *et al.*, 1992).

Initial experiments using exoglycosidases probed the effects of the removal of the terminal sialic acid from hCG, equine CG, and ovine

al., 1975). N, NG, NGA, and NGAM-hCG are neuraminidase-, β-galactosidase-, β-*N*-acetylglucosaminidase-, and α-mannosidase-treated hCG analogs, respectively. (B) Progesterone production in response to hCG and its derivatives. Granulosa cells were incubated in triplicate with hCG and its derivatives *in vitro*. The average picomoles of progesterone present in each flask over 12 days at 37°C minus the amount present at the start of the incubation (0.129 ± 0.12) are shown. The average SEM was less than 7.5% of the mean.

FSH by digestion with neuraminidase (Dufau *et al.*, 1971; Moyle *et al.*, 1975, 1978). The desialyated hCG showed a stronger binding to the receptor and reduced postreceptor activity (Fig. 4) (Moyle *et al.*, 1975), whereas equine CG showed increased receptor binding activity and in addition displayed an increase in cAMP stimulation and steroidogenesis (Fig. 5) (Bahl *et al.*, 1984). These effects were recently postulated to be due to changes in structured water around the hormonal microenvironment and the movement of ions caused by alterations in the negative charges (Renwick and Wiggins, 1992).

Earlier studies were concerned with the contamination of protease activity and of other exoglycosidases (Lee *et al.*, 1986; Thotakura *et al.*, 1990b; Wittwer *et al.*, 1989). Commercially available preparations are purer and are claimed to be sequencing grade; some of the enzymes are recombinant and free of contaminating activities. Initial studies were done by removing terminal sialic acid in PMSG (Moyle *et al.*, 1975) and later sequentially removing the terminal saccharides by exoglycosidase digestions (Moyle *et al.*, 1978). Recent reports have also utilized neuraminidase to remove terminal sialic acids in recombinant hTSH and bovine LH expressed in Chinese hamster ovary cells (Thotakura *et al.*, 1991; Smith *et al.* 1990). The availability of a specific sulfatase made it possible to remove terminal galNAc-linked sulfate moieties to study their role in the action of bovine LH. The removal of sulfate affected neither the receptor binding activity nor the postreceptor *in vitro* activity of the hormone, in contrast to the sialic acid residues in recombinant hormones, the removal of which decreased or increased the *in vitro* bioactivity depending on the hormone but not on the receptor binding activity of the hormone (Thotakura *et al.*, 1991; Smith *et al.*, 1990; Baenziger *et al.*, 1992). Following further sequential enzymatic removal of galactose and *N*-acetylgalactosamine, recombinant hTSH showed similarly increased *in vitro* bioactivity (Thotakura *et al.*, unpublished results). However, the removal of either the sulfate or the sialic acid residues from hCG, hTSH, or bLH dramatically increased their metabolic clearance rate, probably through the galactose receptors in liver, and consequently the hormones displayed a very low *in vivo* bioactivity (Baenziger *et al.*, 1992; Bahl *et al.*, 1984; Szkudlinski and Thotakura, unpublished results) (Fig. 6).

3. BIOSYNTHETIC

The N-linked glycan chains of glycoproteins are transferred cotranslationally as a $Glc_3(Man)_9$–$GlcNAc_2$ unit *en bloc* to the Asn residue and are processed posttranslationally in the endoplasmic reticulum and Golgi by a series of processing enzymes to the high-mannose- or complex-type final glycan chains. Modifications of the glycan chains

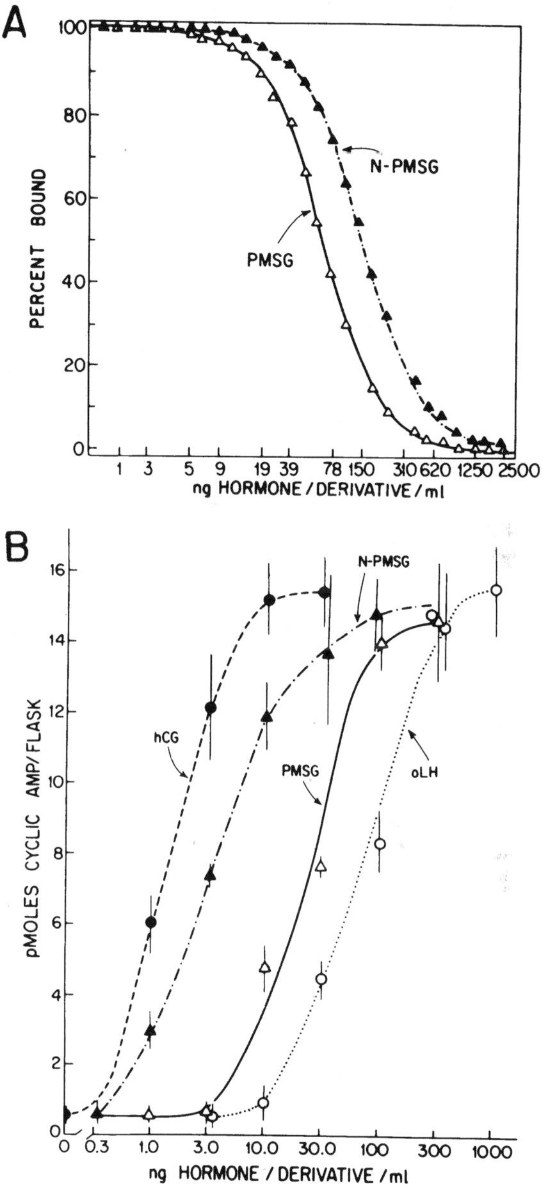

FIG. 5. Receptor-binding and immunological activities of PMSG and N-PMSG. (A) PMSG and N-PMSG at various concentrations were incubated with ^{125}I-PMSG and anti-PMSG serum in 50 m*M* borate and 0.5% BSA, pH 8.0, for 2 hr at 37°C and overnight at 4°C. (B) ^{125}I-PMSG (100,000 cpm) and unlabeled PMSG and N-PMSG were incubated with aliquots of rat testis homogenate for 1 hr at 37°C in 50 m*M* Tris–HCl, pH 7.2, 1 m*M* CaCl, and 0.1% BSA. In the absence of unlabeled hormones, 13,000 cpm of ^{125}I-PMSG was bound, which represents 100% binding.

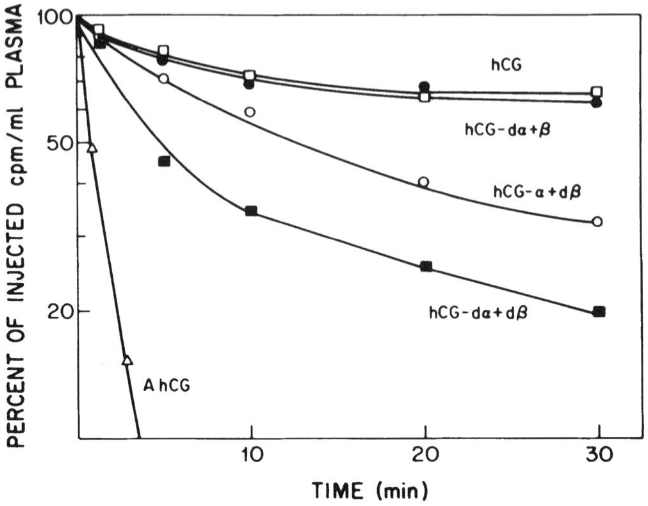

FIG. 6. Plasma survival curves of hCG and derivatives in rats. Mature female rats in the diestrous phase were cannulated through the femoral vein and artery. About 0.5 × 10^6 cpm of ^{125}I-labeled hCG or derivatives was injected through the vein, and blood samples were collected from the artery at the indicated time intervals in heparinized tubes. The radioactivity in 100 μl plasma was counted and expressed as a percentage of the injected cpm in plasma. Each point is the average from three animals and vertical bars extend to the limit of SEM. (●) hCG, (■) dhCG (Bahl *et al.*, 1984).

have been achieved by using inhibitors of one or more of the processing enzymes. Detailed structures and the mechanism action of these inhibitors are described elsewhere in this volume. Inhibition of oligosaccharide transfer using tunicamycin produced glycosylated subunits of TSH, but these subunits were unstable and were rapidly degraded inside the cell (Weintraub *et al.*, 1980), suggesting that glycosylation is essential for proper folding and/or stability of the nascent glycoproteins. Once the protein is folded and secreted, the oligosaccharides are not necessary for the secondary structure of the glycoprotein. Studies using the free α-subunit secreted in the presence of swainsonine demonstrated that a particular type of oligosaccharide structure is essential for hCG subunit combination to form a dimer (Blithe, 1990). Ovine and bovine LH and mouse TSH were biosynthetically produced to contain high-mannose-type oligosaccharides for deglycosylation using endo H and for studying the effects of different types of oligosaccharide structures on the secretion and bioactivity of the hormones (Thotakura and Bahl, unpublished data; Stannard *et al.*, 1989). The stability and secretion, but not the *in vitro* bioactivity, of the TSH were affected by alterations of the oligosaccharide chains (Standdard *et al.*, 1989). As discussed later, glycoproteins lacking in or with altered oligosaccharide chains are produced by expression in prokaryotic systems or in eukaryotic cells lacking one or more of the oligosaccharide-processing enzymes. It was shown that high-mannose-type glycan-containing FSH was rapidly cleared from circulation and thus showed very low *in vivo* bioactivity (Galway *et al.*, 1990), suggesting that the oligosaccharide structures affect the activity of the hormone at several steps of the hormone action.

4. TRANSGLYCOSYLATION

Oligosaccharide chains of hCG and other glycoprotein hormones can also be modified enzymatically using glycosyl transferases in addition to glycosidases. The naturally occurring NeuAcα2 $\rightarrow$ 3Gal linkages of hCG were altered to NeuAcα2 $\rightarrow$ 6Gal by resialylating a desialylated hCG preparation, using the specific sialyl transferase, and it was found that the *in vitro* activity of the hormone was not affected (Amano *et al.*, 1989). Similarly, no differences in the *in vitro* activity of recombinant hTSH expressed in Chinese hamster ovary cells were observed if the NeuAcα2 $\rightarrow$ 6Gal linkages were converted to NeuAcα2 $\rightarrow$ 3Gal by de- and resialyation reactions (Thotakura, unpublished results; Amano *et al.*, 1989). It is also possible to modulate the *in vivo* bioactivity and metabolic clearance of recombinant hTSH by increasing its sialic acid content using sialyl transferases (Thotakura *et al.*, unpublished data). Thus, by engineering the oligosaccharide structures of natural as well as recombinant glycoprotein hormones, glycoforms of the hormones possessing desired *in vivo* bioactivity can be obtained.

B. Modification of Carbohydrates by Recombinant DNA Methodology

The advent of recombinant DNA technology offers a powerful tool to examine the role of the carbohydrate chains in the biological function of glycoproteins. The carbohydrate can be modified in a variety of ways by gene transfer technology. (a) The carbohydrate can be modified by the expression of glycoproteins. This is accomplished by constructing expression vectors with various transcription promoters followed downstream by the cDNA or a gene and subsequently introducing it in one of the several types of available host cells, including those of bacterial, plant, insect, amphibian, and mammalian origin. Because the nature of the carbohydrate is determined by the conformation of the protein matrix and the glycosylation capability of the host cells, recombinant glycoproteins with different carbohydrate variants are likely to be produced. (b) Carbohydrate modification can also be performed by using mammalian cell lines defective in the glycosylation processing pathway. (c) Finally, another way of modifying carbohydrates is site-directed mutagenesis combined with DNA expression techniques. This last method involves the alteration of the asparagine-linked glycosylation consensus sequence, Asn–*X*–Ser/Thr at the individual glycosylation sites, by site-directed mutagenesis in the cDNA, giving upon expression glycoproteins lacking some or all carbohydrate chains. Thus, by this approach it is possible to assess the role of individual N-linked carbohydrate chains in glycoproteins with multiple carbohydrate units.

1. Expression in Unglycosylated System (Prokaryotic Cells)

The expression of the bovine glycoprotein α-subunit in prokaryotic cell *Escherichia coli* was first reported by Pierce *et al.* (as cited in Strickland *et al.*, 1985). The recombinant α-subunit produced from bacteria was completely devoid of all the carbohydrate chains and was found to be present as an insoluble aggregate. The protein was solubilized in guanidine–HCl/mercaptoethanol solution and purified by DEAE and CM–cellulose chromatography in the presence of urea. The purified α-subunit protein had an amino acid composition comparable to that of the bovine LH α-subunit, but only 0.2% of the protein had the native conformation as assessed by a conformationally sensitive radioimmunoassay. However, the yield of the nativelike structure of the α-subunit was increased 18 times upon refolding by reduction and reoxidation. In contrast, under similar conditions the chemically deglycosylated LH α-subunit containing only the chitobiose core could be refolded to almost 100% native conformation. The data were interpreted to indicate that the carbohydrates facilitated the refolding of the α-subunit. The refolded bacteria-derived α-subunit could recombine with the LH β-subunit and the resulting dimer showed nearly 1.75 times the receptor binding potency compared with that of native bovine LH. The steroidogenesis potency of the dimer was evaluated using murine Leydig tumor MA-10 cells and the dimer was found to have only 37% of the steroidogenic activity compared with that of native bovine LH, a result consistent with that obtained with the chemically deglycosylated glycoprotein hormones (Fig. 7).

2. Expression in Heterologous Cells (Eukaryotic Cells)

Expression of hCG in eukaryotic cells such as insect cells and mammalian cells has been reported (Chen *et al.*, 1991, Chen and Bahl, 1991a; Lustbader *et al.*, 1987; Matzuk and Boime, 1988). Insect cells have been found to possess glycosylation machinery similar to that of mammalian cells but with a different processing capability. Although

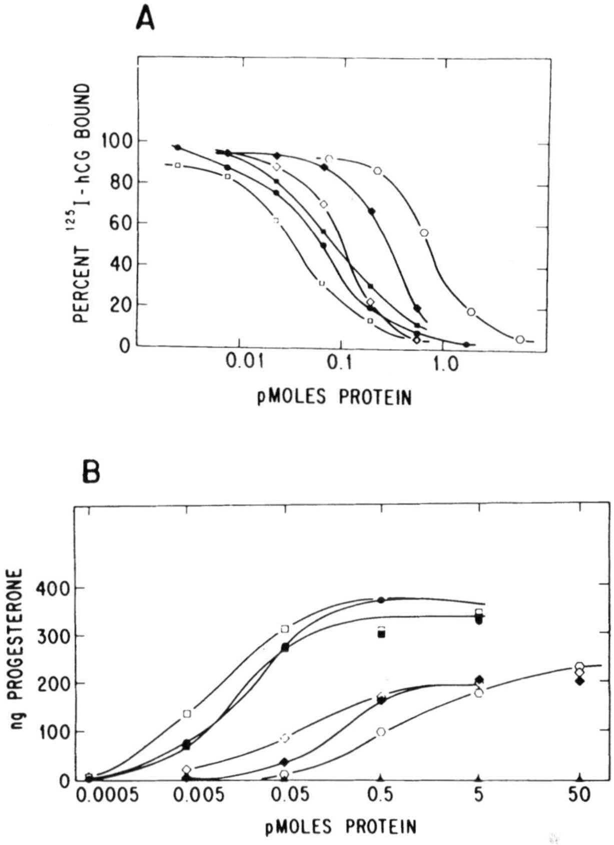

FIG. 7. Radioligand receptor assay (A) and steroidogenic assay (B) or bacterial-expressed bovine LHα recombinant. The total recombinant mixture resulting from incubation of the various α forms with the hCGβ-subunit, with no separation of the $\alpha\beta$ dimer from free subunits, was tested for its ability to inhibit ^{125}I-hCG binding to a rat testicular homogenate and its ability to stimulate steroid production in mouse Leydig tumor cells (Strickland *et al.*, 1985). ●, hCG; □, α + hCGβ; ■, ref α + hCGβ; ◇, dgα + hCGβ; ◆, ref dgα + hCGβ; ○, ref bac α + hCGβ; ▲, hcGβ.

glycoproteins expressed in insect sf9 cells, derived from fall armyworm, have high-mannose-type carbohydrates as shown by endoglycosidase cleavage and by carbohydrate composition analysis, human tissue plasminogen activator was found to have the complex-type structure (Davidson *et al.*, 1990). We have expressed the hCG β-subunit in the baculovirus insect cell expression system by infecting the cells with recombinant viruses with hCG β-cDNA (Chen *et al.*, 1991. The recombinant hCG β-subunit derived from the insect cells had two N-linked high-mannose-type carbohydrates instead of the complex-type carbohydrate chains present in the urinary hCG β-subunit, indicated by the carbohydrate composition and endoglycosidase digestion. The proposed N-linked carbohydrate structure (Chen and Bahl, 1992) in rhCG is shown in Fig. 8. In the same way it was shown that the four O-linked mucin-type oligosaccharide chains in the hCG β-subunit were substituted by simple disaccharide chains in the recombinant hCG β-subunit. Despite the carbohydrate substitution, the recombinant hCG β-subunit could readily recombine with the native hCG α-subunit to form an hCG hybrid $\alpha r\beta$. This hybrid hCG, therefore, would have both the complex-type and the high-mannose-type structures, the complex type being present in the native hCG α-subunit. Another hCG carbohydrate variant with only high-mannose-type carbohydrates was prepared by coinfecting the insect cells with both recombinant viruses with hCG α-subunit and β-subunit cDNAs (Chen and Bahl, 1991a). Since both the recombinant hCG β-subunit and the recombinant hCG were secreted into the medium, they were purified by immunoaffinity chromatography using an hCG monoclonal antibody. Studies with the resulting purified preparation indicated that the carbohydrate substitution had no effect on the secretion, assembly, and receptor binding activities of the hormone. Further evaluation of the *in vitro* biological activities of the hybrid hCG $\alpha r\beta$ or the recombinant hCG revealed that these carbohydrate variants of hCG have a comparable ability to stimulate cAMP accumulation and progesterone production in murine Leydig MA-10 cells (Fig. 9). Thus, it seems that there is no carbohydrate specificity required for the subunit assembly, hormone receptor binding, and the *in vitro* biological activities of hCG. The only common structure among these carbohydrate variants of hCG is the pentasaccharide core.

Expression of the biologically active hCG in mouse C127 cells was achieved by transfection with expression vectors containing hCG α- and β-subunit coding sequences (Lustbader *et al.*, 1987). Recombinant hCG was secreted into the medium accompanied by a 10-fold or more excess β-subunit. The dimeric hormone and the excess hCG β-subunit were found to have carbohydrate structures similar to the native hor-

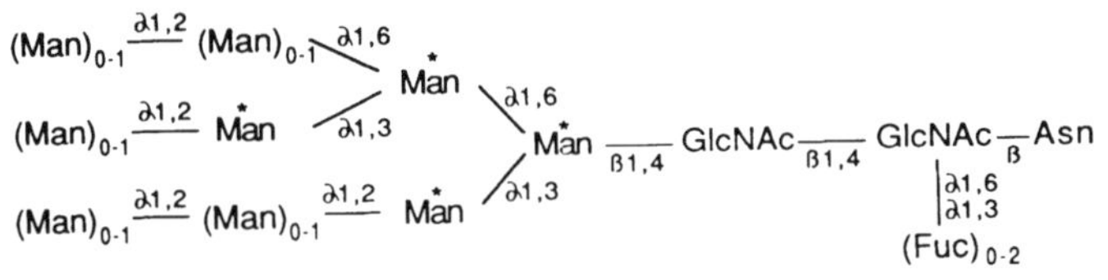

FIG. 8. Tentative structure of high-mannose-type carbohydrate chains in rhCGβ from insect cells. The presence of four mannose residues marked by an asterisk is established by the peroxidate oxidation data (Chen and Bahl, 1992).

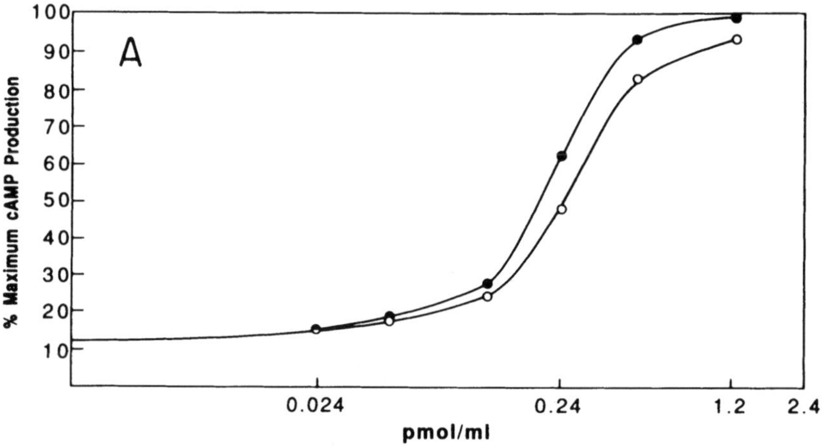

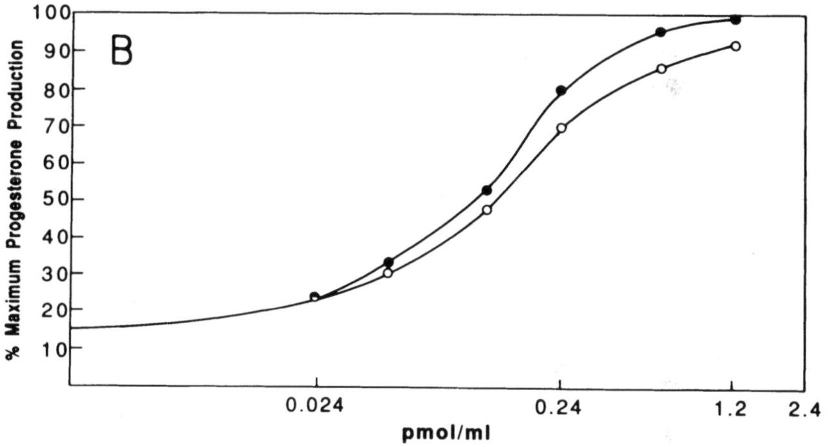

FIG. 9. Cyclic AMP production and steroidogenesis of rhCG from insect cells in mouse Leydig cells. (A) The concentration-dependent stimulation of cAMP in MA-10 cell by native hCG (●) or rhCG (○) was performed by incubation of the cells with various concentrations of hCG or rhCG at 37°C for 1 hr. The cAMP accumulation was determined by cAMP binding assay with [^{3}H]-cAMP and cAMP binding proteins. (B) The effect of rhCG on the steroidogenesis in MA-10 cells was determined by exposing cells to various concentrations of native hCG (●) or rhCG (○) at 37°C for 4 hr. The progesterone in the medium was estimated by radioimmunoassay using ^{125}I-progesterone and anti-progesterone antibodies (Chen and Bahl, 1991a).

mone and were similar to the urinary counterparts as judged by chemical and biological criteria (Fig. 10). The HCG α-subunit alone was also expressed in mouse C127 cells and purified by affinity chromatography using a monoclonal antibody. The free α-subunit was found to have a higher molecular weight. The peptide mapping of the free α-subunit

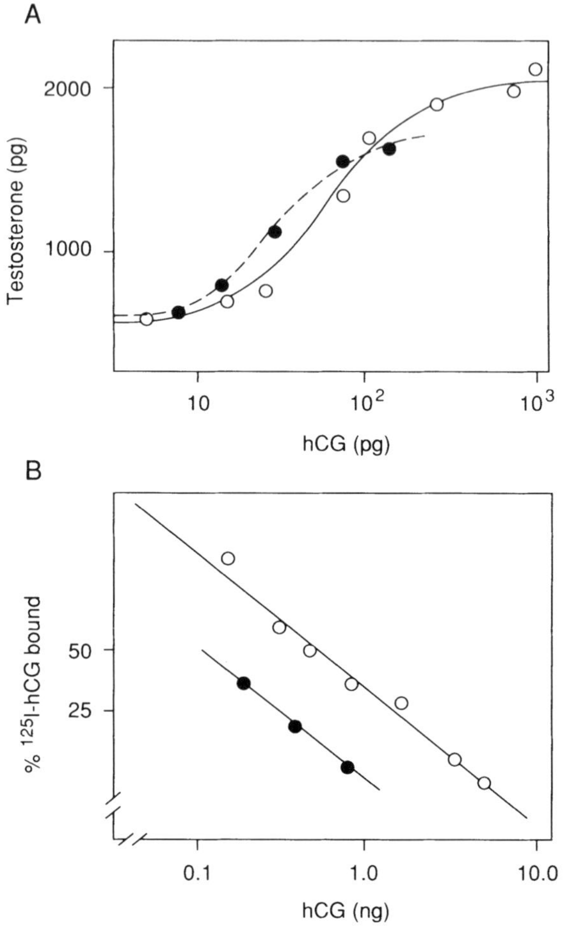

FIG. 10. *In vitro* biological activity of recombinant hCG from mouse C127 cells. (A) Stimulation of testosterone in rat Leydig cells; (B) competitive receptor binding assay. (○) Urinary hCG standard; (●) recombinant hCG (Lustbader *et al.*, 1987).

showed that the difference in the molecular weight resulted from residues 52–91 of the hCG α-subunit. Further studies indicated that the N-linked carbohydrate chains on the free hCG α-subunit were triantennary instead of biantennary as are those present in the urinary hCG α-subunit. However, the α-subunit prepared from recombinant dimeric hCG did not seem to have such hyperglycosylation because its molecular weight was similar to that of the urinary α-subunit. In addition to the hyperglycosylation on the N-linked carbohydrate chains, abnormal O-linked glycosylation was found in the free hCG α-subunit as shown by *in vivo* labeling techniques. This hyperglycosylated form of the

α-subunit lost its ability to recombine with the β-subunit to form a heterodimer. The free hCG α-subunit with hyperglycosylation may have a different conformation compared with the one in the dimeric hormone as demonstrated by studies using the free hCG α-subunit from a tumor JAR cell. The free hCG α-subunit from tumor cells had an extra O-linked glycosylation site at position threonine 39 (Corless *et al.*, 1987).

3. Expression in Mutant Cells

An example of using the glycosylation mutant cell lines for the preparation of carbohydrate variants is a carbohydrate variant of hCG that was prepared from a CHO mutant cell line, IdID (Matzuk *et al.*, 1987). This cell is deficient in UDP–Gal/UDP–GalNAC 4-epimerase and was used to prepare hCG without O-linked oligosaccharides. IdID cells have no ability to synthesize UDP–Gal and UDP–GalNAC under normal culture conditions when glucose is the sole sugar source. This results in the lack of Gal and GalNAC incorporation in N-linked or O-linked carbohydrates in glycoproteins. By supplying Gal or GalNAC in the medium the ability of cells to process the N- or O-linked carbohydrates respectively can be restored (Fig. 11). The recombinant hCG β-subunit devoid of O-linked oligosaccharides was obtained by growing the transfected cells in the culture medium containing Gal. The pulse–chase experiment with [^{35}S]methionine indicated that the hCG β-subunit without O-linked oligosaccharides was secreted at the same rate as the O-glycosylated hCG β-subunit in this cell. When the hCG α- and β-subunits were coexpressed, the loss of O-linked glycosylation had no effect on hormone secretion and subunit assembly (Matzuk *et al.*, 1987). The effect of O-linked oligosaccharides on the *in vitro* biological activities of hCG was examined using the murine Leydig tumor MA-10 cell line in a separate study (Matzuk *et al.*, 1990). Recombinant hCG deficient in the O-linked oligosaccharides was found to have the same binding to the mouse LH receptor and displayed a similar ability to stimulate cAMP accumulation and steroidogenesis in MA-10 cells. These studies indicate that O-linked glycosylation in hCG has little effect on its *in vitro* biological properties. Its importance, however, in its *in vivo* behavior remains to be investigated.

4. Site-Directed Mutagenesis

Site-directed mutagenesis is a powerful tool to investigate the role of the carbohydrates. By disruption of the glycosylation consensus sequence Asn–*X*–Thr/Ser through site-directed mutagenesis in hCG cDNA and its subsequent expression in CHO cells, the effect of the individual N-linked carbohydrate chains on hCG secretion, assembly,

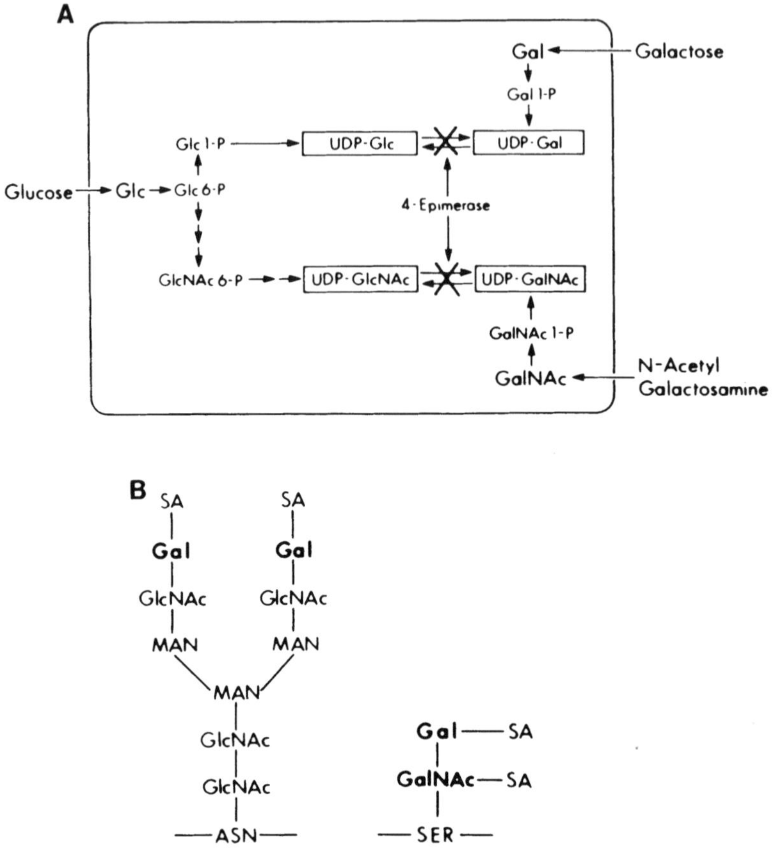

FIG. 11. Reversible glycosylation pathway in CHO mutant IdID cells. (A) Under normal culture conditions with glucose as the sole sugar in the medium, Gal and GalNAC cannot be added to glycoproteins in IdID cells because their UDP–Gal/UDP–GalNAC 4-epimerase deficiency (indicated by ×) prevents the synthesis of UDP–Gal and UDP–GalNAC from glucose. The 4-epimerase deficiency may be bypassed by the direct addition of Gal and GalNAC to the culture medium. (B) Typical N-linked and O-linked carbohydrate structures. The positions of Gal and GalNAC are indicated in boldface. Glc, glucose; GlcNAC, *N*-acetylglucosamine; Man, Mannose; SA, sialic acid (Matzuk *et al.*, 1987).

and biological functions has been examined (Matzuk and Boime, 1988; Bielinska *et al.*, 1989; Matzuk *et al.*, 1989). The substitution of Asn by Asp and/or Thr by Ala in the glycosylation consensus sequences in hCG was used to selectively remove the N-linked carbohydrate chains. While the absence of one or both N-linked carbohydrate chains at Asn13 and Asn30 in the hCG β-subunit had no significant effect on hormone function, the carbohydrate chains in the hCG α-subunit were found to have biological significance. The HCG α-subunit lacking the glycosylation at Asn78 was unstable and degraded rapidly, with less than 20% secreted. However, by coexpression of the Asn78 mutant and the hCG α-subunit, about 45% of the mutated subunit could be secreted into the medium as a dimer. The secretion and stability of the hCG α-subunit without oligosaccharide on Asn52 were identical to those of the wild-type protein, although the removal of this carbohydrate chain did affect dimer formation and the secretion of the dimer. In addition, the glycosylation on this site was shown to be processed in the monomer differently from that in the dimer form. The α-subunit with a carbohydrate chain on Asn52 from the dimer was found to bind essentially Con A while the free hCG α-subunit was processed into a form that failed to bind Con A. Furthermore, this mutated subunit, after recombining with the hCG β-subunit, had much less *in vitro* biological activity such as cAMP production and steroidogenesis in MA-10 cells. The results of these studies are summarized in Table II. This is an elegant example of a molecular biological approach to investigate the role of the carbohydrates in glycoproteins.

Because the hCG β-subunit has four O-linked glycosylation clusters on the 30-amino-acid carboxy terminus, which is unique to the hCG β-subunit and is not present in any other β-subunit of glycoprotein hormones, the coding sequence for this region was specifically deleted from the hCG β-subunit cDNA and expressed in both insect sf9 and CHO cells in order to investigate the biological functions of this carboxy

TABLE II

RECEPTOR BINDING AND STEROIDOGENIC ACTIVITIES OF hCG GLYCOSYLATION MUTANTS IN MURINE LEYDIG MA-10 CELLS[a]

Hormone	Receptor binding (%)	Steroidogenesis (%)
HCG,Wt[b]	100	100
DGα-β[c]	116	58
α-DGβ	102	97
DGα(Asn52)-β	75	69
DGα(Asn78)-β	124	102
DGα(Asn52)-DGβ	83	13
DGα(Asn78)-DGβ	137	99
DGα-DGβ	129	12

Note. Data on the biological properties of recombinant hCG and various mutants expressed in CHO cells are from Matzuk *et al.* (1989).

[a] Modified from Sairam (1989).

[b] Recombinant hCG.

[c] Deglycosylated α or β.

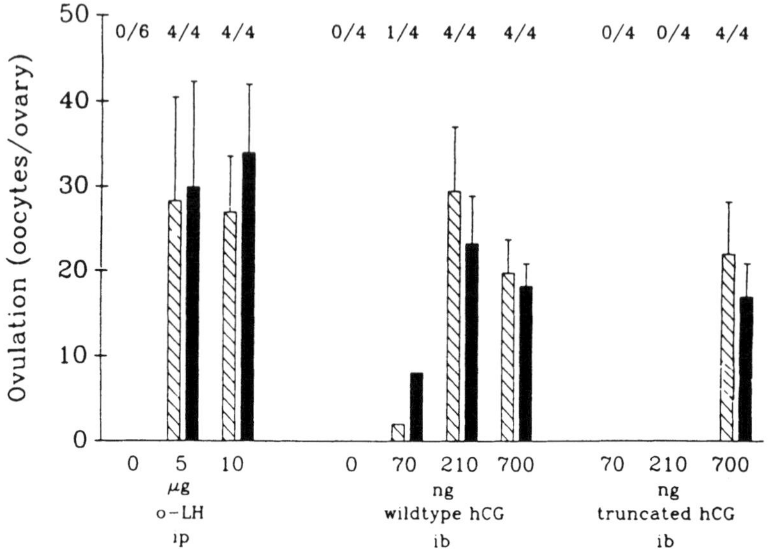

FIG. 12. Number of ovulated oocytes recovered from the oviducts of PMSG-primed female rats after ip injection of ovine LH or ib injection of wild-type hCG or truncated hCG. The number of rats that ovulated per total number of animals used is listed at the top over each column. Ib injections were into left ovaries. ■, oocytes recovered from left ovaries; ▧, oocytes recovered from right ovaries (Matzuk *et al.*, 1990).

terminus and these four O-linked oligosaccharides (Chen and Bahl, 1991b; Matzuk *et al.*, 1990). Both the hCG β-subunit deletion mutant from insect cells with N-linked high-mannose-type carbohydrates and the one from CHO cells with N-linked complex-type carbohydrates were found to be remarkably similar to the intact hCG β-subunit in all their *in vitro* biological aspects examined, including receptor binding, cAMP production, and progesterone accumulation in MA-10 cells. However, *in vivo* studies using the CHO-cell-derived hCG β-subunit mutant, after recombining with the hCG α-subunit, had low biological potency as judged by rat uterine weight assay and by its ability to stimulate mouse ovulation (Fig. 12). Obviously the O-linked oligosaccharides and the carboxy-terminal peptide on the hCG β-subunit might contribute to its long plasma survival time because the dimer comprising this deletion mutant and the hCG α-subunit was found to have a much faster plasma clearance rate compared with that of the urinary hCG.

IV. CONCLUSIONS

The methods for the modification of oligosaccharide chains in glycoproteins can be broadly classified into two groups, biochemical methods and those based on gene transfer methodology. The biochemical methods constitute chemical, enzymatic, and metabolic methods. The most widely used chemicals for deglycosylation are HF and TFMS. Enzymatic deglycosylation or modification can be achieved by endo- or exoglycosidases. The combined use of exoglycosidases and specific transferases can also be made to effect modification of the oligosaccharide chain. The metabolic methods involve the use of carbohydrate processing inhibitors. By this approach, one can obtain completely unglycosylated protein by inhibition of glycosylation by tunicamycin or partially unglycosylated protein by using inhibitors of other glycosidases involved in the various processing steps. The methods based on recombinant DNA methodology include site-directed mutagenesis and the expression of cDNAs or genes of glycoproteins in heterologous cells or their mutants. These are potent methods for the modification of oligosaccharide chains. An added advantage of site-directed mutagenesis is the possibility of the specific deletion of one or more oligosaccharide chains. This approach allows the evaluation of individual glycosyl chains in the biological function of a glycoprotein. The limitations of this approach, however, are the alteration of one or more amino acids in the polypeptide chain, which may or may not have any significant effect on its conformation, and the failure of expression of the mutant gene in the heterologous cell system. Another promising approach to modify the oligosaccharide chain by gene transfer techniques is the coinsertion of a specific glycosyl transferase and glycoprotein cDNAs or genes in a suitable vector for expression in a cell or its mutant lacking the transferase. In this way, it is possible to incorporate a particular sugar in a particular linkage in the oligosaccharide chain. Such approaches will become common place as more and more cDNAs or genes of glycosyl transferases become available. Thus, it is clear that a host of potential options are presently available for the modification of oligosaccharide chains in glycoproteins.

References

Amano, J., Sato, S., Nishimura, R., Mochizuki, M., and Kobata, A. (1989). Sialic acids, but not their linkage to galactose residues, are required for full expression of the biological activity of human chorionic gonadotropin. *J. Biochem.* **105,** 339–340.

Amr, S., Menezes-Ferreira, M. M., Shimohigashi, Y., Chen, H. C., Nisula, B. C., and Weintraub, B. D. (1985). Activities of deglycosylated thyrotropin at the thyroid membrane receptor–adenylate cyclase system. *J. Endocrinol. Invest.* **8,** 537–541.

Baenziger, J. J., Kumar, S., Brodheck, R. M., Smith, P. L., and Beranek, M. C. (1992). Circulatory half-life but not interaction with the lutropin/chorionic gonadotropin receptor is modulated by sulfation of bovine lutropin oligosaccharides. *Proc. Natl. Acad. Sci. USA* **89,** 334–338.

Bahl, O. P., Thotakura, N. R., and Anumula, K. R. (1984). "Biological Role of Carbohydrates in Gonadotropins in Hormone Receptors in Growth and Reproduction" (B. B. Saxena *et al.*, eds.), pp. 165–183. Raven Press, New York.

Blithe, D. L. (1990). N-linked oligosaccharides on free α interfere with its ability to combine with human chorionic gonadotropin-β subunit. *J. Biol. Chem.* **265,** 21951–21956.

Browne, E. D., Flasch, M. V., Sairam, M. R., and Bhalla, V. K. (1990). Is deglycosylated human chorionic gonadotropin an antagonist to human chorionic gonadotropin? Characterization of deglycosylated human chorionic gonadotropin action in two testicular interstitial cell fractions. *Biochim. Biophys. Acta* **1033,** 226–234.

Calvo, F. O., Keutman, H. T., Bergert, E. R., and Ryan, R. J. (1986). Deglycosylated human follitropin: Characterization and effects on adenosine cyclic 3′, 5′-phosphate production in porcine granulosa cells. *Biochemistry* **25,** 3938–3943.

Chen, W., and Bahl, O. P. (1991a). Recombinant carbohydrate variant of human choriogonadotropin β subunit des C-terminal (115–145). *J. Biol. Chem.* **266,** 6246–6251.

Chen, W., and Bahl, O. P. (1991b). Recombinant carbohydrate and selenomethionyl variants of human choriogonadotropin. *J. Biol. Chem.* **266,** 8192–8197.

Chen, W., Shen, Q.-X., and Bahl, O. P. (1991). Carbohydrate variant of the recombinant β-subunit of human choriogonadotropin expressed in baculovirus expression system. *J. Biol. Chem.* **266,** 4081–4087.

Chen, W., and Bahl, O. P. (1992). Polyclonal antibodies against the polypeptide and carbohydrate epitopes of recombinant human choriogonadotropin β-subunit. *Mol Cell. Endocrinol.* **86,** 57–66.

Cole, L. A., Birken, S., and Perini, F. (1985). The structures of the serine-linked sugar chains on human chorionic gonadotropin. *Biochem. Biophys. Res. Commun.* **126,** 333–339.

Cole, L. A., Metsch, L. A., and Grotjan, H. E., Jr. (1987). Significant steroidogenic activity of luteinizing hormone is maintained after enzymatic removal of oligosaccharide. *Mol. Endocrinol.* **1,** 621–627.

Corless, C. L., Bielinska, M., Ramabhadran, T. V., Daniels-McQueen, S., Otani, T., Reitz, B. A., Tiemeier, D. D., and Boime, I. (1987). Gonadotropin α subunit, differential processing of free and combined forms in human trophoblast and transfected mouse cells. *J. Biol. Chem.* **262,** 14197–14203.

Davison, D. J., Fraser, M. J., and Castellino, F. J. (1990). Oligosaccharide processing in the expression of human plasminogen cDNA by lepidopteran insect cells. *Biochemistry* **29,** 5584–5590.

Dufau, M. L., Catt, K. J., and Tsuruhara, T. (1971). Retention of *in vitro* biological activities by desialylated human luteinizing hormone and chorionic gonadotropin. *Biochem. Biophys. Res. Commun.* **44,** 1022–1026.

Edge, A. B., Faltynek, C. R., Hof, L., Reichert, L. E., Jr., and Weber, P. (1981). Deglycosylation of glycoproteins by trifluoromethane sulfonic acid. *Anal. Biochem.* **118,** 131–137.

Galway, A. B., Hsueh, A. J. W., Keene, J. L., Yamoto, M., Fauser, B. C. J. M., and Boime, I. (1990). *In vitro* and *in vivo* bioactivity of recombinant human follicle-stimulating hormone and partially deglycosylated variants secreted by transfected eukaryotic cell lines. *Endocrinology* **127,** 93–100.

Goverman, J. M., Parsons, T. F., and Pierce, J. G. (1982). Enzymatic deglycosylation of the subunits of chorionic gonadotropin. *J. Biol. Chem.* **257,** 15059–15064.

Kalyan, N. K., and Bahl, O. P. (1981). Effect of deglycosylation on the subunit interactions and receptor binding activity of human chorionic gonadotropin. *Biochem. Biophys. Res. Commun.* **102,** 1246–1253.

Kessler, M. J., Reddy, M. S., Shah, R. H., and Bahl, O. P. (1979a). Structures of N-glycosidic carbohydrate units of human chorionic gonadotropin. *J. Biol. Chem.* **254,** 7901–7908.

Kessler, M. J., Mise, T., Ghai, R. D., and Bahl, O. P. (1979b). Structure and location of the O-glycosidic carbohydrate units of human chorionic gonadotropin. *J. Biol. Chem.* **254,** 7909–7914.

Lee, K. O. Gesundheit, N., Chen, H. C., and Weintraub, B. D. (1986). Enzymatic deglycosylation of thyroid-stimulating hormone with peptide *N*-glycosidase F and endo-β-*N*-acetyl glucosaminidase F. *Biochem. Biophys. Res. Commun.* **138,** 230–237.

Liu, L., Southers, J. L., Banks, S. M., Blithe, D. L., Wehman, R. E., Brown, J. H., Chen, H. C., and Nisula, B. C. (1989). Stimulation of testosterone production in the cynomolgus monkey *in vivo* by deglycosylated and desialylated human choriogonadotropin. *Endocrinology* **124,** 175–180.

Lustbader, J., Birken, S., Pollack, S., Levinson, L., Bernstine, E., Hsiung, N., and Canfield, R. (1987). Characterization of the expression products of recombinant human choriogonadotropin and subunits. *J. Biol. Chem.* **262,** 14204–14212.

Matzuk, M. M., Hsueh, A. J. W., Lapolt, P., Tsafriri, A., Keen, J. L., and Boime, I. (1990). The biological role of the carboxyl-terminal extension of human chorionic gonadotropin β subunit. *Endocrinology* **126,** 276–383.

Matzuk, M. M., Keene, J. L., and Boime, I. (1989). Site-specificity of the chorionic gonadotropin N-linked oligosaccharides in signal transduction. *J. Biol. Chem.* **264,** 2409–2414.

Matzuk, M. M., and Boime, I. (1988). The role of the asparagine-linked oligosaccharides of the α-subunit in the secretion and assembly of human chorionic gonadotropin. *J. Cell. Biol.* **106,** 1049–1059.

Matzuk, M. M., Kreiger, M., Corless, C. L., and Boime, I. (1987). Effects of preventing O-glycosylation on the secretion of human chorionic gonadotropin in Chinese hamster ovary cells. *Proc. Natl. Acad. Sci. USA* **84,** 6354–6358.

Merz, W. E. (1988). Evidence for impaired subunit interaction in chemically deglycosylated human choriogonadotropin. *Biochem. Biophys. Res. Commun.* **156,** 1271–1278.

Mizuochi, T., Nishimura, R., Derappe, C., Taniguchi, T., Hamamoto, T., Mochizuki, M., and Kobata, A. (1983). Structure of the asparagine-linked sugar chains of human chorionic gonadotropin produced in choriocarcinoma. *J. Biol. Chem.* **258,** 14126–14129.

Moyle, W. M., Bahl, O. P., and Marz, L. (1975). Role of carbohydrate of human chorionic gonadotropin in the mechanism of hormone actions. *J. Biol. Chem.* **250,** 9163–9169.

Moyle, W. M., Erickson, G. F., Bahl, O. P., Christacos, S., and Gutowski, J. (1978). Action of PMSG and asialo-PMSG on rat Leydig and granulosa cells. *Am. J. Physiol.* **235,** E218–E226.

Nissim, M., Lee, K. O., Petrick, P. A., Dahlberg, P. A., and Weintraub, B. D. (1987). A sensitive thyrotropin (TSH) bioassay based on iodide uptake in rat FRTL-5 thyroid cells: Comparison with the adenosine 3′,5′-monophosphate response to human serum RSH and enzymatically deglycosylated bovine and human TSH. *Endocrinology* **121,** 1278–1287.

Pierce, J. G., and Parsons, T. F. (1981). Glycoprotein hormones: Structure and function. *Annu. Rev. Biochem.* **50,** 465–495.

Rebois, R. V., and Fishman, P. H. (1984). Antibodies against human chorionic gonadotropin convert the deglycosylated hormone from an antagonist to an agonist. *J. Biol. Chem.* **259,** 8087–8090.

Reichert, L. E., Jr., Dattatreyamurthy, B., Grasso, P., and Santa-Coloma, T. A. (1991). Structure–function relationships of the glycoprotein hormones and their receptors. *Trends Pharmacol. Sci.* **12,** 199–203.

Ronin, C., Papandreou, M. J., Canonne, C., and Weintraub, B. D. (1987). Carbohydrate chains of human thyrotropin are differentially susceptible to endoglycosidase removal on combined and free polypeptide subunits. *Biochemistry* **26,** 5848–5853.

Renwick, A., and Wiggins, P. (1992). An antipodean perception of the mode of action of glycoprotein hormones. *FEBS Lett.* **297,** 1–3.

Sairam, M. R. (1989). Role of carbohydrates in glycoprotein hormone transduction. *FASEB J.* **3,** 1915–1926.

Smith, P. L., Kaetzel, D., Nilson, J., and Baenziger, J. U. (1990). The sialylated oligosaccharides of recombinant bovine lutropin modulate hormone bioactivity. *J. Biol. Chem.* **265,** 874–881.

Sojar, H. T., and Bahl, O. P. (1987). A chemical method for the deglycosylation of proteins. *Arch. Biochem. Biophys.* **259,** 52–57.

Sojar, H. T., and Bahl, O. P. (1987). Chemical deglycosylation of glycoproteins. *In* "Methods in Enzymology" (V. Ginsburg, ed.), Vol. 138, pp. 341–350. Academic Press, San Diego.

Stannard, B. S., Gesundheit, N., Thotakura, N. R., Gyves, P. W., Ronin, C., and Weintraub, B. D. (1989). Differential effect of inhibitors of oligosaccharide processing on the secretion of thyrotropin from dispersed rodent pituitary cells. *Biochem. Biophys. Res. Commun.* **165,** 788–794.

Strickland, T. W., Thomason, A. R., Nilsen, J. H., and Pierce, J. G. (1985). The common α subunit of bovine glycoprotein hormones limited formation of native structure by the totally nonglycosylated polypeptide chain. *J. Cell. Biochem.* **29,** 235–237.

Tarentino, A. L. (1989). Enzymatic approaches for studying the structure, synthesis and processing of glycoproteins. *Methods Cell Biol.* **32,** 111–139.

Tarentino, A. L., Quinones, G. T., Shrader, W. P., Changchein, L. M., and Plumer, T. H., Jr. (1992). Multiple endoglycosidase (Endo) F activities expressed by *Flavobacterium meningosepticum. J. Biol. Chem.* **267,** 3868–3872.

Thotakura, N. R., and Bahl, O. P. (1982). Role of carbohydrate in human chorionic gonadotropin: deglycosylation uncouples hormone-receptor complex and adenylate cyclase system. *Biochem. Biophys. Res. Commun.* **108,** 399–405.

Thotakura, N. R., and Bahl, O. P. (1987). Enzymatic deglycosylation of glycoproteins. *In* "Methods in Enzymology" (V. Ginsburg, ed.), Vol. 138, pp. 350–359. Academic Press, San Diego.

Thotakura, N. R., Weintraub, B. D., and Bahl, O. P. (1990a). The role of carbohydrate in human choriogonadotropin (hCG) action: Effects of N-linked carbohydrate chains from hCG and other glycoproteins on hormonal activity. *Mol. Cell. Endocrinol.* **70,** 263–272.

Thotakura, N. R., LiCalzi, L., and Weintraub, B. D. (1990b). The role of carbohydrate in thyrotropin action assessed by a novel approach using enzymatic deglycosylation. *J. Biol. Chem.* **265,** 11527–11534.

Thotakura, N. R., Desai, R. K., Bates, L. G., Cole, E. S., Pratt, B. M., and Weintraub, B. D. (1991). Biological activity and metabolic clearance of a recombinant human thyrotropin produced in Chinese hamster ovary cells. *Endocrinology* **128,** 341–348.

Thotakura, N. R., Desai, R. K., Szkudlinski, M. W., and Weintraub, B. D. (1992). The role of the oligosaccharide chains of thyrotropin α and β subunits in hormone action. *Endocrinology* **131,** 82–88.

Wittwer, A. J., Howard, S. C., Carr, L. S., Harakas, N. K., Feder, J., Parekh, R. B., Rudd, P. M., Dwek, R. A., and Rademacher, T. W. (1989). Effects of N-glycosylation on *in vitro* activity of Bowes melanoma and human colon fibroblast derived tissue plasminogen activator. *Biochemistry* **28,** 7662–7669.

The Role of Glycosylation in Adhesive Protein Function

Marvin L. Tanzer,* Martin S. Giniger,*
John W. Dean III,† and S. Chandrasekaran*

**Department of BioStructure and Function, School of Dental Medicine, University of Connecticut Health Center, Farmington, Connecticut 06030-3705 and †Department of Periodontology, University of Tennessee College of Dentistry, Memphis, Tennessee 38163*

I. Introduction

In multicellular organisms, all cells are surrounded by an extracellular matrix. This is a complex milieu composed of some or all of the following: structural components; attachment factors for both cellular and noncellular components; signals for cells to spread, migrate, proliferate, or synthesize products, a selective filter system; and possibly even signals telling specific cells to die or radically change their phenotype. The content of the extracellular matrix varies with its role and location and may change with developmental stages, aging, or neoplastic transformation. Many of the minor components have yet to be identified and even some of the better known constituents have not been fully

Cell Surface and Extracellular Glycoconjugates

characterized. Glycoproteins, including laminin, fibronectin, tenascin, thrombospondin, and other cell adhesion molecules, have been localized to a variety of extracellular matrices. They are large, complex, multidomain molecules with numerous biological activities, one of the most important of which involves their adherence to a variety of cell types through cell surface membrane receptors. Until recently, the ways in which the asparagine-linked carbohydrate moieties present on extracellular matrix glycoproteins might influence their cellular interactions have not been extensively studied.

In contrast to proteins and nucleic acids, carbohydrates are capable of encoding considerably more information per monomer unit, due to a large repertoire of positional and anomeric linkages. This feature is both a strength and a weakness. Many unique structures can be generated and, indeed, are found. However, such diversity of structure has sometimes been viewed as implying a lack of specificity. This point, coupled with the fact that the structures must be generated by posttranslational pathways, has often tempered interest in the biological roles of glycoconjugates. The recent studies indicating the potential importance of animal cell surface lectins in attachment to other cells or the extracellular matrix have revealed that some systems do make use of environmental carbohydrate signals. The reasons why there should be two recognition systems in many instances, namely carbohydrate and protein, through which cells can receive signals from their environment are presently unclear. Possibly the need for a fine-tuning process before substrate-bound cells commit to additional responses, such as spreading or neurite extension, or a requirement for redundancy of systems may be essential. The well-documented alterations of the oligosaccharides found on glycoproteins produced by tumor cells with different metastatic potentials serve to highlight the potential importance of the study of the role of glycosylation in adhesive protein function.

II. Protein Glycosylation

The enzymes involved in N-glycosylation, their intracellular locations, and the components of the oligosaccharide chains at various stages of maturation (which varies for different positions in the glycoprotein; i.e., not all N-linked oligosaccharides on a mature glycoprotein are identical) have been elucidated (Kornfeld and Kornfeld, 1985). In glycoproteins potential N-glycosylation sites contain the sequence asparagine–*X*–serine (threonine), where *X* can be any amino acid except proline (Bause and Hettkamp, 1979). It remains unclear why all of these sites are not glycosylated in a given glycoprotein. The lipid-linked precursor for all N-linked glycosylation, glucose$_3$–mannose$_9$–(N-acetylglucosamine)$_2$, is synthesized stepwise in the rough endoplasmic reticulum (RER) (Fig. 1). The terminal N-acetylglucosamine (GlcNAc), which ultimately will be attached to an asparagine residue in the glycosylated protein, is coupled to the dolichol carrier via a pyrophosphate linkage. The glycosylation is catalyzed by an oligosaccharyltransferase in a cotranslational manner as the nascent polypeptide is vectorially transported across the RER membrane (Kornfeld and Kornfeld, 1985; Hirschberg and Snider, 1987).

The first step in processing the core oligosaccharide is the removal of the terminal glucose residue by α-1,2 glucosidase I, followed by the removal of the other two glucose residues by α-1,3 glucosidase II. The final step in the RER is removal of one mannose residue by a specific α-mannosidase (see Fig. 1). The immature glycoprotein is then transported in vesicles that bud off from the RER, moving to the cis-Golgi for further processing. The next enzyme is Golgi α-mannosidase I, which removes three more mannose residues from high-mannose oligosaccharides, resulting in a carbohydrate chain of Man$_5$–GlcNAc$_2$. Next a GlcNAc residue is added to one of the remaining Man residues by N-acetylglucosaminyltransferase followed by the removal of two other Man residues by Golgi mannosidase II, yielding Man$_3$–GlcNAc$_3$. N-Acetylglucosaminyltransferase II adds either one or two GlcNAc residues to the available mannose residue. Subsequently, fucosyltransferase, galactosyltransferase, and sialyltransferase add their specific products to the carbohydrate chain, resulting in a mature, complex, N-linked oligosaccharide that is classified by its number of antennae. The mature glycoprotein now leaves the trans-Golgi and becomes localized intracellularly or is secreted (Kornfeld and Kornfeld, 1985).

Why the oligosaccharide moieties at some sites on a glycoprotein are left in a high-mannose state while others become more fully processed is unclear but several mechanisms may be involved. Differences in the glycosyltransferases and hydrolases, either their presence or their absence, or their level of activity, may affect carbohydrate processing. It has been reported that the substrate specificities of some processing enzymes can vary between tissues and among species (Kornfeld and Kornfeld, 1985). Also, the conformation of a glycoprotein may account for which oligosaccharides continue to be processed into more mature forms. Oligosaccharide moieties that are not readily accessible during transit through the endoplasmic reticulum and Golgi might be unavailable for processing. Chaperon proteins may bind and affect conformation and oligosaccharide addition or structure (Gething and Sambrook, 1992). Furthermore, sites of carbohydrate–carbohydrate interaction between oligosaccharide chains coupled to proteins during transit

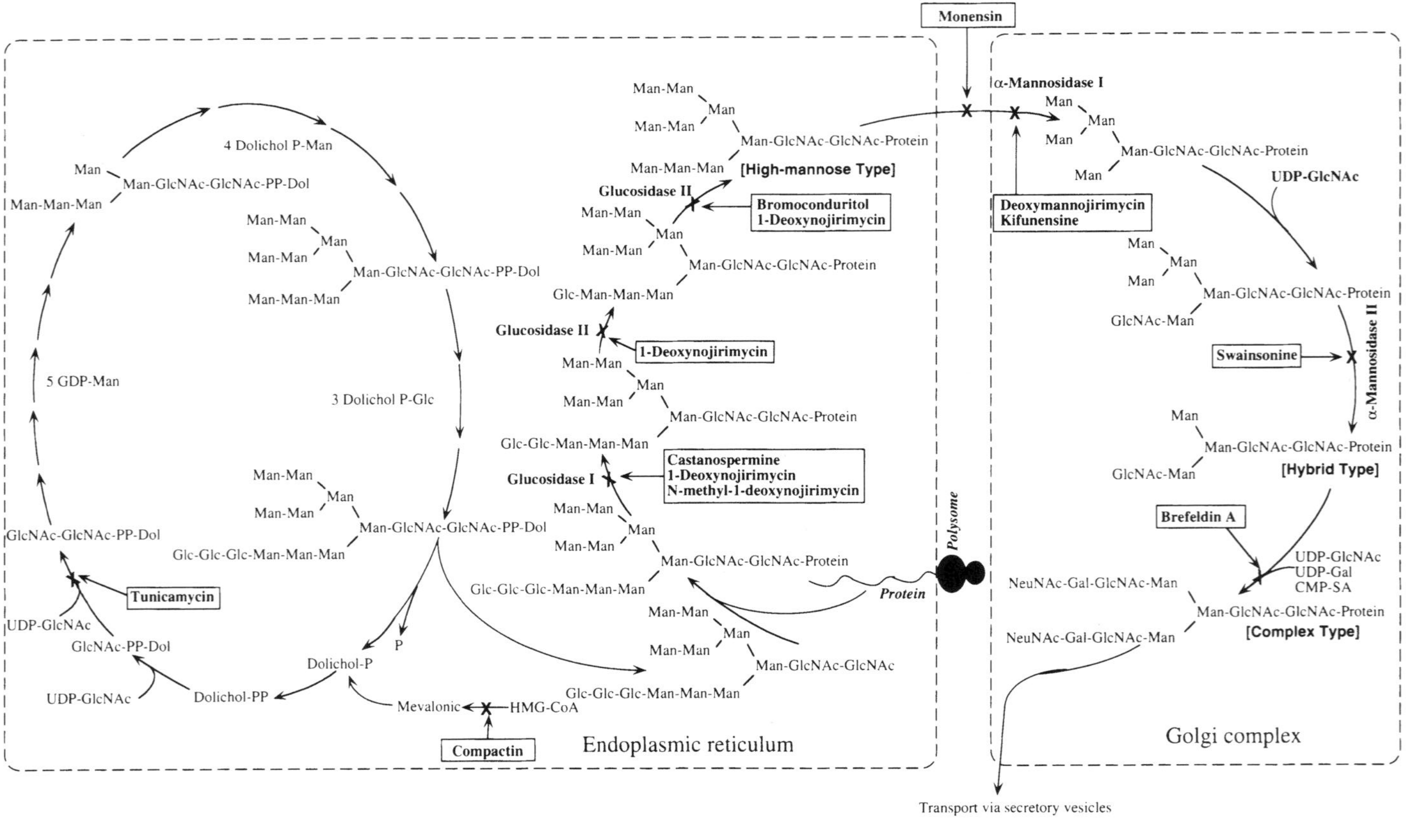

FIG. 1. Pathway of N-glycosylation of proteins. The left side shows the portion of the reactions that occur in the rough endoplasmic reticulum, yielding a core structure that then is added to the emerging nascent protein. Following incorporation, the oligosaccharide chain undergoes modifications before it is exported to the Golgi apparatus, where further processing takes place. In the case of extracellular matrix glycoproteins, the fully glycosylated molecules are then secreted. Specific inhibitors of the glycosylation process are also indicated.

through the RER and Golgi, as has been found for IgG heavy chains, may be less available for processing (Sutton and Phillips, 1983).

III. N-Glycosylation Inhibitors

There are many, primarily naturally occurring compounds that can selectively inhibit specific enzymes of the N-glycosylation pathway. These compounds, which tend to be noxious, can be very useful for producing glycoproteins with defined oligosaccharide chains. These inhibitors are generally used on cells in culture with dosages experimentally determined to balance deleterious side effects against efficacy of specific N-glycosylation pathway enzyme inhibition.

Tunicamycin, a highly potent protein synthesis (Tamura and Ito, 1982) and glycosylation (Elbein, 1987) inhibitor, was originally isolated by Takatsuki *et al.* while screening the culture broth of *Streptomyces lysosuperficus* for antiviral antibiotics (Takatsuki and Tamura, 1971, 1982). Tunicamycin acts by inhibiting the biosynthesis of polyisoprenol–carbohydrate conjugates, the most notable of which, dolichol pyrophosphate *N*-acetylglucosamine, is produced in the RER (see Fig. 1). The formation of this molecule is the initial step in the synthesis of the nascent oligosaccharide Glc_3–Man_9–$GlcNAc_2$, which is coupled via its terminal GlcNAc to dolicholpyrophosphate and is transferred to an appropriate asparagine residue in a newly synthesized protein in the RER. The resultant dolichol phosphate is then rephosphorylated and reused as a carrier molecule.

Castanospermine (1,6,7,8-tetrahydroxyindolidizine), a plant alkaloid isolated from the nut of an Australian chesnut tree, impairs complex oligosaccharide formation by acting on glucosidases I and II in the RER (Elbein, 1987). This compound, a competitive enzyme inhibitor, when administered to cells in culture, leads to the accumulation of glycoproteins containing the immature structure Glc_3–Man_9–$GlcNAc_2$. 1-Deoxynojirimycin, a glucose analog, also leads to the accumulation of glycoproteins that have carbohydrate chain structures of Glc_3–Man_9–$GlcNAc_2$. The actinomycete *Kitasatosporia kifunense* produces a potent inhibitor of Golgi α-mannosidase I known as kifunensine (Elbein *et al.*, 1990). This recently discovered alkaloid compound is a cyclic oxamide derivative of 1-amino mannojirimycin that results in the production of glycoproteins with high-mannose oligosaccharides of the Man_9–$GlcNAc_2$ structure (Elbein *et al.*, 1990). 1-Deoxymannojirimycin is also an inhibitor of Golgi α-mannosidase I and is frequently used to produce glycoproteins with high-mannose oligosaccharides similar to those found when kifunensine is employed (Elbein, 1987). Swainsonine (1,2,8-trihydroxyoctahydroindolizidine), a poisonous alkaloid from the locoweed plant, is a reversible, active site-directed inhibitor of Golgi α-mannosidase II that acts almost at the end of the N-glycosylation pathway, causing cells in culture to produce glycoproteins with oligosaccharide chains with the structure Man_5–$GlcNAc_3$ (Schwarz and Datema, 1984; Elbein, 1987).

IV. Principles of Cell Adhesion

The adherence of cells to the extracellular matrix or homologous cells during development, migration, wound healing, and morphogenesis is vital for these processes to proceed properly. Attachment occurs via cell surface receptors, many of which span the membrane and also serve as sites of communication between the cytoplasm and the extracellular matrix. There is increasing evidence that these molecular receptors are important sites of information transfer (from molecular signals in the extracellular matrix to the cell cytoplasm). These can influence processes such as cell movement, proliferation, and neurite outgrowth, not only via reorganization of the cytoskeleton, but also by modification of gene expression.

Adherence to a favorable substratum is a critical first step for many types of cells when newly placed in culture, or *in vivo* during embryogenesis or metastasis. Substrate attachment-dependent effects on cell growth, morphology, and behavior are now believed to all be mediated through transmembrane receptors. For example when a fibroblast-like cell binds to a favorable substratum via receptors it begins to make flattened extensions, lamellipodia. These attach to the substrate via specific structures, adhesion plaques, or focal adhesions, which have integrin molecules as key receptor components (Burridge *et al.*, 1988). Adhesion plaques, first described in cultured fibroblasts (Abercrombie *et al.*, 1971), were found to be areas of close contact between the cell and its substratum. The lamellipodia contain no organelles and are a meshwork of actin fibers inside the cell membrane. A cell that spreads and then remains stationary develops highly organized actin "stress fibers" that become visible in the cytoplasm. Focal adhesions apparently are anchors to attach stress fibers to the cell membrane as well as potentially having a role in the maintenance and assembly of the actin cytoskeleton.

As a migratory cell moves in a given direction, it extends lamellipodia or smaller microspikes, some of which adhere to the substratum while others move back in waves over the cell, the phenomenon of ruffling. It is thought that cell membrane is retrieved via this ruffling process over the Golgi apparatus so that forward migration of the cell can continue. New membrane is inserted at the leading edge from the Golgi vesicles as the cell proceeds on its way and a tail, or retraction fiber, extends behind the cell. This eventually breaks, leaving a tiny piece of cell membrane behind (a footprint), with the rest retracting into the main body of the cell (Singer and Kupfer, 1986).

Neurite extension is similar in many respects to cell spreading and migration except that the cell body remains in one place and only processes are extended (Singer and Kupfer, 1986). A growth cone is present at the distal end of the advancing neurite and has many similarities to the leading edge of a migrating cell. However, the appearance and behavior of these structures varies markedly with both cellular origin and substratum (Bray and Hollenbeck, 1988). These growth cones can retain their navigational ability even when they are severed from the cell body and are composed mainly of actin, actin-associated proteins, and microtubules (Bray and Hollenbeck, 1988). There do not seem to be any ruffling phenomena or a membrane retrieval system in neurons as has been found in migratory cells, resulting in the continuous addition of new membrane (from a pool of membrane saccules that is similar to, and may be a product of, the Golgi apparatus) to the axonal process as the growth cone proceeds (Bray and Hollenbeck, 1988). Therefore, the total amount of cell membrane increases as neurites are extended and the cell body remains stationary.

V. Adhesive Proteins

A. *Neural Cell Adhesion Molecule*

Neural cell adhesion molecule (NCAM) (Fig. 2) is a cell surface glycoprotein that is expressed by all major cell types in the nervous system. It is the most abundant and widespread cell surface adhesion molecule and was the first to be fully characterized (Rutishauser, 1991). It is a multifunctional protein that possibly mediates both cell–cell and cell–matrix adhesion (Nybroe and Bock, 1989). This Ca^{2+}-independent adhesion molecule occurs in the adult central nervous system in three molecular forms: 180, 140, and 120 kDa; all are derived from alternative RNA splicing from a single structural gene (Pollersberg *et al.*, 1985). While these three forms of NCAM appear to contain a single polypeptide chain, there is much structural heterogeneity in both the protein portion and the carbohydrate content of this molecule. Each form can be identified on the basis of the apparent molecular weight of the backbone polypeptide chain or on the basis of its relatively high- or low-sialic-acid content (Rutishauser, 1991).

NCAM has seven potential asparagine-linked glycosylation sites (Fig. 2), of which only five are actually glycosylated (Lyles *et al.*, 1984;

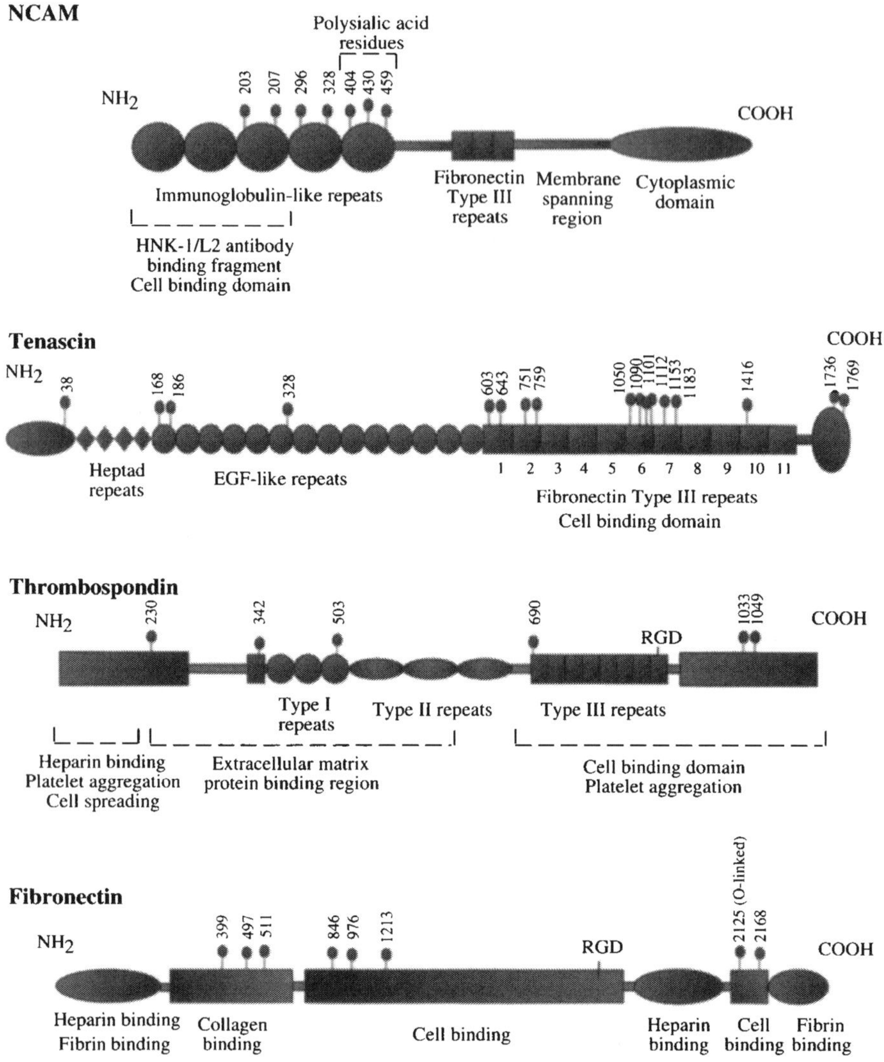

FIG. 2. Domain structure and potential sites of N-linked carbohydrate addition for NCAM, tenascin, thrombospondin, and fibronectin. Domains are denoted by circles, ovals, diamonds, and rectangles and are appropriately labeled but not drawn to scale. Various potential N-glycosylation sites are marked as lollipops and the numbers above them indicate the amino acid position for each site. The NCAM domain structure is representative of NCAM-180 (Cunningham *et al.*, 1987). Tenascin is representative of the tenascin splice variant Tn230 (Spring *et al.*, 1989). The thrombospondin domain structure is representative of endothelial cell-derived thrombospondin (Lawler and Hynes, 1986). The basic fibronectin domain structure was derived from the longest polypeptide chain of human plasma fibronectin (Skorstengaard *et al.*, 1986).

Cunningham *et al.*, 1987). The high polysialic acid form of this molecule is expressed predominantly by fetal tissue and is approximately 30% sialic acid by weight, whereas the low-sialic-acid form accounts for only about 10%. Most of the sialic acid in NCAM appears to be in the form of unbranched a-2,8 polymers (Finne, 1982) and is located in the central part of the molecule.

The highly sialylated form of NCAM is the less adhesive of the two forms, probably because the large numbers of polysialic acid chains can create a hydrated domain around the molecule, thus causing steric hinderance and interfering with adhesion (Rutishauser *et al.*, 1988). Several investigators have demonstrated this negative regulatory effect on cell adhesion. Sadoul *et al.* (1983) have demonstrated that cell membranes that contain the embryonic, highly sialylated NCAM on its surface exhibit decreased binding and aggregation properties. Hoffman and Edelman (1983) have found an inverse relationship between NCAM's sialic acid content and its adhesiveness. Furthermore, Rutishauser *et al.* (1985) have shown that partial removal of polysialic acid from embryonic brain membranes enhanced adhesion properties.

Besides polysialic acid, other carbohydrate moieties may play an important role in NCAM's cell adhesion properties. An additional carbohydrate epitope important for cell adhesion has been revealed by monoclonal antibodies against L2 and HNK-1 (Schachner, 1989). This epitope contains a 3-sulfated, glucuronal-substituted lacto-series oligosaccharide (Ariga *et al.*, 1987). Other glycoproteins that express this HNK-1 epitope include tenascin (Künemund *et al.*, 1988) and human neuroectoderm-derived fibronectin (Harper *et al.*, 1990). The exact mechanism by which this carbohydrate functions in cell adhesion is still unclear; however, several experiments demonstrate clearly that these kinds of sugars play an important role in cell adhesion. Cole and Schachner (1987) have demonstrated that incubation of retinal cells on an NCAM substratum with L2 monoclonal antibody inhibits cell attachment to this substratum. Similarly, Künemund *et al.* (1988) have demonstrated that the L2/HNK-1 carbohydrate of extracellular matrix glycoproteins acts as a ligand in cell adhesion in microexplant cultures of early postnatal mouse cerebellum.

B. Tenascin

Tenascin (Fig. 2) is a large, multisubunit glycoprotein of the extracellular matrix (Bourdon and Ruoslahti, 1989). It was first described as a glial-mesenchymal extracellular matrix glycoprotein, GMEM, by Bourdon *et al.* (1983). After that, several other investigators isolated it from a variety of sources and independently named this molecule myotendenous antigen (Chiquet and Fambrough, 1984) , cytotactin (Grumet *et al.*, 1985) , hexabrachion (Erickson and Iglesias, 1984), and J1 protein (Kruse *et al.*, 1985). The name "tenascin" was ultimately proposed by Chiquet-Ehrismann *et al.* (1986). It was derived from a combination of two latin verbs, "tenere" (to hold) and "nasci" (to be born), and describes one of its prominent features and its developmental expression.

A major form of tenascin, isolated from embryonal chicken fibroblasts, is a large (>1000 kDa), disulfide-linked hexamer composed of three different kinds of polypeptide subunits that migrate at 230, 200, and 190 kDa, respectively (Spring *et al.*, 1989). These three polypeptides are known to be closely related since all monoclonal antibodies raised against one chain will always recognize the other two variants (Chiquet and Fambrough, 1984; Chiquet-Ehrismann *et al.*, 1988; Pearson *et al.*, 1988). The "hexabrachion" structure of tenascin is composed of two triplets that are covalently linked within its central globule. The polypeptides show homology to heptad repeats, EGF, fibronectin type III, and fibrinogen and appear to be mosaics of repetitive structural building blocks arranged through evolution by alternative splicing and exon shuffling. The central region seems to be the only region unique to tenascin (Pearson *et al.*, 1988).

More than half of this glycoprotein is composed of fibronectin type III repeats, all being homologous to each other and centrally located. At the amino terminus, the type III repeats are linked to 13.5 EGF-like repeats. At the COOH terminus, a Ca^{2+} binding fibrinogen-like sequence is found (Chiquet-Ehrismann, 1991). The larger 230-kDa subunit has 17 potential N-linked glycosylation sites (Fig. 2), while the smaller subunits have 11 each. The extra six N-linked glycosylation sites are found within two additional type III repeats specific for the larger 230-kDa subunit (Pearson *et al.*, 1988). Approximately two-thirds of potential N-linked glycosylation sites in tenascin are found within the type III repeats (Table 1).

To date there is very little evidence as to the functional importance of the carbohydrate side chains of this glycoprotein. In fact, there is much controversy as to the exact role of this protein in adhesive functions. Most investigators have characterized this molecule as being an antiadhesive molecule, causing the inhibition of cell adhesion and spreading (Chiquet-Ehrismann, 1991). It has been reported that tenascin inhibits cell adhesion to fibronectin (Chiquet-Ehrismann *et al.*, 1988) and that this glycoprotein acts as a repulsive substrate for central nervous system neurons (Faissner and Kruse, 1990). However, other investigators have shown that tenascin can promote cell adhesion in an RGD-dependent manner that is distinct from fibronectin

TABLE I

POTENTIAL *N*-LINKED GLYCOSYLATION SITES OF VARIOUS GLYCOPROTEIN CELL ADHESION DOMAINS

Cell adhesion molecule	Number of potential N-linked sites	Number of potential N-linked sites contained within cell binding domains	Percentage of total potential N-linked sites contained within cell binding domains
NCAM	7	2	29
Tenascin	17	11	65
Thrombospondin	6	3	50
Fibronectin	7	3	43
Laminin	68	30[a]	44

[a] Combines potential N-linked glycosylation sites from the two major cell binding domains of laminin: E1 (Fujiwara, 1988) and E8 (Deutzmann, 1990; Sasaki *et al.*, 1987, 1988; Sasaki and Yamada, 1987).

in that little cell spreading is shown on tenascin substrata (Bourdon and Ruoslahti, 1989).

Further studies using recombinant tenascin fragments showed that type III repeats of the tenascin arms promoted cell adhesion, whereas the fragments containing the cysteine-rich EGF-like repeats were anti-adhesive (Spring *et al.*, 1989). Since the majority of potential N-linked glycosylation sites reside within the type III repeats, it is conceivable that the oligosaccharides of tenascin may play a role in cellular adhesion. One piece of evidence that points to such a potential role comes from Kruse *et al.* (1985), who demonstrated that a polyclonal antibody of the L2/HNK-1 family reacted with a J1 glycoprotein and interfered with astrocyte–oligodendrocyte cell adhesion.

C. *Thrombospondin*

Thrombospondin (Fig. 2) is a large, multifunctional glycoprotein secreted from activated platelets and many mesenchymal and epithelial cells (Mosher, 1990). It is a disulfide-joined, trimeric, 450-kDa, cellular and extracellular matrix glycoprotein that interacts via specific domains with a wide variety of molecules (Asch and Nachman, 1989). The protein contains globular carboxy-terminal and amino-terminal domains (Dixit *et al.*, 1985; Lawler *et al.*, 1985; Frazier, 1987). Three types of repeating amino acid sequence are present: type I repeats show homology to the circumsporozoite protein from *Plasmodium falciparum,* type II repeats show homology with epidermal growth factor, and type III repeats have structural homology with calmodulin and other calcium binding sequences. There is an RGD sequence in the last type III repeat that is most likely the site for thrombospondin–cell adhesion.

There are six potential N-linked glycosylation sites (Fig. 2), of which two are located within type III repeats (Lawler and Hynes, 1986). Structural studies of the oligosaccharides of human platelet thrombospondin revealed that this glycoprotein contains mono-, bi-, tri-, and tetraantennary complex-type structures. The majority (70%) of these sugar chains have a fucosylated trimannosyl core. A bisected *N*-acetylglucosamine core was present in 19% of biantennary structures. An extremely small number (0.5%) of these complex-type sugar chains were of the high-mannose type, which is quite different from other platelet membrane glycoproteins such as GPIIb and GPIIa, which have 52% high-mannose-type chains (Furukawa *et al.*, 1989). Evidence also exists for the existence of O-linked sugar chains consisting of glucose and xylose in bovine thrombospondin within one of its three EGF-like domains (Nishimura *et al.*, 1992).

Thrombospondin has demonstrated an affinity for a variety of molecules: sulfated glycolipids (Roberts, 1986; Ginsburg and Roberts, 1988; Roberts and Ginsburg, 1988), collagen (Lahav *et al.*, 1982; Mumby *et al.*, 1984; Galvin *et al.*, 1987), fibronectin (Lahav *et al.*, 1982), fibrinogen (Leung and Nachman, 1982; Lahav *et al.*, 1984), laminin (Frazier, 1987; Asch and Nachman, 1989), heparan sulfate proteoglycans, and integrins (Prater *et al.*, 1991). It has been demonstrated that the C-terminal domain is required for melanoma cell and platelet adhesion (Dixit *et al.*, 1985; Roberts *et al.*, 1987). On the other hand, the N-terminal sulfatide and heparin binding domain is required for cell spreading, but not initial cell binding (Roberts *et al.*, 1987; Ginsburg and Roberts, 1988).

There has been some controversy as to the nature and consequences of thrombospondin interaction with cells. This molecule has been described as an adhesive protein for keratinocytes, melanoma cells, and platelets (Roberts *et al.*, 1987; Varani *et al.*, 1988; Asch and Nachman, 1989; Sage and Bornstein, 1991), whereas it has been shown to reduce focal adhesions in endothelial cells and fibroblasts (Murphy-Ullrich and Höök, 1989). The last two cell types attach themselves to a thrombospondin substratum, but fail to spread (Lawler *et al.*, 1988). Among the proadhesion proponents, plausible arguments have been made that cell adhesion to thrombospondin is mediated by an integrin receptor (Lawler *et al.*, 1988; Tuszynski *et al.*, 1989), heparan sulfate proteoglycan (Roberts *et al.*, 1987; Kaesberg *et al.*, 1989), or cell membrane glycolipids containing sulfate esters (Ginsburg and Roberts, 1988).

It is possible that the carbohydrate moieties of thrombospondin play a role in adhesion since anionic polysaccharides and neutral and anionic monosaccharides can inhibit the binding of thrombospondin to cell membrane sulfated glycolipids. The sulfated fucan fucoidan is the best inhibitor (Roberts *et al.*, 1986). Mannose-6 phosphate and fucoidan have both been shown to inhibit binding of lymphocytes to the extracellular matrix of endothelial venules. Since multiple modes of interactions between adhesive glycoproteins and cells are common, it is quite likely that different sites on the thrombospondin molecule and several distinct receptors may play a role in thrombospondin-mediated cell adhesion.

D. Fibronectin

The fibronectins (Fig. 2) are large, multifunctional glycoproteins found in the blood plasma and extracellular matrix (Ruoslahti, 1988). Although intensive research has made it the prototype cell adhesion protein, the role of its carbohydrate side chains in this process is not yet fully understood. Besides mediating adhesive interactions between the cell and the extracellular matrix, it can also interact with a number of biologically active macromolecules such as collagen, fibrin, and glycosaminoglycan (Yamada, 1989).

The complete amino acid sequence of human fibronectin has been determined from its cloned cDNA sequence (Kornblihtt *et al.*, 1985). In addition, the sequences of bovine, rat, and chicken fibronectins are also known (Hynes, 1985; Schwarzbauer *et al.*, 1985; Skorstengaard *et al.*, 1986). The fibronectin molecule is made up of two polypeptide chains, one slightly longer than the other, that are associated through two disulfide bonds near the carboxy terminus. The dimer has an approximate molecular weight of 550 kDa. In general, each polypeptide is composed of three different kinds of repeats, types I, II, and III (Petersen and Skorstengaard, 1985). The type I repeat is about 45 amino acids long and found at both termini of the molecule. It has homology to domains found in tissue plasminogen and factor XII (Patthy *et al.*, 1984; Patel *et al.*, 1987). Two 60-amino-acid type II repeats are found at the amino-terminal end, interrupting a row of type I repeats, while 15–17 type III segments make up the middle of each polypeptide. Type II homologies are present in factor XII, prothrombin, plasminogen, and urokinase (Patthy *et al.*, 1984; Patel *et al.*, 1987); however, the type III repeats have few similarities with any other protein domain (Ruoslahti, 1988).

Fibronectin contains 4–9% carbohydrate, depending on the tissue or cell origin (Petersen *et al.*, 1989). The fibronectin from adult human skin fibroblasts has low carbohydrate content, whereas fibronectin derived from tumor cells has a higher content (Ruoslahti, 1988). In general, each polypeptide chain has six or seven N-linked oligosaccharides and one or two O-linked sugar moieties (Ruoslahti, 1988; Petersen *et al.*, 1989). The oligosaccharides of plasma and fibroblast fibronectins comprise complex type chains, primarily containing mannose and sialic acid.

Although the physiological functions of the carbohydrates of fibronectin are unresolved, it is known that they protect the molecule from proteolysis (Bernard *et al.*, 1982). In addition, it appears that high-molecular-weight N-linked polylactosamine containing chains found covalently attached to placental fibronectin weakens binding of this glycoprotein to collagen (Zhu and Laine, 1985, 1990). The significance of this is not known, but perhaps it is analogous to the negative regulatory role of polysialic acid in N-CAM function observed in avian and rat embryological tissues (Finne *et al.*, 1983). It has been hypothesized that post-translational modification of fibronectin and N-CAM by these long carbohydrate chains may allow mobility of certain cells through the embryonic extracellular matrix, allowing them to reach their target locations (Zhu and Laine, 1990).

Olden *et al.* (1989) have reported, using tunicamycin as an inhibitor of N-glycosylation, that the carbohydrate moieties of fibronectin were not necessary for fibroblast adhesion, but protected the protein against proteolysis. Although attention has focused on the role of the RGD sequence in fibronectin-mediated cell adhesion, Jones *et al.* (1986) have presented data that suggest that the oligosaccharides of fibronectin may also be important, acting as modulators of fibroblast adhesion and spreading. Harper *et al.* (1990) have recently reported that human tumors of neuroectodermal origin secrete fibronectin that contains the cell-adhesion-related HNK-1 epitope. This is consistent with earlier work that shows that fibronectin appears to be capable of transiently acquiring the HNK-1 carbohydrate domain and thereby mediating activities specific to this epitope, including cell–matrix interaction and adhesion (Bronner-Fraser, 1985; Duband *et al.*, 1986; Harper *et al.*, 1990). Therefore fibronectin now joins the growing list of RGD-containing extracellular matrix glycoproteins that possess the HNK-1 epitope and strengthens the correlation between these two structures. Further study is now required to determine the relative importance of fibronectin sugar chains in cell adhesive function.

E. Laminin

Laminins (Fig. 3), a family of large multidomain glycoproteins (M_r 900,000) of the basal lamina, have been implicated in the development and maintenance of cellular and tissue organization (Martin and Timpl, 1987; Beck *et al.*, 1990). Laminin was originally purified from two systems, an embryonal carcinoma cell line and a mouse sarcoma (Chung *et al.*, 1977; Timpl *et al.*, 1979). Subsequently, isoforms and

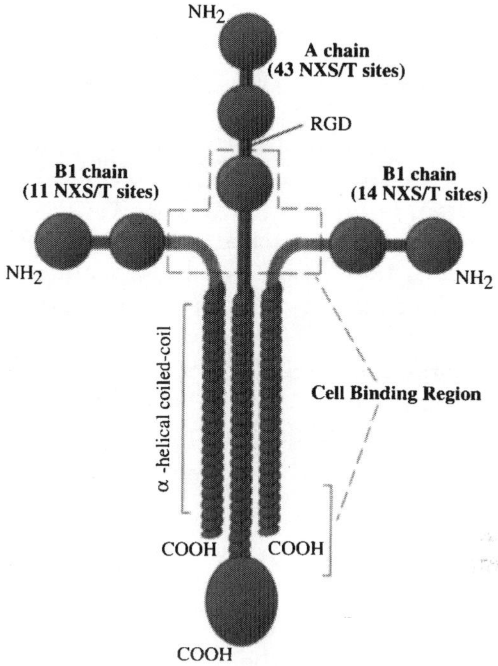

FIG. 3. Schematic diagram of laminin, showing the different domain structures and the number of potential glycosylation sites.

structural variants of laminin have been characterized following their purification from different sources (Hunter *et al.*, 1989; Paulsson and Saladin, 1989; Beck *et al.*, 1990; Ehrig *et al.*, 1990). Laminin is made up of three polypeptide chains, A, B1, and B2, linked by disulfide bonds. The three chains of laminin are arranged in a unique cruciform shape as viewed by electron microscopy. The A chain has a large terminal globule on its long arm, at its carboxy terminus, and three smaller globules, one at the amino terminus and the other two evenly spaced along its shorter arm. The B chains each have a small globular region at their amino termini with another globular region closer to the center of molecule.

1. LAMININ GLYCOSYLATION

Most studies show that laminin contains 12–15% carbohydrate (Arumugham *et al.*, 1986; Fujiwara *et al.*, 1988), although recent data indicate almost twice that amount (Knibbs *et al.*, 1989). Most of the carbohydrate is present in complex-type Asn-linked oligosaccharides encompassing many different structures (Arumugham *et al.*, 1986;

Fujiwara *et al.*, 1988; Knibbs *et al.*, 1989). Prominent features of the oligosaccharides are the presence of terminal α-galactose moieties, terminal *N*-acetylneuraminic acid residues, and poly-*N*-acetyllactosamine containing chains. The last have both linear blood group i and branched blood group I structures (Knibbs *et al.*, 1989). A total of 68 potential glycosylation sites have been determined by molecular cloning studies for mouse EHS tumor laminin (Sasaki *et al.*, 1987; Sasaki and Yamada, 1987; Sasaki *et al.*, 1988) , of which about 40 sites are actually glycosylated (Fujiwara *et al.*, 1988). Lectin binding studies indicated that the carbohydrate moieties present on laminin differ in various regions of the molecule (Rao *et al.*, 1983). The carbohydrate substituents of laminin may vary with the species and tissue source of the molecule, similar to other glycoproteins (Feizi, 1987). The structural studies of laminin carbohydrates carried out to date suggest that laminin contains only N-linked oligosaccharides (Arumugham *et al.*, 1986; Fujiwara *et al.*, 1988; Knibbs *et al.*, 1989).

2. Role of Glycosylation on the Adhesive Function of Laminin

Studies related to laminin carbohydrates and their implications for biological properties have intensified in recent times. In an early study, the carbohydrate moieties of laminin were not found to increase the stability of the molecule during protease digestion (Howe, 1984). The same study also suggests that the carbohydrates of laminin do not enhance heparin binding properties of laminin. Recent studies, however, have shown that the carbohydrate moieties of laminin are implicated in several cellular events (Tanzer *et al.*, 1991). The carbohydrate moieties of basal lamina, where laminin is primarily located, have been shown to play an essential role in attachment and spreading of basal corneal epithelial cells (Trinkaus-Randall *et al.*, 1988). In addition, many of the lectin-like receptors discussed later have been shown to interact with laminin. One of the early studies on cell attachment to laminin and type IV collagen showed that such adhesion can be partially modulated by the expression of specific cell surface-associated oligosaccharide structures (Dennis *et al.*, 1984). Codogno *et al.* (1988) have recently described a cell surface receptor responsible for concanavalin A-induced inhibition of fibroblast spreading on laminin. The finding that this carbohydrate containing receptor is operant in cell spreading on laminin but not on fibronectin highlights the diversity and specificity of these systems. The importance of carbohydrates in development has been suggested by Bronner–Fraser, who was able to disrupt avian neural crest cell migration *in vitro* on a laminin substrate by introducing a carbohydrate binding antibody (HNK-1).

In our laboratory, several studies have been conducted to examine the structure–function relationships of laminin carbohydrates. Laminin was examined for cell binding and spreading activities in two cell culture systems: mouse B16 F1 melanoma cells and rat PC-12 pheochromocytoma cells, which develop neurite extensions (Dean *et al.*, 1988). Under suitable circumstances the melanoma cells will adhere and spread in the cell culture wells while the pheochromocytoma cells will respond by extending neurites. To initially test the hypothesis that the carbohydrates of laminin are informational for such cellular responses, specific lectins were used to adhere to sugar ligands contained within the oligosaccharides of laminin. The lectins used were wheat germ agglutinin (WGA), concanavalin A (Con A), and *Griffonia simplicifolia* agglutinin I (GSA I). Both WGA and GSA I inhibited cell adhesion in a concentration-dependent fashion when either melanoma cells or pheochromocytoma cells were tested. In contrast, Con A allowed cells to adhere, but in the case of the melanoma cells, spreading did not occur whereas the pheochromocytoma cells failed to extend neurite processes. Similar results have also been reported by Bouzon *et al.* (1990), who studied the binding and spreading of mouse melanoma cells in the presence or absence of various lectins. These results implied that specific oligosaccharide moieties may play a role in cell adhesion, cell spreading, and neurite outgrowth.

In a later study, unglycosylated laminin was purified from lysates of a murine embryonal carcinoma-derived cell line treated with the glycosylation inhibitor tunicamycin (Dean *et al.*, 1990). This unglycosylated laminin supported adhesion of melanoma cells equivalent to glycosylated laminin, but cell spreading was minimal in the absence of the carbohydrates. When the unglycosylated laminin was used as a substrate for pheochromocytoma cells, neurite outgrowth was markedly impaired; no effect was noted on the number of cells bound, compared with glycosylated laminin. Thus, the results from these studies show that once cells become bound to laminin the carbohydrate residues of that glycoprotein must be available to enable the cells to spread or to extend neurite processes.

In other studies, laminins containing immature oligosaccharides were produced using the glycosylation inhibitors swainsonine and castanospermine (Fig. 1) (Chandrasekaran *et al.*, 1991). When such laminins were used in the biological assays, laminin containing immature high-mannose-type oligosaccharides was as effective as laminin containing fully processed oligosaccharides. In contrast, laminin with hybrid-type oligosaccharides had incomplete activity. These results suggest that large, high-mannose oligosaccharide chains impart the same information for cellular responses as do laminin molecules that contain mature sugar chains of similar size. Furthermore, the size of the oligosaccharide chain itself may also be an important factor in the cellular response. Interestingly, a variety of candidate mono- and disaccharides have failed to elicit cell spreading or neurite outgrowth on unglycosylated laminin (Chandrasekaran *et al.*, 1991).

Several reconstitution experiments have been carried out to determine if the spreading/neurite outgrowth response could be restored to cells adhering to unglycosylated laminin (Chandrasekaran *et al.*, 1991). When a mixture of glycosylated and unglycosylated laminins was used as a substratum, some cells began to spread when 30% glycosylated laminin was present. At least 65% glycosylated laminin was required to elicit a spreading response by the majority of the cells. Cell spreading has also been found to be restored completely by a pronase digest of glycosylated laminin; a similar digest of unglycosylated laminin had no effect (Chandrasekaran *et al.*, 1991). These particular results not only indicate that laminin carbohydrates, rather than polypeptide sequences, are responsible for cell spreading but also suggest that substrate attachment of the carbohydrate moieties is not essential. These composite reconstitution experiments show that laminin carbohydrates provide essential information to responsive cells, enabling them to progress from an adherent state to a spread form or to extend neurite processes.

VI. Adhesion Receptors

Several adhesion receptor families have been described, which include: (a) integrins; (b) the adhesion molecules of the immunoglobulin superfamily; (c) cadherins, calcium-dependent cell–cell adhesion receptors; (d) selectins; and (e) lymphocyte homing receptors (Albelda and Buck, 1990). Receptors are, in general, composed of an assembly of glycoproteins that are embedded in the plasma membrane lipid bilayer and that transduce biological information conferred in signals. The carbohydrate moieties of these glycoproteins are usually clustered on the periphery of the molecule so that this hydrophilic area is exposed to the external surface of the cell membrane. Parenthetically, we note that carbohydrate structures also serve as important and unique signals or markers on glycoproteins that are in the soluble form as well as those situated on the cell surface (Sairam, 1985).

A. *Cell Surface Carbohydrates*

A layer of carbohydrates or glycocalix arising from glycoproteins, glycolipids, and glycosaminoglycans covers the entire surface of most cells. The oligosaccharide chains of these carbohydrates, directed toward the aqueous environment of the cell, can significantly modulate cellular interactions. The cell surface carbohydrate moieties are known to take part in many biological events, including cell–cell and cell–matrix interactions (Elbein, 1991), transport of glycoproteins (Varki, 1992), malignant transformation and tumor progression (Dennis, 1992), antigenic determination (Feizi, 1989), and mammalian fertilization (Wasserman, 1992).

Carbohydrates of glycoproteins have the structural complexity to act as informational molecules. Just a few monosaccharide units can generate a bewildering array of oligosaccharide structures. These highly complex carbohydrate structures arise due to the variations in the linkage positions and anomeric configurations of monosaccharides. Different branching patterns within the oligosaccharide side chains and their potential to carry various charged substituents such as sulfates, phosphates, and other groups further complicate the structure of the carbohydrates of glycoproteins. The diverse oligosaccharide structures on the cell surface participate in the regulation of cell growth and differentiation during embryogenesis and oncogenesis (Feizi, 1987). Several tumor-associated carbohydrate antigens have been described recently (Hakomori, 1991). *In vitro* studies have shown that alterations in cell surface carbohydrates affect the invasion process (Schallier *et al.,* 1988). Tumor cell invasion has been correlated with a gradient in the level of sialylation (Schallier *et al.,* 1988), although sialic acid itself has been reported to have a negative influence on the adhesive forces between cells, as mediated by CAMs (Brackenbury, 1985). Dennis *et al.* (1987) have reported that the expression of the metastatic phenotype of certain tumor cell lines is directly related to cell surface carbohydrates. In a recent study, Miyake and Hakomori (1991) have demonstrated the inhibition of metastatic deposition of a highly metastatic variant of mouse melanoma cells using monoclonal antibodies that specifically recognize the disaccharide structure Fucα1–1Galβ1–R; the structures of the R group itself had no effect. These studies and others emphasize the potential role of carbohydrates in the metastatic process of cells.

B. Lectin Receptors

The importance of lectins has been highlighted by various investigators (Sharon and Lis, 1989). In general, lectins are defined as carbohydrate binding proteins other than enzymes or antibodies (Barondes, 1988). While lectins are best known in plants, they have been found in all species, with those present in animals receiving recent emphasis (Barondes, 1986, 1988; Drickamer, 1988; Zhou and Cummings, 1992).

Lectins always have at least two sugar binding sites, generally require divalent cations for activity, and are usually defined in terms of binding specificity to monosaccharides or short oligosaccharides. Vertebrate cells produce a wide variety of lectins that are known to take part in various biological functions, including lymphocyte recirculation (Rosen and Stoolman, 1987), immune regulation (White, 1987; Kery, 1991), transport of glycoproteins (Newton *et al.*, 1987) , intercellular signaling (Krusius *et al.*, 1987), gamete recognition (Macek and Shur, 1988), and cell–cell and cell–matrix interactions (Sharon and Lis, 1989). Drickamer (1988) has classified animal lectins into two distinct classes: C-type lectins, which require Ca^{2+} for carbohydrate binding activity, and S-type lectins, which require free sulfhydryl groups for activity. C-type lectins, have so far been found only extracellularly while S-type lectins, which are mostly soluble lectins, have been found in intracellular as well as in extracellular environments.

The potential modes of binding of receptors during cell–cell and cell–matrix interactions are shown in Fig. 4. Cell surface lectins interact with other cells by binding to complementary sugar moieties. These molecules also use their carbohydrate binding abilities to interact with the extracellular matrix by directly binding to the carbohydrate moie-

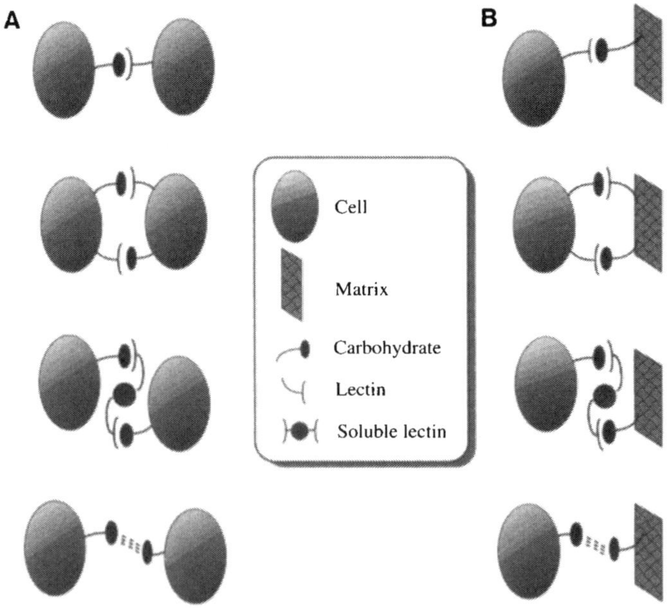

FIG. 4. The potential modes of binding of carbohydrate-mediated receptors during cell–cell (A) and cell–matrix interactions (B).

ties or through the carbohydrates of soluble glycoproteins as mediators (Sharon and Lis, 1989). There has been evidence suggesting the possibilities of carbohydrate–carbohydrate interactions during cell binding (Eggens *et al.,* 1989; Kojima and Hakomori, 1989).

Ashwell *et al.* (1982) found a hepatic receptor that recognized terminal galactoses of circulating serum glycoproteins. As these glycoproteins age, they lose their terminal sialic acid residues, exposing galactose residues, which then leads to the binding and removal of these proteins from the circulation by the liver. Lectins present on the surface of sperm have been shown to exhibit high specificity and affinity with complex glycoconjugates of the egg's extracellular coat during fertilization in a number of mammalian and nonmammalian species (Macek and Shur, 1988). Specific cell adhesion of neural retina cells to substratum-bound gangliosides, a class of neuronal cell surface glycoconjugates, was reported by Blackburn *et al.* (1986). A mannose receptor expressed on the surface of macrophages contains multiple carbohydrate-recognition domains (Taylor *et al.,* 1992).

Carbohydrate binding protein 35 (CBP35) is a galactose-specific lectin originally found in mouse (Roff and Wang, 1983) and human (Roff *et al.,* 1983) fibroblasts. Subsequently, several other proteins with either identical or homologous amino acid sequences have been reported (Anderson and Wang, 1992). Mouse Mac-2, a galactose binding lectin expressed by thioglycolate-induced inflammatory macrophages (Ho and Springer, 1982), is identical to CBP35 (Cherayil *et al.,* 1989). Mac-2, synthesized *in vitro* by human and mouse macrophages, binds to laminin in a galactose-dependent manner, suggesting the role of lectin receptors in cell–matrix interaction (Cherayil *et al.,* 1990). Mac-2 has been described as a major nonintegrin receptor (Woo *et al.,* 1990) whose extracellular properties include the ability to agglutinate cells and to bind to laminin (Woo *et al.,* 1991) and to other glycoproteins (Rosenberg *et al.,* 1991). A number of endogenous galactose binding lectins involved in tumor cell adhesion and metastasis have been reported (Raz and Lotan, 1987). Bao *et al.* (1992) found a developmentally regulated galactoside binding lectin in chicken tissues that specifically recognizes laminin.

Many vertebrate tissues contain soluble lactose binding lectins (Barondes, 1984, 1986; Drickamer, 1988; Zhou and Cummings, 1992). Most of these soluble lectins are S-type lectins (Zhou and Cummings, 1992), which are known to bind extracellularly to laminin (Zhou and Cummings, 1990) and to other glycoproteins containing poly-*N*-acetyllactosamine sequences (Do *et al.,* 1990; Zhou and Cummings, 1992). Extracts of human lung contain three major β-galactoside binding proteins (Sparrow *et al.,* 1987) homologous to the CBP35-related lectins discussed earlier (Anderson and Wang, 1992). Similarly, nine soluble lactose binding lectins were purified from rat intestine (Leffler *et al.*, 1989). The metastatic adhesion of tumor cells has been found to be mediated through S-type lectins expressed in many tumor cell lines (Lotan and Raz, 1988; Raz and Lotan, 1988). It has been found that expression of these lectins is positively correlated to the metastatic potential in a number of tumor cell line variants (Raz *et al.*, 1984). Although a definitive function for soluble lectins is yet to be established, they may be mediating cell–cell and cell–matrix interactions through complimentary sugars on the cell surface and on extracellular matrix glycoproteins (Sharon and Lis, 1989). All of these studies of lectin-like receptors center around the question of whether the carbohydrate binding properties of cells have biological significance. Further studies should define the precise mechanisms underlying the cellular response(s) toward the carbohydrate moieties of other cells and of the extracellular matrix.

C. 67-kDa Elastin/Laminin Receptor

A 67-kDa elastin binding protein was originally isolated from cultures of bovine ligamentum nuchae fibroblasts (Wrenn *et al.*, 1988). It was subsequently found to be a galactose binding lectin (Hinek *et al.*, 1988). In a later study, 67-kDa proteins isolated from fetal bovine auricular chondroblasts and human melanoma cells were shown to share structural and functional similarities (Mecham *et al.*, 1989). Both of these proteins bound to elastin and laminin and their interaction to laminin was inhibited by elastin peptides (Mecham *et al.*, 1989). The lectin properties of these peripheral membrane proteins are important in modulating their association with the cell surface (Mecham *et al.*, 1991). Of special interest is the finding that these proteins contain a galactoside binding domain that shows immunological similarity to a 14.5-kDa rat lung lectin and apparently has a regulatory function (Hinek *et al.*, 1988). Several other 67-kDa proteins showing high affinity toward laminin have been reported (Mecham, 1991b). Although none of these proteins possess elastin binding properties, it has been suggested that they are the same as the 67-kDa elastin/laminin receptor (Mecham, 1991a). Mecham (1991a) has proposed that the receptor contains a protein binding site, a carbohydrate binding site, and a membrane binding site. Attachment of a sugar moiety to the carbohydrate binding site lowers the affinity of the protein binding site, which causes the release of bound protein (elastin or laminin) from the receptor. Thus, the occupancy of the lectin domain somehow modulates protein binding to the receptor.

D. Cell Surface Galactosyltransferases

Glycosyltransferases are conventionally considered to be constituents of intracellular membranes. However, studies suggest that cell surface glycosyltransferases participate in biological phenomena (Shur, 1989). A cell surface β-1,4-galactosyltransferase has recently been shown to be associated with the cytoskeleton of migrating mesenchymal cells (Eckstein and Shur, 1992). The best-established role of cell surface glycosyltransferases is their active participation in cell–cell and cell–matrix interactions during development and cell migration. Galactosyltransferases (GalTase) represent a diverse and heterogeneous family of glycosyltransferases, known to mediate a variety of interactions through complimentary oligosaccharide substrates on adjacent cell surfaces and in the extracellular matrix (Shur, 1984). The receptor function of GalTase has been demonstrated in many biological processes. During fertilization GalTase on the sperm surface mediates sperm binding to the egg by recognizing and binding to its substrate on the zona pellucida (Shur and Hall, 1982; Lopez *et al.*, 1985; Shur and Neely, 1988). GalTase may play an important role during F9 embryonal carcinoma cell differentiation (Lopez *et al.*, 1989). Control of normal, transformed, and metastatic cell behavior has also been shown to be associated with GalTase activity (Shur, 1989). Elevated levels of cell surface GalTase activity have been documented in a number of malignant tissues (Chatterjee and Kim, 1977; Plotkin *et al.*, 1979; Marchase *et al.*, 1988). Increased activities of cell surface GalTase have been reported to be associated only with metastatic variants of cells compared with the nonmetastatic variants of the same cells (Plotkin *et al.*, 1979). These studies provide insights concerning mechanisms underlying cell migration and the potential importance of glycosyltransferases during metastasis.

Runyan *et al.* (1986) reported that neural crest cells used cell surface GalTase during their migration and that laminin, with its many carbohydrate residues, was the preferred substrate *in vitro*. There has been some evidence indicating that there are two separate populations of β-1,4-GalTase molecules that are similar yet distinct, one for Golgi processing of glycoproteins and the other for cell surface attachment interactions (Lopez *et al.*, 1989; Youakim and Shur, 1990). Eckstein and Shur (1989) found that laminin containing matrices induced the stable expression of this enzyme on the surface of migratory cells, suggesting the active participation of GalTase in cell migration. By using antibodies to this enzyme, they found that the GalTase molecules were localized primarily to the active lamellipodia of the cells and could be simultaneously associated with actin containing microfilaments.

Also, cell surface GalTase seems to participate in the initiation of neurite outgrowth from PC12 cells on laminin substrates (Begovac and Shur, 1990; Begovac *et al.*, 1991). These results highlight the regulatory role of the oligosaccharide chains present on extracellular matrix molecules and present intriguing possibilities for model systems in which cell surface GalTase binds to exposed GlcNAc residues, attaching galactose residues as the enzyme releases and the cell migrates.

Although most studies related to cell glycosyltransferases have focused upon various forms of GalTases, there have been a few reports on other glycosyltransferases involved in cellular interactions (Shur, 1989). A prototypical example is the involvement of *N*-acetylgalactosaminyltransferase (GalNAcTase) in adhesive function of CAMs (Balsamo *et al.*, 1986). Paietta *et al.* (1987) suggest that a surface sialyltransferase acts as a hemagglutinin receptor on lymphocytes.

E. Selectins

In recent years a new family of adhesion receptors critical to interactions of circulating cells within the vasculature has been the subject of intensive research. This family has been designated the lectin cell adhesion molecules (LEC-CAMs), or selectins (Bevilacqua, 1989). Selectins are classified into three groups: L-selectin (LECAM-1, LAM-1, $gp90^{MEL}$, Leu-8, TQ-1, and DREG), E-selectin (LECAM-2 and ELAM-1), and P-selectin (GMP-140, PADGEM, LECAM-3, and CD62) (Stoolman, 1992). On the basis of a defined sequence homology, all selectins known so far have a common structure containing five contiguous domains (Fig. 5). A short carboxy-terminal region linked to the transmembrane domain is the only intracellular domain of the molecule. The hydrophobic transmembrane domain extends through the plasma membrane lipid bilayer and links extracellularly to a set of consensus repeats that are homologous to complement regulatory proteins. The number of repeats of the consensus sequence differs among the three groups of selectins. A short EGF-like sequence links the complement regulatory domains to a lectin-like domain located at the N terminus of the molecule. The lectin-like domains have so far been identified to be Ca^{2+} dependent and therefore fall into the category of C-type lectins (Drickamer, 1988). Recent studies suggest that this lectin-like domain encodes the major cell binding sites in selectins (Stoolman, 1992).

The selectin family of cell adhesion molecules initiates attachment to the microvasculature during leukocyte recruitment and lymphocyte recirculation (Stoolman, 1992). L-selectin is constitutively expressed on lymphocytes, monocytes, neutrophils, and eosinophils (Stoolman, 1989) and is highly dependent on calcium for adhesion (Yednock and

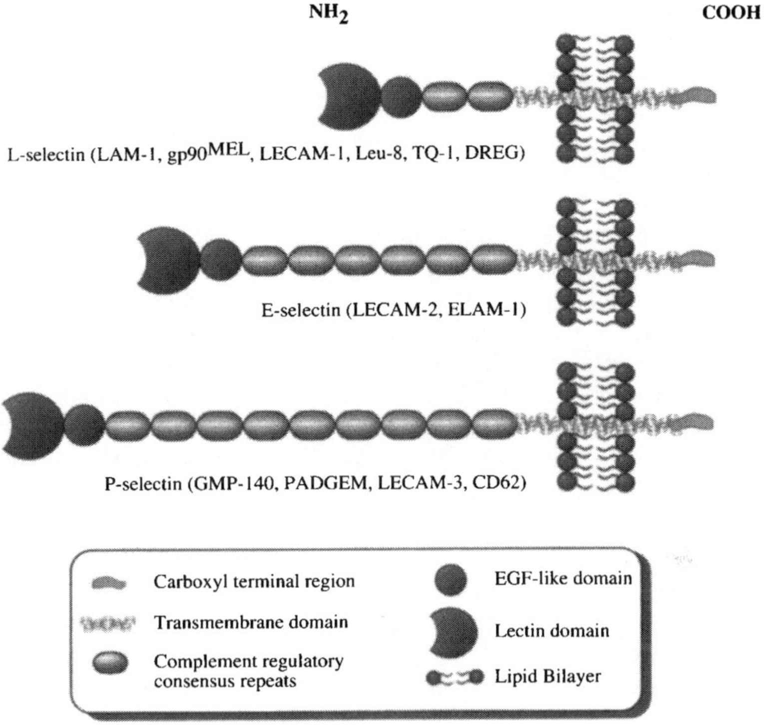

FIG. 5. Domain structures of the selectin family of cell adhesion molecules.

Rosen, 1989). In an early study, carbohydrate structures based on L-fucose and D-mannose were shown to react with one or more receptors on the lymphocyte surface, selectively inhibiting its adhesion (Stoolman and Rosen, 1983). More recently, Yednock *et al.* (1987) have reported a carbohydrate-dependent adhesion of lymphocytes mediated by the lectin-like domain of L-selectin.

E-selectin is an adhesive protein expressed on the surface of the vascular endothelium and it has been implicated in the initial events of neutrophil extravasation. E-selectin has been reported to recognize ligands containing sialylated lactosamine structures with fucose linked to the 3-hydroxyl group of at least one GlcNAc residue (Brandley *et al.*, 1990; Moore *et al.*, 1991). Kameyama *et al.* (1991) have reported that fucose and sialic acid residues are important for recognition. Tyrrell *et al.* (1991) have recently proposed a model of functional groups necessary for recognition by E-selectin. By employing direct binding and inhibition studies they have reported that tetrasaccharides such as Sia(α2–3)Gal(β1–4)[Fuc(α1–3)] GlcNAc and Sia(αa2–3)Gal(β1–4)-[Fuc(α1–3)]Glc are the smallest structures required by E-selectins to recognize their target ligands. E-selectin has also been shown to bind to a pentasaccharide epitope called VIM-2, with the following structure: Sia(α2–3)Gal(β1–4)GlcNAc(β1–3)Gal(β1–4)[Fuc(α1–3)]GlcNAc (Stoolman, 1992).

P-selectin, the third member of the selectin family, is a component of Weibel–Palade bodies of endothelial cells (Bonfanti *et al.*, 1989). It is reported to be an integral membrane protein located in secretory granules of platelets and endothelium (McEver *et al.*, 1989). Evidence to date suggests that sialylated oligosaccharides interact with P-selectin (Stoolman, 1992). Recently, Zhou *et al.* (1991) have reported that GMP-140 binds to sialylated, fucosylated lactosaminoglycans on both myeloid and nonmyeloid cells. The carbohydrate recognition properties of E-selectin and P-selectin have been shown to have striking similarity, although the amino acid sequences of their lectin domains are not identical (Zhou *et al.*, 1991). In general, sialic acid has been reported to be one of the essential components of the ligands for all three groups of selectins (Rosen *et al.*, 1985; True *et al.*, 1990; Zhou *et al.*, 1991). Kojima *et al.* (1992) have recently demonstrated that application of benzyl-α-*N*-acetyl galactosamine, an inhibitor of O-linked carbohydrate chain extension, resulted in reduced adhesion of tumor cells to activated endothelial cells or platelets mediated by E-selectin or P-selectin. These authors suggest the possibility that targeting of O-glycosylation inhibitors or modifiers to tumor cells may suppress metastatic potential.

F. Integrins

Integrins are a diverse group of noncovalently linked heterodimers that form a gene superfamily of membrane-spanning, cell surface receptors (Albelda and Buck, 1990). Integrins bind numerous ECM components, including fibronectin, laminin, vitronectin, types I, IV, and VI collagen, fibrinogen, and complement cascade components. These dimeric glycoprotein receptors are located on the cell surface and have a short membrane spanning region and a small cytoplasmic domain. They are composed of a larger subunit designated a plus a smaller disulfide-rich subunit designated β. This receptor complex is thought to provide a link between the microfilaments of the cell cytoskeleton, via the molecules vinculin and talin, and the external environment. Studies by several groups have contributed considerable information on integrin interaction mediated through the tetrapeptide sequence, arginine–glycine–aspartic acid–serine (RGDS). Since the discovery of the RGDS site in fibronectin (Pierschbacher and Ruoslahti, 1984) , the RGD tripeptide has been found in almost all other extracellular matrix molecules (Humphries *et al.*, 1991). Laminin contains one RGD sequence in its A-chain (Sasaki *et al.,* 1988) and at least six integrin receptors bind to laminin (Mecham, 1991a). Most of the studies of laminin–integrin interaction are non-RGD mediated (Mecham, 1991a). Goodman *et al.* (1991) have described an integrin–laminin interaction mediated through the short arm of laminin which contains the RGD sequence. Although numerous reports have shown integrin binding to proteins and peptides, studies of integrin interactions with carbohydrate moieties of extracellular matrix molecules have been sparse.

Studies at this laboratory (Dean *et al.,* 1988, 1990; Chandrasekaran *et al.,* 1991) and elsewhere (Bouzon *et al.,* 1990) have demonstrated laminin carbohydrate-dependent adhesion and spreading of the mouse melanoma cell line B16F1. A highly metastatic variety of these melanoma cells, B16F10, has been shown to bind to laminin through a 120- to 140-kDa cell surface glycoprotein (gp120/140) (Chammas *et al.,* 1991). Polyclonal and monoclonal antibody binding studies showed that gp120/140 was an $\alpha6/\beta1$ integrin. This study also suggested that the carbohydrate moieties of both laminin and gp120/140 may be implicated in their mutual interactions. On the basis of these observations, two alternative mechanisms have been postulated: (a) carbohydrate moieties from each ligand interact with the protein backbones of the complementary molecule, as in lectin–lectin interactions (Sharon and Lis, 1989); and/or (b) reciprocal, specific carbohydrate–carbohydrate interactions occur (Eggens *et al.,* 1989; Kojima and Hakomori, 1989). Recently, a 190-/220-kDa laminin binding lectin has been purified from chicken tissues (Bao *et al.,* 1992). Although it copurified with integrin, this protein has been found to be a nonintegrin lectin with galactose binding specificity.

Adhesion of leukocyte integrins LFA-1 (lymphocyte function-associated antigen-1) and Mac-1 to immunoglobulin has been shown to be regulated by glycosylation of the immunoglobulin molecule (Diamond *et al.,* 1991). This report indicates that mutations at two sites in the third domain of intercellular adhesion molecule 1 (ICAM-1), a member of the immunoglobulin superfamily, destroy N-glycosylation sites, which enhance the binding of Mac-1 to immunoglobulin. The size of the N-linked oligosaccharide affects the adhesion properties of ICAM-1 (Diamond *et al.,* 1991). Adhesion depends on LFA-1 if ICAM-1 has a larger N-linked side chain, but on Mac-1 if ICAM-1 has a smaller carbohydrate side chain.

VII. Epilogue

Information storage, retrieval, and processing are essential functions for biological systems. Although DNA is the central repository for all primary information, transfer of secondary information (mRNA), tertiary information (proteins), and quaternary information (oligosaccharides and glycosaminoglycans) is also critical for cell processes and behavior. We have summarized the current knowledge concerning the participation of structural carbohydrate in cell–cell and cell–matrix recognition. That body of knowledge indicates that carbohydrate recognition is intrinsic to specific cellular responses. The challenge for the future is to determine which specific molecular structures, receptors, and mechanisms underlie the various biological phenomena. The only credible prediction is that there will be plenty of surprises awaiting those who have ventured forth in this exciting area.

Acknowledgments

The experimental work of the authors was partially supported by NIH Grants AR-17720, AR-07430, and DE-00157.

References

Abercrombie, M., Heaysman, J., and Pegrum, S. M. (1971). The locomotion of fibroblasts in culture. *Exp. Cell Res.* **67,** 359–367.

Albelda, S. M., and Buck, C. A. (1990). Integrins and other cell adhesion molecules. *FASEB J.* **4,** 2868–2880.

Anderson, R. L., and Wang, J. L. (1992). Carbohydrate binding protein 35. *Trends Glycosci. Glycotechnol.* **4,** 43–52.

Ariga, T., Kohriyama, T., Freddo, L., Latov, N., Megumi, S., Kon, K., and Yu, R. K. (1987). Characterization of sulfated glucuronic acid containing glycolipids reacting with IgM M-proteins in patients with neuropathy. *J. Biol. Chem.* **262,** 848–854.

Arumugham, R. G., Hsieh, T. C., Tanzer, M. L., and Laine, R. A. (1986). Structures of the asparagine-linked sugar chains of laminin. *Biochim. Biophys. Acta* **883,** 112–126.

Asch, A. S., and Nachman, R. L. (1989). Thrombospondin: Phenomenology to function. *Prog. Hemostasis Thromb.* **9,** 157–176.

Ashwell, G., and Harford, J. (1982). Carbohydrate-specific receptors of the liver. *Annu. Rev. Biochem.* **51,** 531–554.

Balasmo, J., Pratt, R. S., and Lilian, J. (1986). Chick neural retina *N*-acetylgalactosaminyltransferase/acceptor complex: Catalysis involves transfer of *N*-acetylgalactosamine phosphate to endogenous acceptors. *Biochemistry* **25,** 5402–5407.

Bao, Z. Z., Muschler, J., and Horwitz, A. F. (1992). LBL, a novel, developmentally regulated, laminin-binding lectin. *J. Biol. Chem.* **267,** 4974–4980.

Barondes, S. H. (1984). Soluble lectins: A new class of extracellular proteins. *Science* **223,** 1259–1264.

Barondes, S. H. (1986). Vertebrate lectins: properties and function. *In* "The Lectins—Properties, Functions and Applications in Biology and Medicine" (I. E. Liener, N. Sharon, and I. J. Goldstein, eds.), pp. 437–466.

Barondes, S. H. (1988). Bifunctional properties of lectins: Lectins redefined. *Trends Biochem. Sci.* **13,** 480–482.

Bause, E., and Hettkamp, H. (1979). Primary structural requirements for N-glycosylation of peptides in rat liver. *FEBS Lett.* **108,** 341–344.

Beck, K., Hunter, I., and Engel, J. (1990). Structure and function of laminin: Anatomy of a multidomain glycoprotein. *FASEB J.* **4,** 148–160.

Begovac, P. C., Hall, D. E., and Shur, B. D. (1991). Laminin fragment E8 mediates PC12 cell neurite outgrowth by binding to cell surface beta-1,4-galactosyltransferase. *J. Cell. Biol.* **113,** 637–44.

Begovac, P. C., and Shur, B. D. (1990). Cell surface galactosyltransferase mediates the initiation of neurite outgrowth from PC12 cells on laminin. *J. Cell. Biol.* **110,** 461–70.

Bernard, B. A., Yamada, K. M., and Olden, K. (1982). Carbohydrates selectively protect a specific domain of fibronectin aginst proteases. *J. Biol. Chem.* **257,** 8549–8554.

Bevilacqua, M. P., Stengelin, S., Gimbrone, M. A., Jr., and Seed, B. (1989). Endothelial leukocyte adhesion molecule 1: An inducible receptor for neutrophils related to complement regulatory proteins and lectins. *Science* **243,** 1160–1165.

Blackburn, C. C., Swank-Hill, P., and Schnaar, R. L. (1986). Gangliosides support neural retina cell adhesion. *J. Biol. Chem.* **261,** 2873–2881.

Bonfanti, R. B., Furie, B. C., Furie, B., and Wagner, D. D. (1989). PADGEM (GMP140) is a component of Weibel–Palade bodies of human endothelial cells. *Blood* **73,** 1109–1112.

Bourdon, M. A., and Ruoslahti, E. (1989). Tenascin mediates cell attachment through an RGD-dependent receptor. *J. Cell Biol.* **108,** 1149–1155.

Bourdon, M. A., Wikstrand, C. J., Furthmayr, H., Matthews, T. J., and Bigner, D. D. (1983). A human glioma–mesenchymal extracellular matrix antigen defined by monoclonal antibody. *Cancer Res.* **43,** 2796–2806.

Bouzon, M., Dussert, C., Lissitzky, J.-C., and Martin, P.-M. (1990). Spreading of B16 F1 cells on laminin and its proteolytic fragments P1 and E8: Involvement of laminin carbohydrate chains. *Exp. Cell Res.* **190,** 47–56.

Brackenbury, R. (1985). Molecular mechanisms of cell adhesion in normal and transformed cells. *Cancer Metastasis Rev.* **4,** 41–58.

Brandley, B. K., Swiedler, S. J., and Robbins, P. W. (1990). Carbohydrate ligands of the LEC cell adhesion molecules. *Cell* **63,** 861–863.

Bray, D., and Hollenbeck, P. J. (1988). Growth cone motility and guidance. *Annu. Rev. Cell Biol.* **4,** 43–61.

Bronner-Fraser, M. (1985). Alteration in neural crest migration by a monoclonal antibody that affects cell adhesion. *J. Cell Biol.* **101,** 610–617.

Bronner-Fraser, M. (1986). An antibody to a receptor for fibronectin and laminin perturbs cranial neural crest development *in vivo*. *Dev. Biol.* **117,** 528–536.

Burridge, K., Fath, K., Kelly, T., Nuckolls, G., and Turner, C. (1988). Focal adhesions: Transmembrane junctions between the extracellular matrix and the cytoskeleton. *Annu. Rev. Cell Biol.* **4,** 487–525.

Chammas, R., Veiga, S. S., Line, S., Potocnjak, P., and Brentani, R. R. (1991). Asn-linked oligosaccharide dependent interaction between laminin and gp 120/140. *J. Biol. Chem.* **266,** 3349–3355.

Chandrasekaran, S., Dean, J. W., Giniger, M. S., and Tanzer, M. L. (1991). Laminin carbohydrates are implicated in cell signaling. *J. Cell. Biochem.* **46,** 115–124.

Chatterjee, S. K., and Kim, U. (1977). Galactosyltransferase activity in metastasizing and nonmetastasizing rat mammary carcinomas and its possible relationship with tumor cell surface antigen shedding. *J. Natl. Cancer Inst.* **58,** 273–280.

Cherayil, B. J., Chaitovitz, S., Wong, C., and Pillai, S. (1990). Molecular cloning of a human macrophage lectin specific for galactose. *Proc. Natl. Acad. Sci. USA* **87,** 7324–7328.

Cherayil, B. J., Weiner, S. J., and Pillai, S. (1989). The mac-2 antigen is a galactose-specific lectin that binds IgE. *J. Exp. Med.* **170,** 1959–1972.

Chiquet, M., and Fambrough, D. M. (1984). Chick myotendinous antigen II: A novel extracellular glycoprotein complex consisting of large disulfide-linked subunits. *J. Cell Biol.* **98,** 1937–1946.

Chiquet-Ehrismann, R. (1991). Anti-adhesive molecules of the extracellular matrix. *Curr. Opin. Cell Biol.* **3,** 800–804.

Chiquet-Ehrismann, R., Kalla, P., Pearson, C., Beck, K., and Chiquet, M. (1988). Tenascin interferes with fibronectin action. *Cell* **53,** 383–390.

Chiquet-Ehrismann, R., Mackie, E. J., Pearson, C. A., and Sakakura, T. (1986). Tenascin: An extracellular matrix protein involved in tissue interactions during fetal development and oncogenesis. *Cell* **47,** 131–139.

Chung, A. E., Freeman, I. L., and Brainski, J. E. (1977). A novel extracellular membrane elaborated by a mouse embryonal carcinoma-derived cell line. *Biochem. Biophys. Res. Commun.* **79,** 859–868.

Codogno, P., Doyennette-Moyne, M. A., Botti, J., and Aubery, M. (1988). Concanavalin A-induced impairment of fibroblast spreading on laminin but not on fibronectin. *J. Cell. Physiol.* **136,** 463–470.

Cole, G. J., and Schachner, M. (1987). Localization of the L2 monoclonal antibody binding site on chicken neural cell adhesion molecule (NCAM) and evidence for its role in NCAM-mediated cell adhesion. *Neurosci. Lett.* **78,** 227–232.

Cunningham, B. A., Hemperly, J. J., Murray, B. A., Prediger, E. A., Brackenberry, R., and Edelman, G. M. (1987). Neural cell adhesion molecule: Structure, immunoglobulin-like domains, cell surface modulation and alternative RNA splicing. *Science* **236,** 799–806.

Dean, J. W., Chandrasekaran, S., and Tanzer, M. L. (1988). Lectins inhibit cell binding and spreading on a laminin substrate. *Biochem. Biophys. Res. Commun.* **156,** 411–416.

Dean, J. W., Chandrasekaran, S., and Tanzer, M. L. (1990). A biological role of the carbohydrate moieties of laminin. *J. Biol. Chem.* **265,** 12553–12562.

Dennis, J. W. (1992). Changes in glycosylation associated with malignant transformation and tumor progression. *In* "Cell Surface Carbohydrates and Cell Development" (M. Fukuda, ed.), pp. 25–69. CRC Press, Boca Raton, FL.

Dennis, J. W., and Laferte, S. (1987). Tumor cell surface carbohydrates and metastatic phenotype. *Cancer Metastasis Rev.* **5,** 185–204.

Dennis, J. W., Waller, C. A., and Schirrmacher, V. (1984). Identification of asparagine-linked oligosaccharides involved in tumor cell adhesion to laminin and type IV collagen. *J. Cell. Biol.* **99,** 1416–1423.

Deutzmann, R., Aumailley, M., Wiedemann, H., Pysny, W., Timpl, R., and Edgar, D. (1990). Cell adhesion, spreading and neurite stimulation by laminin fragment E8 depends on maintenance of secondary and tertiary structure in its rod and globular domain. *Eur. J. Biochem.* **191,** 513–522.

Diamond, M. S., Staunton, D. E., Marlin, S. D., and Springer, T. A. (1991). Binding of the integrin Mac-1 (CD11b/CD18) to the third immunoglobulin-like domain of ICAM-1 (CD54) and its regulation by glycosylation. *Cell* **65,** 961–971.

Dixit, V. M., Haverstick, D. M., O'Rourke, K. M., Hennessy, S. W., Grant, G. A., Santoro, S. A., and Frazier, W. A. (1985). A monoclonal antibody against human thrombospondin inhibits platelet aggregation. *Proc. Natl. Acad. Sci. USA* **82,** 3472–3476.

Do, K. Y., Smith, D. F., and Cummings, R. D. (1990). LAMP-1 in CHO cells is a primary carrier of poly-*N*-acetyllactosamine chains and is bound preferentially by a mammalian S-type lectin. *Biochem. Biophys. Res. Commun.* **173,** 1123–1128.

Drickamer, K. (1988). Two distinct classes of carbohydrate-recognition domains in animal lectins. *J. Biol. Chem.* **263,** 9557–9560.

Duband, J. L., Roder, S., Chen, W. T., Yamada, K. M., and Thiery, J. P. (1986). Cell adhesion and migration in the early vertebrate embryo: Location and possible role of the putative fibronectin receptor complex. *J. Cell Biol.* **102,** 160–178.

Eckstein, D. J., and Shur, B. D. (1989). Laminin induces the stable expression of surface galactosyltransferase on lamellipodia of migrating cells. *J. Cell. Biol.* **108,** 2507–2517.

Eckstein, D. J., and Shur, B. D. (1992). Cell surface beta-1,4-galactosyltransferase is associated with the detergent-insoluble cytoskeleton on migrating mesenchymal cells. *Exp. Cell. Res.* **201,** 83–90.

Eggens, I., Fenderson, B., Toyokuni, T., Dean, B., Stroud, M., and Hakomori, S. (1989). Specific interaction between Le^x and Le^x determinants. *J. Biol. Chem.* **264,** 9476–9484.

Ehrig, K., Leivo, I., Argraves, W. S., Ruoslahti, E., and Engvall, E. (1990). Merosin, a tissue-specific basement membrane protein, is a laminin-like protein. *Proc. Natl. Acad. Sci. USA* **87,** 3264–3268.

Elbein, A. D. (1987). Inhibitors of the biosynthesis and processing of N-linked oligosaccharide chains. *Annu. Rev. Biochem.* **56,** 497–534.

Elbein, A. D. (1991). The role of N-linked oligosaccharides in glycoprotein function. *Trends Biotechnol.* **9,** 346–352.

Elbein, A. D., Trpoea, J. E., Mitchell, M., and Kaushal, G. P. (1990). Kifunensine, a potent inhibitor of the Golgi processing mannosidase I. *J. Biol. Chem.* **265,** 15599–15605.

Erickson, H. P., and Iglesias, J. L. (1984). A six-armed oligomer isolated from cell surface fibronectin preparations. *Nature* **311,** 267–269.

Faissner, A., and Kruse, J. (1990). J1/Tenascin is a repulsive substrate for central nervous system neurons. *Neuron* **5,** 627–637.

Feizi, T. (1987). Significance of carbohydrate components of cell surfaces. *In* "Autoimmunity and Autoimmune Diesease," Ciba Foundation Symposium, Vol. 129, pp. 43–58. Wiley, Chichester.

Feizi, T. (1989). Glycoprotein oligosaccharides as recognition structures. *In* "Carbohydrate Recognition in Cellular Function," Ciba Foundation Symposium, Vol. 145, pp. 62–74. Wiley, Chichester.

Finne, J. (1982). Occurrence of unique polysialosyl carbohydrate units in glycoproteins of developing brain. *J. Biol. Chem.* **257,** 11966–11970.

Finne, J., Finne, U., Deagostini-Bazin, H., and Gordis, C. (1983). Occurrence of α2-8 linked polysialosyl units in a neural cell adhesion molecule. *Biochem. Biophys. Res. Commun.* **112,** 482–487.

Frazier, W. A. (1987). Thrombospondin: A modular adhesive glycoprotein of platelets and nucleated cells. *J. Cell Biol.* **105,** 625–632.

Fujiwara, S., Shinkai, H., Deutzmann, R., Paulsson, M., and Timpl, R. (1988). Structure and distribution of N-linked oligosaccharide chains on various domains of mouse tumour laminin. *Biochem. J.* **252,** 453–461.

Furukawa, K., Roberts, D.D., Endo, T., and Kobata, A. (1989). Structural study of the sugar chains of human platelet thrombospondin. *Arch. Biochem. Biophys.* **270,** 302–312.

Galvin, N. J., Vance, P. M., Dixit, V. M., Fink, B., and Frazier, W. A. (1987). Interaction of human thrombospondin with types I–V collagen: Direct binding and electron microscopy. *J. Cell Biol.* **104,** 1413–1422.

Gething, M. J., and Sambrook, J. (1992). Protein folding in the cell. *Nature* **355,** 33–45.

Ginsburg, V., and Roberts, D. D. (1988). Glycoconjugates and cell adhesion: The adhesive proteins laminin, thrombospondin and von Willebrand's factor bind specifically to sulfated glycolipids. *Biochimie* **70,** 1651–1659.

Goodman, S. L., Aumailley, M., and von der Mark, H. (1991). Multiple cell surface receptors for the short arms of laminin: $\alpha1\beta1$ integrin and RGD-dependent proteins mediate cell attachment only to domains III in murine tumor laminin. *J. Cell. Biol.* **113,** 931–941.

Grumet, M., Hoffman, S., Crossin, K. L., and Edelman, G. M. (1985). Cytotactin, an extracellular matrix protein of neural and non-neural tissues that mediates glia–neuron interaction. *Proc. Natl. Acad. Sci. USA* **82,** 8075–8079.

Hakomori, S. (1991). Possible functions of tumor-associated carbohydrate antigens. *Curr. Opin. Immunol.* **3,** 646–653.
Harper, J. R., Perry, S. K., Davis, R. M., and Laufer, D. M. (1990). Human neuroectoderm-derived cell line secretes fibronectin that shares the HNK-1/10C5 carbohydrate epitope with neural cell adhesion molecules. *J. Neurochem.* **54,** 395–401.
Hinek, A., Wrenn, D. S., Mecham, R. P., and Barondes, S. H. (1988). The elastin receptor: A galactoside binding protein. *Science* **239,** 1539–1541.
Hirschberg, C. B., and Snider, M. D. (1987). Topography of glycosylation in the rough endoplasmic reticulum and Golgi apparatus. *Annu. Rev. Biochem.* **56,** 63–87.
Ho, M. K., and Springer, T. A. (1982). MAC-2, a novel 32,000 M_r mouse macrophage subpopulation-specific antigen defined by monoclonal antibodies. *J. Immunol.* **128,** 1221–1228.
Hoffman, S., and Edelman, G. M. (1983). Kinetics of homophilic binding by embryonic and adult forms of the neural cell adhesion molecule. *Proc. Natl. Acad. Sci. USA* **80,** 5762–5766.
Howe, C. C. (1984). Functional role of laminin carbohydrate. *Mol. Cell. Biol.* **4,** 1–7.
Humphries, M. J., Paul Mould, A., and Yamada, K. M. (1991). Matrix receptors in cell migration. *In* "Receptors for Extracellular Matrix" (J. A. McDonald and R. P. Mecham, eds.), pp. 195–253. Academic Press, San Diego.
Hunter, D. D., Shah, V., Merlie, J. P., and Sanes, J. R. (1989). A laminin-like adhesive protein concentrated in the synaptic cleft of the neuromuscular junction. *Nature* **338,** 229–234.
Hynes, R. (1985). Molecular biology of fibronectin. *Annu. Rev. Cell Biol.* **1,** 67–90.
Jones, G. E., Arumugham, R., and Tanzer, M. L. (1986). Fibronectin glycosylation modulates fibroblast adhesion and spreading. *J. Cell Biol.* **103,** 1663–1670.
Kaesberg, P. R., Ershler, W. B., Esko, J. D., and Mosher, D. F. (1989). Chinese hamster ovary cell adhesion to platelet thrombospondin is dependent on cell surface heparan sulfate proteoglycan. *J. Clin. Invest.* **83,** 994–1001.
Kameyama, A., Ishida, H., Kiso, M., and Hasegawa, A. (1991). Total synthesis of sialyl Lewis X. *Carbohydr. Res.* **209,** c1–c4.
Kery, V. (1991). Lectin–carbohydrate interactions in immunoregulation. *Int. J. Biochem.* **23,** 631–640.
Knibbs, R. N., Perini, F., and Goldstein, I. J. (1989). Structure of the major concanavalin A reactive oligosaccharides of the extracellular matrix component laminin. *Biochemistry* **28,** 6379–6392.
Kojima, N., and Hakomori, S. (1989). Specific interaction between gangliotriaosylceramide (Gg3) and sialosyllactosylceramide (Gm3) as a basis for specific cellular recognition between lymphoma and melanoma cells. *J. Biol. Chem.* **264,** 20159–20162.
Kojima, N., Handa, K., Newman, W., and Hakomori, S. (1992). Inhibition of selectin-dependent tumor cell adhesion to endothelial cells and platelets by blocking O-glycosylation of these cells. *Biochem. Biophys. Res. Commun.* **182,** 1288–1295.
Kornblihtt, A. R., Umezawa, K., Vibe-Pedersen, K., and Baralle, F. E. (1985). Primary structure of human fibronectin: Differential splicing may generate at least 10 polypeptides from a single gene. *EMBO J.* **4,** 1755–1759.
Kornfeld, R., and Kornfeld, S. (1985). Assembly of asparagine linked oligosaccharides. *Annu. Rev. Biochem.* **54,** 631–664.
Kruse, J., Keilhauer, G., Faissner, A., Timpl, R., and Schachner, M. (1985). The J1 glycoprotein—A novel nervous system cell adhesion molecule of the L2/HNK-1 family. *Nature* **316,** 146–148.
Krusius, T., Gehlsen, K. R., and Ruoslahti, E. (1987). A fibroblast chondroitin sulfate

proteoglycan core protein contains lectin-like and glowth factor-like properties. *J. Biol. Chem.* **262,** 13120–13125.

Künemund, V., Jungalwala, F. B., Fischer, G., Chou, D. K. H., Keilhauer, G., and Schachner, M. (1988). The L2/HNK-1 carbohydrate of neural cell adhesion molecules is involved in cell interactions. *J. Cell Biol.* **106,** 213–223.

Lahav, J., Lawler, J., and Gimbrone, M. A. (1984). Thrombospondin interactions with fibronectin and fibrinogen: Mutual inhibition in binding. *Eur. J. Biochem.* **145,** 151–156.

Lahav, J., Schwartz, M. A., and Hynes, R. O. (1982). Analysis of platelet adhesion with a radioactive chemical crosslinking agent: Interaction of thrombospondin with fibronectin and collagen. *Cell* **31,** 253–262.

Lawler, J., Derick, L. H., Connolly, J. E., Chen, J. H., and Chao, F. C. (1985). The structure of human platelet thrombospondin. *J. Biol. Chem.* **260,** 3762–3772.

Lawler, J., and Hynes, R. O. (1986). The structure of human thrombospondin, an adhesive glycoprotein with multiple calcium-binding sites and homologies with several different proteins. *J. Cell. Biol.* **103,** 1635–1648.

Lawler, J., Weinstein, R., and Hynes, R. O. (1988). Cell attachment to thrombospondin: The role of ARG-GLY-ASP, calcium, and integrin receptors. *J. Cell. Biol.* **107,** 2351–2361.

Leffler, H., Masiarz, F. R., and Barondes, S. H. (1989). Soluble lactose-binding vertebrate lectins: A growing family. *Biochemistry* **28,** 9222–9229.

Leung, L. L., and Nachman, R. L. (1982). Complex formation of platelet thrombospondin with fibrinogen. *J. Clin. Invest.* **70,** 542–549.

Lopez, L. C., Bayna, E. M., Litoff, D., Shaper, N. L., Shaper, J. H., and Shur, B. D. (1985). Receptor function of mouse sperm surface galactosyltransferase during fertilization. *J. Cell. Biol.* **101,** 1501–1510.

Lopez, L. C., Maillet, C. M., Oleszkowicz, K., and Shur, B. D. (1989). Cell surface and Golgi pools of beta-1,4-galactosyltransferase are differentially regulated during embryonal carcinoma cell differentiation. *Mol. Cell. Biol.* **9,** 2370–2377.

Lotan, R., and Raz, A. (1988). Endogenous lectins as mediators of tumor cell adhesion. *J. Cell. Biochem.* **37,** 107–117.

Lyles, J. M., Linnemann, D., and Bock, E. (1984). Biosynthesis of D2-cell adhesion molecule: Post-translational modifications, intracellular transport and developmental changes. *J. Cell Biol.* **99,** 2082–2091.

Macek, M. B., and Shur, B. (1988). Protein–carbohydrate complementarity in mammalian gamete recognition. *Gamete Res.* **20,** 93–109.

Marchase, R. B., Kidd, V. J., Rivera, A. A., and Humphreys-Beher, M. G. (1988). Cell surface expression of 4b-galactosyltransferase accompanies rat parotid gland acinar cell transition to growth. *J. Cell. Biochem.* **36,** 453–465.

Martin, G. R., and Timpl, R. (1987). Laminin and other basement membrane components. *Annu. Rev. Cell Biol.* **3,** 57–85.

McEver, R. P., Beckstead, J. H., Moore, K. L., Marshall-Carlson, L., and Bainton, D. F. (1989). GMP-140, a platelet alpha granule membrane protein, is also synthesized by vascular endothelial cells and is localized in Weibel–Palade bodies. *J. Clin. Invest.* **84,** 92–99.

Mecham, R. P. (1991a). Laminin receptors. *Annu. Rev. Cell Biol.* **7,** 71–91.

Mecham, R. P. (1991b). Receptors for laminin on mammalian cells. *FASEB J.* **5,** 2538–2546.

Mecham, R. P., Hinek, A., Griffin, G. L., Senior, R. M., and Liotta, L. A. (1989). The elastin receptor shows structural and functional similarities to the 67-kDa tumor cell laminin receptor. *J. Biol. Chem.* **264,** 16652–16657.

Mecham, R. P., Whitehouse, L., Hay, M., Hinek, A., and Sheetz, M. P. (1991). Ligand affinity of the 67-kD elastin/laminin binding protein is modulated by the protein's lectin domain: Visualization of elastin/laminin-receptor complexes with gold-tagged ligands. *J. Cell Biol.* **113,** 187–194.

Miyake, M., and Hakomori, S. I. (1991). A specific cell surface glycoconjugate controlling cell motility: Evidence by functional monoclonal antibodies that inhibit cell motility and tumor cell metastasis. *Biochemistry* **30,** 3328–3334.

Moore, K. L., Varki, A., and McEver, R. P. (1991). GMP-140 binds to a glycoprotein receptor on human neurtophils: Evidence for a lectin-like interaction. *J. Cell. Biol.* **112,** 491–499.

Mosher, D. F. (1990). Physiology of thrombospondin. *Annu. Rev. Med.* **41,** 85–97.

Mumby, S. M., Raugi, G. J., and Bornstein, P. (1984). Interactions of thrombospondin with extracellular marix proteins: Selective binding to type V collagen. *J. Cell Biol.* **98,** 646–652.

Murphy-Ullrich, J. E., and Höök, M. (1989). Thrombospondin modulates focal adhesions in endothelial cells. *J. Cell Biol.* **109,** 1309–1319.

Newton, S. A., Yeo, K. T., Yeo, T. K., Parent, J. B., and Olden, K. (1987). Possible involvement of intracellular lectins in intracellular transport of glycoproteins. *In* "Vertebrate Lectins" (K. Olden and J. B. Parent, eds.), pp. 211–226. Van Nostrand–Reinhold, New York.

Nishimura, H., Yamashita, S., Zeng, Z., Walz, D.A., and Iwanaga, S. (1992). Evidence for the existence of O-linked sugar chains consisting of glucose and xylose in bovine thrombospondin. *J. Biochem.* **111,** 460–464.

Nybroe, O., and Bock, E. (1989). Structure and function of the neural cell adhesion molecules NCAM and L1. *In* "Molecular Aspects of Development and Aging of the Nervous System" (J. M. Lauder, ed.), Vol. 265, pp. 185–196. Plenum, New York.

Olden, K., Pratt, R. M., and Yamada, K. M. (1979). Role of carbohydrate in biological function of the adhesive glycoprotein fibronectin. *Proc. Natl. Acad. Sci. USA* **76,** 3343–3347.

Paietta, E., Hubbard, A. L., Wiernik, P. H., Diehl, V., and Stockert, R. J. (1987). Hodgkin's cell lectin: An ectosialyltransferase and lymphocyte agglutinant related to the hepatic asialoglycoprotein receptor. *Cancer Res.* **47,** 2461–2467.

Patel, R. S., Odermatt, E., Schwarzbauer, J. E., and Hynes, R. O. (1987). Organization of the fibronectin gene provides evidence for exon shuffling during evolution. *EMBO J.* **6,** 2565–2572.

Patthy, L., Trexler, M., Váli, Z., Bányai, L., and Váradi, A. (1984). Homology of the gelatin-binding region of fibronectin with the kringle structures of proteases. *FEBS Lett.* **171,** 131–136.

Paulsson, M., and Saladin, K. (1989). Mouse heart laminin: Purification of the native protein and structural comparison with Engelbreth–Holm Swarm tumor laminin. *J. Biol. Chem.* **264,** 18726–18732.

Pearson, C. A., Pearson, D., Shibahara, S., Hofsteenge, J., and Chiquet-Ehrismann, R. (1988). Tenascin: cDNA cloning and induction by TGF-β. *EMBO J.* **7,** 2977–2982.

Petersen, T. E., and Skorstengaard, K. (1985). Primary structure of fibronectin. *In* "Plasma Fibronectin: Structure and Function" (J. McDonagh, ed.), Vol. 5, pp. 7–30. Dekker, New York.

Petersen, T. E., Skorstengaard, K., and Vibe-Pedersen, K. (1989). Primary structure of fibronectin. *In* "Fibronectin" (D. R. Mosher, ed.), pp. 1–45. Academic Press, San Diego.

Pierschbacher, M. D., and Ruoslahti, E. (1984). Cell attachment activity of fibronectin can be duplicated by small synthetic fragments of the molecules. *Nature* **309,** 30–33.

Plotkin, G. M., Wides, R. J., Gilbert, S. L., Wolf, G., Hagen, I. K., and Prout, G. R. (1979).

Galactosyltransferase activity in human transitional cell carcinoma lines and in benign and neoplastic human bladder epithelium. *Cancer Res.* **39,** 3856–3860.

Pollersberg, E. G., Sadoul, R., Goridis, C., and Schachner, M. (1985). Selective expression of the 180-kD component of the neural cell adhesion molecule N-CAM during development. *J. Cell Biol.* **101,** 1921–1929.

Prater, C. A., Plotkin, J., Jaye, D., and Frazier, W. A. (1991). The properdin-like type I repeats of human thrombospondin contain a cell attachment site. *J. Cell Biol.* **112,** 1031–1040.

Rao, C. N., Goldstein, I. J., and Liotta, L. A. (1983). Lectin-binding domains of laminin. *Arch. Biochem. Biophys.* **227,** 118–124.

Raz, A., and Lotan, R. (1987). Endogenous galactoside-binding lectins: A new class of functional tumor cell surface molecules related to metastasis. *Cancer Metastasis Rev.* **6,** 433–452.

Raz, A., and Lotan, R. (1988). Tumor cell surface lectins and metastasis. *Tum. Progr. Metastasis,* 237–244.

Raz, A., Meromsky, L., Carmi, P., Prakash, R., Lotan, D., and Lotan, R. (1984). Monolclonal antibodies to endogenous galactose-specific tumor cell lines. *EMBO J.* **3,** 2979–2983.

Roberts, D. D. (1986). Sulfatide-binding proteins. *Chem. Phys. Lipids* **42,** 173–183.

Roberts, D. D., and Ginsburg, V. (1988). Sulfated glycolipids and cell adhesion. *Arch. Biochem. Biophys.* **267,** 405–415.

Roberts, D. D., Rao, C. N., Liotta, L. A., Gralnick, H. R., and Ginsburg, V. (1986). Comparison of the specificities of laminin, thrombospondin, and von Willebrand factor for binding to sulfated glycolipids. *J. Biol. Chem.* **261,** 6872–6877.

Roberts, D. D., Sherwood, J. A., and Ginsburg, V. (1987). Platelet thrombospondin mediates attachment and spreading of human melanoma cells. *J. Cell Biol.* **104,** 131–139.

Roff, C. F., Rosevear, P. R., Wang, J. L., and Barker, R. (1983). Identification of carbohydrate-binding proteins from mouse and human fibroblasts. *Biochem. J.* **211,** 625–629.

Roff, C. F., and Wang, J. J. (1983). Endogenous lectins from cultured cells: Isolation and characterization of carbohydrate-binding proteins from 3T3 fibroblasts. *J. Biol. Chem.* **258,** 10657–10663.

Rosen, S. D., Singer, M. S., Yednock, T. A., and Stoolman, L. M. (1985). Involvement of sialic acid on endothelial cells in organ-specific lymphocyte recirculation. *Science* **228,** 1005–1007.

Rosen, S. D., and Stoolman, L. M. (1987). Potential role of cell surface lectin in lymphocyte recirculation. *In* "Vertebrate Lectins" (K. Olden and J. B. Parent, eds.), pp. 152–181. Van Nostrand–Reinhold, New York.

Rosenberg, I., Cherayil, B. J., Isselbacher, K. J., and Pillai, S. (1991). Mac-2 binding glycoproteins: Putative ligands for a cytosolic beta galactoside lectin. *J. Biol. Chem.* **266,** 18731–18736.

Runyan, R. B., Maxwell, G. D., and Shur, B. D. (1986). Evidence for a novel enzymatic mechanism of neural crest cell migration on extracellular glycoconjgate matrices. *J. Cell. Biol.* **102,** 432–441.

Ruoslahti, E. (1988). Fibronectin and its receptors. *Annu. Rev. Biochem.* **57,** 375–413.

Rutishauser, U., Acheson, A., Hall, A. K., Mann, D. M., and Sunshine, J. (1988). The neural cell adheson molecule (NCAM) as a regulator of cell–cell interactions. *Science* **240,** 53–57.

Rutishauser, U., Watanabe, M., Silver, J., Troy, F. A., and Vimr, E. R. (1985). Specific alteration of NCAM-mediated cell adhesion by an endoneuraminidase. *J. Cell Biol.* **101,** 1842–1849.

Rutishauser, U. S. (1991). Neural cell adhesion molecule and polysialic acid. *In* "Receptors for Extracellular Matrix" (J. A. McDonald and R. P. Mecham, eds.), pp. 131–156. Academic Press, San Diego.

Sadoul, R., Hirn, M., Deagostini-Bazin, H., Rougon, G., and Goridis, C. (1983). Adult and embryonic mouse neural cell adhesion molecules have different binding properties. *Nature* **304,** 347–349.

Sage, E. H., and Bornstein, P. (1991). Extracellular proteins that modulate cell–matrix interactions: SPARC, tenascin, and thrombospondin. *J. Biol. Chem.* **266,** 14831–14834.

Sairam, M. R. (1985). Glycosylation and receptor–ligand interaction. *In* "The Receptors" (P. M. Conn, ed.), Vol. II, pp. 307–340. Academic Press, Orlando, FL.

Sasaki, M., Kato, S., Kohno, K., Martin, G. R., and Yamada, Y. (1987). Sequence of the cDNA encoding the laminin B1 chain reveals a multidomain protein containing cysteine-rich repeats. *Proc. Natl. Acad. Sci. USA* **84,** 935–939.

Sasaki, M., Kleinman, H. K., Huber, H., Deutzmann, R., and Yamada, Y. (1988). Laminin, a multidomain protein: The A chain has a unique globular domain and homology with the basement membrane proteoglycan and the laminin B chains. *J. Biol. Chem.* **263,** 16536–16544.

Sasaki, M., and Yamada, Y. (1987). The laminin B2 chain has a multidomain structure homologous to the B1 chain. *J. Biol. Chem.* **262,** 17111–17117.

Schachner, M. (1989). Families of neural cell adhesion molecules. In "Carbohydrate Recognition in Cellular Function," Ciba Foundation Symposium, Vol. 145, pp. 156–188. Wiley, Chichester.

Schallier, D. C. C., Bolscher, J. G. M., and Smets, L. A. (1988). Alterations in cell surface carbohydrates affecting invasion *in vitro*. *In* "Lectins and Glycoconjugates in Oncology" (H. J. Gabius and G. A. Nagel, eds.), pp. 87–96. Springer-Verlag, Berlin.

Schwarz, R. T., and Datema, R. (1984). Inhibitors of trimming: New tools in glycoprotein research. *Trends Biol. Sci.,* 32–34.

Schwarzbauer, J. E., Paul, J. I., and Hynes, R. O. (1985). On the origin of species of fibronectin. *Proc. Natl. Acad. Sci. USA* **82,** 1424–1428.

Sharon, N., and Lis, H. (1989). Lectins as cell recognition molecules. *Science* **246,** 227–246.

Shur, B. D. (1984). The receptor function of galactosyltransferase during cellular interactions. *Mol. Cell. Biochem.* **61,** 143–158.

Shur, B. D. (1989). Expression and function of cell surface galactosyltransferase. *Biochim. Biophys. Acta* **988,** 389–409.

Shur, B. D., and Hall, N. G. (1982). A role for mouse sperm surface galactosyltransferase in sperm-binding to the egg zona pellucida. *J. Cell. Biol.* **95,** 574–579.

Shur, B. D., and Neely, C. A. (1988). Plasma membrane association, purification and partial characterization of mouse sperm β1,4-galactosyltransferase. *J. Biol. Chem.* **263,** 17706–17714.

Singer, S. J., and Kupfer, A. (1986). The directed migration of eukaryotic cells. *Annu. Rev. Cell Biol.* **2,** 337–365.

Skorstengaard, K., Jensen, M. S., Petersen, T. E., and Magnusson, S. (1986). Complete primary stucture of bovine plasma fibronectin. *Eur. J. Biochem.* **161,** 441–453.

Sparrow, C. P., Leffler, H., and Barondes, S. H. (1987). Multiple soluble beta-galactoside-binding lectins from human lung. *J. Biol. Chem.* **262,** 7383–7790.

Spring, J., Beck, K., and Chiquet-Ehrismann, R. (1989). Two contrary functions of tenascin: Dissection of the active sites by recombinant tenascin fragments. *Cell* **59,** 324–334.

Stoolman, L. M. (1989). Adhesion molecules controlling lymphocyte migration. *Cell* **56,** 907–910.

Stoolman, L. M. (1992). Selectins (LEC-CAMs): Lectin-like receptors involved in lymphocyte recirculation and leukocyte recruitment. *In* "Cell Surface Carbohydrates and Cell Development" (M. Fukuda, ed.), pp. 71–97. CRC Press, Boca Raton, FL.

Stoolman, L. M., and Rosen, S. D. (1983). Possible role for cell-surface carbohydrate-binding molecules in lymphocyte recirculation. *J. Cell Biol.* **96,** 722–729.

Sutton, B. J., and Phillips, D. C. (1983). IgG heavy chain carbohydrate interactions. *Biochem. Soc. Trans.* **11,** 130–132.

Takatsuki, A., and Tamura, G. (1971). Tunicamycin production. *J. Antibiot.,* 224–231.

Takatsuki, A., and Tamura, G. (1982). Inhibitors of glycoconjugate biosynthesis by tunicamycin. *In* "Tunicamycin" (G. Tamura, ed.), pp. 38–47. Japan Scientific Societies, Tokyo.

Tamura, G., and Ito, T. (1982). Discovery and chemistry of tunicamycin. *In* "Tunicamycin" (G. Tamura, ed.), pp. 3–29. Japan Scientific Societies Press, Tokyo.

Tanzer, M. L., Dean, J. W., and Chandrasekaran, S. (1991). Cell signalling: A role for laminin carbohydrates. *Trends Glycosci. Glycotechnol.* **3,** 302–314.

Taylor, M. E., Bezouska, K., and Drickamer, K. (1992). Contribution to ligand binding by multiple carbohydrate-recognition domains in the macrophage mannose receptor. *J. Biol. Chem.* **267,** 1719–1726.

Timpl, R., Rohde, H., Robey, P. G., Rennard, S. I., Foidart, J. M., and Martin, G. R. (1979). Laminin—A glycoprotein from basement membranes. *J. Biol. Chem.* **254,** 9933–9937.

Trinkaus-Randall, V., Newton, A. W., Gipson, I. K., and Franzblau, C. (1988). Carbohydrate moieties of the basal lamina: Their role in attachment and spreading of basal corneal epithelial cells. *Cell. Tissue Res.* **251,** 315–323.

True, D. D., Singer, M. S., Lasky, L. A., and Rosen, S. D. (1990). Requirement for sialic acid on the endothelial ligand of a lymphocyte homing receptor. *J. Cell Biol.* **110,** 2757–2764.

Tuszynski, G. P., Karczewski, T., Smith, L., Murphy, A., and Rothman, V. (1989). The GPIIb/IIIa-like complex may function as a human melanoma cell adhesion receptor for thrombospondin. *Exp. Cell Res.* **182,** 473–481.

Tyrrell, D., James, P., Rao, N., Foxall, C., Abbas, S., Dasgupta, F., Nashed, M., Hasegawa, A., Kiso, M., Asa, D., Kidd, J., and Brandley, B. K. (1991). Structural requirements for the carbohydrate ligand of E-selectin. *Proc. Natl. Acad. Sci. USA* **88,** 10372–10376.

Varani, J., Nickoloff, B. J., Riser, B. L., Mitra, R. S., and O'Rourke, K. (1988). Thrombospondin-induced adhesion of human keratinocytes. *J. Clin. Invest.* **81,** 1537–1544.

Varki, A. (1992). Role of oligosaccharides in the intracellular and intercellular trafficking of mammalian glycoproteins. *In* "Cell Surface Carbohydrates and Cell Development" (M. Fukuda, ed.), pp. 25–69. CRC Press, Boca Raton, FL.

Wasserman, P. (1992). Cell surface carbohydrates and mammlian fertilization. *In* "Cell Surface Carbohydrates and Cell Development" (M. Fukuda, ed.), pp. 215–238. CRC Press, Boca Raton, FL.

White, S. L. (1987). Role of lectins in immune regulation. *In* "Vertebrate Lectins" (K. Olden and J. B. Parent, eds.), pp. 182–194 Van Nostrand–Reinhold, New York.

Woo, H. J., Lotz, M. M., Jung, J. U., and Mercurio, A. M. (1991). Carbohydrate-binding protein 35 (Mac-2), a laminin-binding lectin forms functional dimers using cysteine 186. *J. Biol. Chem.* **266,** 18419–18422.

Woo, H. J., Shaw, L. M., Messier, J. M., and Mercurio, A. M. (1990). The major non-integrin laminin binding protein of macrophages is identical to carbohydrate binding protein 35 (Mac-2). *J. Biol. Chem.* **265,** 7097–7099.

Wrenn, D. S., Hinek, A., and Mecham, R. P. (1988). Kinetics of receptor-mediated binding of tropoelastin to ligament fibroblasts. *J. Biol. Chem.* **263,** 2280–2284.

Yamada, K. M. (1989). Fibronectin domains and receptors. *In* "Fibronectin" (D. R. Mosher, ed.), pp. 47–121. Academic Press, San Diego.

Yednock, T. A., Butcher, E. C., Stoolman, L. M., and Rosen, S. D. (1987). Receptors involved in lymphocyte homing: Relationship between a carbohydrate-binding receptor and the MEL-14 antigen. *J. Cell. Biol.* **104,** 725–731.

Yednock, T. A., and Rosen, S. D. (1989). Lymphocyte homing. *Adv. Immunol.* **44,** 313–378.

Youakim, A., and Shur, B. D. (1990). The relationship between Golgi and cell surface beta-1,4-galactosyltransferase. *Glycobiology,* 107–114.

Zhou, Q., and Cummings, D. (1990). The S-type lectin from calf heart tissue binds selectively to the carbohydrate chains of laminin. *Arch. Biochem. Biophys.* **281,** 27–35.

Zhou, Q., and Cummings, R. D. (1992). Animal lectins: A distinct group of carbohydrate binding proteins involved in cell adhesion, molecular recognition, and development. *In* "Cell Surface Carbohydrates and Cell Development" (M. Fukuda, ed.), pp. 99–125. CRC Press, Boca Raton, FL.

Zhou, Q., Moore, K. L., Smith, D. F., Varki, A., McEver, R. P., and Cummings, R. D. (1991). The selectin GMP-140 binds to sialylated, fucosylated lactosaminoglycans on both myeloid and nonmyeloid cells. *J. Cell Biol.* **115,** 557–564.

Zhu, B.C.R., and Laine, R.A. (1985). Polylactosamine glycosylation on human fetal placental fibronectin weakens the binding affinity of fibronectin to gelatin. *J. Biol. Chem.* **260,** 4041–4045.

Zhu, B.C.R., and Laine, R.A. (1990). Novel hyperglycosylated weak gelatin binding fibronectin from human fetal placenta. *Eur. J. Biochem.* **188,** 67–71.

Index